Big Steel

BIG

The First Century of
the United States Steel Corporation
1901–2001

Kenneth Warren

STEEL

UNIVERSITY OF PITTSBURGH PRESS

For Peter, John, and David

Contents

Illustrations

Tables

Preface

At a justly famous dinner given in New York City on Wednesday, 12 December 1900, the commercial advantages that could be gained from the creation of a steel combination of unprecedented size were eloquently laid out before a gathering comprised primarily of distinguished bankers and industrialists. In less than four months the United States Steel Corporation began business. During its first year the new trust produced two-thirds of the nation's raw steel; in 1902 it accounted for almost 30 percent of the steel made throughout the world. At home, fear of competition from the huge organization speeded reconstruction on the part of much smaller rivals; anticipating its invasion of world markets, some foreign iron and steel companies set about consolidation and modernization. By 1920, although its share of world production was still over 25 percent, US Steel made well under half the nation's steel. By the late 1990s, the US Steel Group of the recently formed USX Corporation was producing less than one in every eight tons of steel made in the United States and just over 1.5 percent of global output. Yet, despite these declines, over the whole century US Steel made far more steel than any other company in the world. It disbursed immense sums in dividends and unquestionably played a central role in United States economic development. Nonetheless, in many ways it failed to live up to the expectations of those who listened so eagerly to the only important after-dinner speech delivered on that mid-winter evening so long ago. What follows analyzes the course of this corporation's operations during its first century of existence.

In following this long, complex business history, it has been necessary to be selective. By no means are all plants, products, or personalities covered, nor could they all be in a manageable presentation. There is, for instance, only passing mention of the operations at Worcester or Cleveland, and relatively little about those at Youngstown; while there is a good deal concerning rails, structurals, and thin flat-rolled products, there is much less about bars or wire. As a history of the company, this study is not concerned with the character of leading individuals or with labor history except insofar as these are factors or, as with the latter, leading consequences of corporate policy. My focus has been on themes such as technological and locational change, major shifts in corporate policy, and reaction to new patterns of demand or government pressures of all kinds. I have aimed to chart what happened and, more importantly whenever possible, to explore why the Corporation followed a particular course of action. To a large extent, the first of these aims can be satisfied from a careful examination of published materials: steel industry and government reports or trade journals, the latter of which, particularly in early decades of the twentieth century, have been rich sources of information. There are also many important secondary sources, but a fuller understanding of events involves an evaluation of the vital processes of decision making—an insight into the choices between alternatives—and depends on examination of company archives.

The US Steel Group of USX generously allowed the author access to its extraordinarily abundant and detailed corporate records in the course of three research visits, one in early fall 1991 and the others during 1999. It also provided the financial support that made my 1999 research work possible, at the same time explicitly denying to its officers any oversight of or other influence on the writer's freedom of interpretation. I can confirm that these commitments have been scrupulously followed. The kind welcome and helpfulness of all in the Department of Public Affairs was greatly appreciated. During my last visit in October and November 1999, I benefited from discussing the developments of the last few decades and the prospects of the near future with the present chairman and president, two previous chairmen, one former president, and a former vice chairman and chief finance officer—in all some twelve hours of interviews. Although I am extremely grateful for their recollections and insights and have made full use of them, it would have been inappropriate to make specific references to the sources. In the hope of gaining a rounded view of the events of recent decades, I also wrote to request interviews or comment from three present or recently retired senior officials of the United Steelworkers of America. I received no reply from any of them. Despite this, I have tried throughout to be dispassionate in my judgments of matters in which their opinions and perspectives might have been valuable.

Though both the United States Steel Group and its parent company, USX, are now exceeded in size by many industrial corporations, probably not one of the latter has played a more vital or intriguing part in the economic history of twentieth-century America. I hope that a fuller appreciation of this—and also some of the associated drama—comes through in the following account of the first century of US Steel.

January 2000   *Kenneth Warren*

Preeminent Size

The Economies and Diseconomies of Scale

The large industrial operation typically enjoys many economic benefits compared to smaller rivals. Some of these advantages are technical, resulting from larger units of equipment in which the ratio of inputs such as energy or labor and the outputs of saleable products are favorable in relation to the size, capital, and operating costs of the physical plant. For the big firm, particularly if it is a multiplant operation, there may be other gains, including savings on the management and sales accounts, a greater capacity to support research and to utilize its results, and the ability to allocate orders so as to keep specialized units in fuller production than if they were independent, thereby achieving reductions in the overall costs of distribution. This reality (or hope) of such commercial advantages helps explain the tendency over generations and throughout the industrial world for companies to increase in size. The desire for increased power or for empire building is another, though less rational, consideration but one that nonetheless is also commonly involved. There is, however, also a downside to size. Whatever the initial motivation, giant businesses have to face up to important disadvantages or at least commercial dangers. The century-long career of US Steel illustrates both the gains and the penalties of preeminent size.

When the United States Steel Corporation was formed, it was the largest industrial corporation in the world. At that time in history, exceptional size was expected to confer exceptional commercial advantages. These were specifically spelled out as including economies in administration and sales, opportunities for plant specialization—with consequent scale economies in production at each of them—and reduced cross hauling in making deliveries. It was reckoned to command the best talents in top direction, managers, engineers, and those involved in technical innovation. With such assets, US Steel would become the industry price leader and might dominate trade worldwide. Yet in practice, many of these supposed advantages proved illusory. With a fine irony, this had been anticipated by Andrew Carnegie when he pointed out that large combinations were, like huge, straggling empires, no match for tightly organized single states—or companies.

Throughout the next ninety years, US Steel dominated the American industry in terms of size. To public and government, the Corporation epitomized the power of big business in this most basic of major manufacturing activities. But during this period its share of the industry, judged either by capacity or production, more or less steadily declined. Not only did other companies grow more quickly, but many of them also proved more willing to move into new growth lines of production or to adopt newer technologies. Why in these various ways did US Steel perform less well than the industry as a whole? Technological timidity may have been of major importance, but it in turn was a reflection of entrepreneurial failure. Why, with all the talent the Corporation could command, should there be such a failure? Some have traced the problems of US Steel to its relations with labor. Certainly this relationship was distant, frosty, and woefully lacking in sympathetic understanding over the first third of its corporate history and has often been difficult since, but in these respects US Steel's record has been no worse than that of the other companies in the industry. A central belief of this interpretation of the Corporation's history is that it has been paying a high price for what at the beginning was commonly regarded as its main asset, its preeminent size. This made US Steel uniquely difficult to direct and manage, a huge mass to reorient for meeting new structures and locations of demand, with too much existing plant to make it easy to justify building new ones. Even more subtly, the company could make high profits while failing to keep pace with the best practices of the industry. Its dominating size also exposed it to public scrutiny and to the often unwelcome oversight of government, reaction to which inclined management toward caution more often than to initiative. Time and again US Steel was served and sometimes led by excellent men, but these executives were inhibited and crabbed by corporate unwieldiness, structures, traditions, and defensive attitudes. Only at the end of its first century, when it was subjected to a pruning that reduced its steel capacity by at least two-thirds, did the steel giant begin to produce results that ranked it as a national and world leader in steel-making efficiency rather than merely size.

Examination of the company's earliest years indicates that even then both those who conceived US Steel and those inordinately fearful of its powers were too impressed by possible scale economies and ignored the presence of major diseconomies. There were not only too many plants, widely scattered, but there were also divisions and often disagreements between a distant, finance-dominated headquarters and a system of production units made up of formerly independent companies and of product divisions, resulting in a rather clumsy corporate structure. In an age deeply suspicious of trusts, predominating size required a continuing caution in the exercise of industrial power, an avoidance of cutthroat competition, and a general concern to avoid alienating major customers, the public, and government. Yet this very caution, particularly when reflected in price policies, provided existing competitors and newcomers with a favorable business environment for growth. For its vitally important first quarter-century, a period of large—and in relative terms unexceeded—increase in national steel capacity, US Steel was dominated by a man who was by nature cautious and conservative and had little direct knowledge of steel making. Taken in conjunction with the impediments considered above, this meant that the Corporation fell further behind in its share of production, efficiency, and adjustment to the changing structure of demand. Executives adopted a hostile attitude toward organized labor and were generally harsh in their physical demands on employees, though in some respects these men implemented paternalistic policies. The character and traditions of US Steel, shaped during this period, continued to burden it long afterward. During these years, a considerable number of new, commercially aggressive steel companies were established, and the sluggish corporate response in taking up new products or processes—key examples of which were modern styles and mills for heavy structural shapes, the manufacture of electric furnace steel, and in the 1920s the production of electrical resistance weld pipe—provided prima facie evidence of entrepreneurial failure. Most critical of all, US Steel was slow to adopt the wide continuous hot-strip mill.

In the late 1920s, a new stage of development began following the death of the former chairman. Over the next ten years there were more positive responses to changes in the national economy, marked by the first US Steel involvement in the American West. New top direction; keener competition from companies pioneering in new products, processes, and locations; and the collapse in demand for established staples as a result of the Depression ensured that the Corporation belatedly carried through a large program for the installation of hot-strip mills. During the 1930s, there were wholesale closures or disposals of plants. While this process involved a considerable loss of jobs, a more enlightened attitude was adopted toward labor relations, in the spirit of the age, with the recognition of the Steel Workers Organizing Committee (later reconstituted as the United Steel Workers of America) and the introduction of collective bargaining.

High activity in World War II and a subsequent large and more-or-less continuous expansion of the national economy were accompanied by major extensions by the industry and by US Steel until the early 1960s. Plant was usually almost fully employed during these decades. New sources of ore supply were opened in the traditional areas of the Upper Great Lakes, in the West, and overseas. For the first time in almost thirty years—and for the last time—the Corporation built new works on virgin sites in areas where before it had been scarcely represented, the Atlantic Coast and the West. In the 1950s easier operating conditions once more induced a mood of complacency in top management, and US Steel's share of national output continued to fall. In some respects a more damaging result of these circumstances was that the company again lagged in innovation, especially in bulk steel manufacturing by the oxygen converter. Later it was slow to adopt continuous casting methods.

After a check to the upward movement of product demand and a serious strike that interrupted domestic supplies, steel imports grew rapidly in the early 1960s. While many spokesmen of the American industry alleged that much of this steel was "dumped" on the market by foreign companies, there could be no question about its competitive price or its adverse effect on levels of demand for domestically produced material. At the same time, like other bulk producers, US Steel began to suffer from the challenge of "mini-mills," unintegrated domestic companies that, using electric steel furnaces, continuous casting, and new rolling mills, were able to sharply reduce overhead, energy, and labor costs, thereby undercutting larger, fully-integrated producers. Into the mid-1970s the industry continued to hope that government restraints on imports and national economic growth might together enable big companies to survive and even justify large extensions and plant replacements. Eventually, though, slower growth, periods of sharp recession, and a maturing of the "postindustrial" society (in which steel played a smaller role than in the past), continued heavy inflows of steel, and further increases in mini-mill output brought a frank recognition that the steel industry had entered a state of crisis. Major companies again resorted to mergers and plant closures. This rationalization was nowhere more radical than at US Steel. One outcome of the vigor with which that challenge was faced was that by the 1990s the company's great tally of major works had been reduced to three fully integrated operations, only one of which had existed in 1901. In its tenth decade, the Corporation was at last able to act as a straightforward commercial enterprise, free of both outside opprobrium and, perhaps even more important, its own delusions of grandeur. As a result US Steel became one of the world's most efficient major steel producers.

Part One

The Gary Era

Origins

The Creation of the United States Steel Corporation

On 1 April 1901 the United States Steel Corporation was incorporated under the laws of New Jersey. It was far and away the nation's biggest steel company and the largest industrial organization of any kind worldwide. The capital involved was $1.1 billion, though, as in other trusts of the time, this figure was grossly inflated with "water"—that is, much of its nominal capital was not represented by physical plant. The origins of this gigantic agglomeration of capacity and capital may be traced through a range of forces—some general, many peculiar—to the particular stage reached in national economic development at the time. An early historian of the Corporation distinguished the "enabling" or "desirable" factors for a giant consolidation from the unique circumstances of the particular occasion.[1] These last, which included trade, company, and personal factors, gave particular shape and content to the new organization.

In the course of the 1880s and 1890s, the output of iron and steel had grown massively. As late as 1880, the nation made half as much pig iron as the United Kingdom; by 1900 it turned out more than 150 percent of U.K. output. Over those twenty years U.S. iron production rose from 4 to 14 million metric tons, amounting to almost half the world's increase. American technology and organization became in many respects the best practice. Although it was tempting to ignore the fact that there was much obsolescent equipment, rapid growth in demand and production generally encouraged modernization. As a visiting Englishman found during extensive travels in 1903 and 1904, "Each new plant that is erected is the epitome of the latest practice in its own particular field"; the improved efficiency of the new works made it

necessary to modernize the older ones.[2] Rich national mineral endowments
and the generally large and rapidly growing markets for standardized steel
products differentiated the American experience from that of other leading
industrial nations. But it was also true that enterprise and capital had responded positively to both opportunities and challenges. Writing in 1896,
Pennsylvania Steel Company engineer Henry Huse Campbell, proudly summing up what had been done, recognized the costs as well as the benefits:
"Within the last decade America has made marvellous developments in her
iron industry, until she now leads the world in the quantity of her products,
and bows to none in their quality. This wonderful progress has not been the
unearned harvest of bounteous nature, for it has been accomplished in defiance of mighty obstacles in the enormous distances through which the raw
materials must be carried, and, although the achievement may be a just
source of national pride, it involves inevitable expenses and disadvantages
which may be lessened by energy, but which can never be swept away."[3] This
great system of production was still further extended and expanded during
the next few years. Yet, though the industry's technical and organizational
achievements were universally admired by steel experts, it was in many respects flawed commercially. Large-scale combination was one of the ways by
which financial returns might be made to match excellence of practice.

In economic activity, generally, the trend toward concentration in bigger,
more concentrated units was now at full flood.[4] In steel the technology of the
time provided a strong upward impetus, large furnaces, forges, or rolling mills
usually producing greater output at lower unit costs than smaller units. Further economies gained from linking the various stages in the manufacturing
process, especially in such a heat-intensive industry, meant that iron making,
steel production, and rolling and finishing operations were now commonly
combined at a single site. Ideally, such integrated plants should attempt to secure the most efficient multiples of capacities in each of the various subdivi-
sions of the operations. But growth and rounding out of individual works,
though often accommodated by the rapid extension of steel consumption,
could also mean keen, sometimes cutthroat, competition. In bad times this
meant acceptance of unprofitable operation of plant that had been modernized at high cost. Such conditions encouraged acquisitions or amalgamations,
which might for a time contain the ever-present drive for expansion and lead
to more gradual increases of capacity, better tailored to market growth. The
combination of formerly separate enterprises made easier the maintenance of
"reasonable" prices, elimination of inefficient plants, and specialization to cut
overall costs. It produced ever-bigger companies.

Other forms of organization aiming to ensure "ordered" competition in which all of the important producers could be reasonably prosperous had been tried and found far from satisfactory. Since the late 1880s, leading firms in various product sectors had formed "pools" that allocated production, fixed prices, or both. There were temporary successes, but pooling arrangements proved unstable, often broke down, and were followed by periods in which prices plummeted. Many companies—and dependent communities—suffered severely. It was unfortunate for long-term stability that even under these dire conditions all but the weakest tended to struggle on and survive into the next upward phase.

Some more efficient way of controlling competition was needed. The "trust" style of organization seemed to offer an advantage. By the 1890s there was a well-established trend to combination in a range of industries. Incentive for the processes of merger, rationalization, and the shaping of a leaner industry was always to be found in the promise of a more remunerative future. In addition to realistic or likely promises to the industry of genuine improvements in efficiency, the combination process was attractive to those who carried it through as a source of promotion profits. These seemed justified by the fact that a promoter required not only imagination to see the possibilities of combination but also the ability to secure financial backing and agreement from those controlling the firms in question. As a result, beyond the financial interests of the new group comprising the merging companies, there was added stock to reward the promoter and those underwriting the resultant organization with working capital until it was a going concern. Steel seemed an appropriate field for large-scale application of this process. To understand the necessities and possibilities, some other aspects of the structure of the industry at this time must be considered.

Steel was finished in a wide variety of forms: bars of a range of sizes and shapes for further processing, plate, sheet, tinplate, wire, tubes, structural shapes, rails, and other items as well as castings and forgings. In a nation still building its infrastructure, a few of these products had long dominated the rest. This had been especially the case with rails, but now the economy was diversifying. In 1880 U.S. production of rolled steel was 1.012 million tons, of which rails made up 85 percent. By 1900 the rail tonnage was not far short of three times that of 1880, but this now represented only 31 percent of rolled-steel output. In the 1890s a range of new products grew rapidly. The McKinley Tariff Act of 1890 had been followed by the establishment of an American tinplate industry, which then mushroomed, providing an outlet both for a large increase in "blackplate" (sheet) production and of the steel bars from which this sheet was rolled. Expansion in shipbuilding, both mercantile and naval, meant more need for plate and angles. New construction in bridges and especially in multistory office and apartment buildings called forth a massive expansion in output of the bigger structural shapes. Large-scale growth in wire and in tube and pipe manufacture, for use in fencing and the extension of the oil and natural gas industries, required new wire-rod, skelp, strip, and plate mills. In 1889 the combined production of iron and steel plates, sheets, skelp, and structural shapes was some 400,000 tons less than that of rails; ten years later it was 1.7 million tons greater.[5] Yet, despite the huge scale of national economic growth, the expanding industry was marked by gross overcapacity, and there was an apparently irresistible temptation to extend plant far in excess of likely demand. Between 1887 and 1898, the steel-making ability of the nation was reckoned to have increased from 5.85 to 15.64 million net tons. Both years were ones of record output, but the amount of steel produced went up only from 3.34 to 8.93 million tons. In 1887 about 57 percent of capacity was utilized; eleven years later the proportion was almost exactly the same. The picture was confused by the fact that much obsolete plant was included in the capacity figures as well as plant that was reasonably competitive. The next year posted record production figures for both pig iron and crude steel, but for the first the estimated capacity exceeded the high level of the year's output in iron by 37.3 percent and in steel by 56.3 percent.[6] Even more important, the rounding out of the industry company by company, some needing more primary capacity—pig iron and crude steel—others more plant for finished rolled steel products, threatened vastly greater overcapacities, ever keener competition, and reduced returns. From 1898 to 1900, these problems seemed to be building up to a crisis.

Some of the additional capacity was provided by diversification on the part of existing major companies, some by new operations. Two companies stood out from the list of major integrated companies, partly by virtue of the capacity they controlled and the fact that they owned a number of works, but

TABLE 1.1
Output of various finished iron and steel products, 1890, 1895, and 1900
(thousand gross tons)

Year	Rails	Plates and sheets except nail plate	Structural shapes	Wire rods	All other rolled products	Rails as % all rolled products	Tinplate
1890	1,885	810	incl. in other	457	2,870	31.3	n.a.
1895	1,306	991	518	791	2,583	21.1	114
1900	2,386	1,794	815	846	3,646	25.1	303

Sources: AISA annual reports.

mostly by their efficiency. These were Federal Steel and Carnegie Steel, which together controlled well over one-quarter of the national crude steel capacity. Federal had been formed in late summer 1898 when the ore-mining Minnesota Iron Company, the Lorain Steel Works on Lake Erie, and a few additional operations were merged into the older Illinois Steel Company of Chicago. The industrial zone at the head of Lake Michigan had long been recognized as a highly desirable location both for making iron and steel and for distribution of the finished products. For years, though, the greater efficiency of the Carnegie operations in Pittsburgh had largely cancelled out the natural advantages of their chief rival. Elbert H. Gary, general counsel for Illinois Steel during the 1890s, later recalled that Carnegie Steel once sold steel rails in the Chicago district for eighteen dollars a ton, and that to meet the challenge his own company had to sell at below the costs of production. "The Carnegie Company could do what it came very near to doing at one time, namely force the Illinois Steel Company into the hands of a receiver. In 1896 . . . the papers were drawn, in fact." On the other hand, even though in normal times Illinois Steel could not compete with Carnegie beyond one hundred miles east of Chicago, 95 percent or so of its output could be disposed of where it enjoyed striking advantages of accessibility, its huge natural market area west of the Indiana-Ohio line.[7] The main producers east of the Alleghenies, for whom rail manufacture had also been the staple trade, were less well sheltered from the harsh winds of Carnegie competition. As Carnegie wrote in September 1899: "My view is that sooner or later Harrisburg [Steelton], Sparrows Point, and Scranton will cease to make rails, like Bethlehem. The autumn of last year seemed as good a time to force them out of business as any other. It did not prove so. The boom came and cost us a great deal of money." Even so, whereas in 1888 Allegheny County had produced 12.3 percent of the rail tonnages rolled in Pennsylvania and Illinois combined, by 1897 its share was three times as large.[8]

Although the superiority of Carnegie and to a lesser extent of Federal over other established integrated firms had been proved, there was at the end of the 1890s a new challenge in the form of big, expansion-minded combinations in the fields of the main finished products. Many of the plants of these new companies bought steel in semifinished form from the fully integrated iron, steel, and rolling-mill firms. By the close of the century, a large proportion of the finishing firms were combined in a series of major horizontal mergers, the resulting specializations and range of which were reflected in their new company names. They came in quick succession and in two great waves. The process was helped along by increasing output and rising prices that seemed to promise a rosy future. As an early student of US Steel put it, these compa-

nies were "industrial experiments," but he might have added that they followed a well-tried route and, as a contemporary noted, "hit off the psychological moment."[9] In December 1898 the creation of the American Tin Plate
Company brought together firms operating thirty-nine plants containing 279
mills and making up about 90 percent of national tinplate production. A
month later the American Steel and Wire Company (AS&W), incorporated
under the laws of New Jersey, included most of the wire firms left out of a
smaller combination formed under Illinois corporate law ten months earlier.
In June 1899 the National Tube Company combined twenty-one separate
companies and some nine-tenths of United States wrought-tube capacity. (In
this case, one-quarter of the $80 million capitalization was said to represent
promotion profits.) After this there was a pause until spring 1900, when in
succession the Shelby Steel Tube Company brought together some 90 percent
of the seamless-tube capacity, American Sheet Steel incorporated 70 percent
of sheet making, and American Steel Hoop linked nine firms making bars,
hoops, cotton ties, and skelp. The main aim of those who in those months
formed a constructional steel combine, American Bridge, seems to have been
to yield promoters' profits.

There is no doubt that the product groupings yielded many economies.
John W. Gates of AS&W claimed that reduced cross hauling alone saved a half-
million dollars per year.[10] But the process was also fraught with penalties, restrictions, and outright extortion. For example, one of the stipulations of
contracts for sale of plant to American Tin Plate was that the former owners
could not start tinplate production again for a period of fifteen years within
1,500 miles of the site they had sold. In their quest to control trade, the trusts
incorporated much unsuitable plant, so much of which seemed haphazard.
As one academic of the time put it, the works of AS&W "were sown broadcast
over the whole face of the land. A grant of land, a cash bonus, ten year's exemption from taxation, a local connection, among one of a number of causes
entirely disconnected from considerations of economical production, had determined the original locations of these plants, the burden of whose malad-
justments the steel trusts had now to assume and to carry."[11] Local industrial
communities suffered from the weeding out that followed combination. By
summer 1899 American Tin Plate was dismantling works at Baltimore and
another location in Maryland, in Brooklyn, and in Cleveland. D. G. Reid, Tin
Plate's first president, revealed another aspect of the restraint of trade involved. He told the Stanley Committee, a House of Representatives commission charged with investigating US Steel, that his company had a contract
with roll makers to take their output but could not remember whether they

also had an arrangement with manufacturers not to sell to independent producers.[12] A notorious instance of the ruthless greed of these sales, mergers, and new ventures was revealed years later when John Stevenson Jr. summed up his working philosophy of the time, "'We'll shake the apple tree again' I told my associates."[13]

Most combines, like the smaller finishing companies of which they were composed, bought semifinished steel from the longer-established large companies that had mineral resources, blast furnaces, and steel works as well as rolling mills. Of these there were a considerable number, but their size and location in relation to the distribution of the finishing companies meant the two main suppliers were the Federal and Carnegie companies, for which the increasing outlets for semifinished steel supplemented their trade in such major fully finished lines as plates, structurals, and rails. Federal's Chicago plants delivered wire rods to American Steel and Wire in the Midwest; its Lorain works sold billets to the Ohio plants of National Tube. Carnegie supplied various finishing groups around Pittsburgh and in the Ohio–Lake Erie belt. Two circumstances threatened this commercial arrangement and seemed likely to plunge the industry into unprecedented uncertainty and possible chaos. First, some finishing combines began to contemplate or even to carry through plans to reduce their costs by producing some or more of their own steel. Increasing difficulties in finding adequate markets after spring 1900 pushed them further in this direction. National Tube had iron and steel plant at McKeesport, near Pittsburgh, and in autumn 1900 revealed plans for a steel works at Wheeling, West Virginia. AS&W controlled some Connellsville coking coal, had associated interests in lake ore carriers, and in 1899 began to build what was then expected to be a fully integrated iron and steel operation on Neville Island at the head of the Ohio River; reportedly, the site had been laid out for up to six blast furnaces.[14] At year's end the company announced plans for a Milwaukee steelworks to supply its western plants. Second, the formation of a new company expressly to supply various rolling mill combinations threatened the regional arrangements. National Steel was organized in February 1899 and soon built up a fair-sized mineral resource base. Henry Oliver sold National his New Castle blast furnace, one-sixth of the ore production of the Oliver Iron Mining company for fifty years, the whole output of his coke works near Uniontown for ten years, and reserves of coking coal in Westmoreland County.[15] In its first year, National Steel produced 12 percent of the nation's crude steel, primarily from works established some years before in Ohio. As billets and sheet and tinplate bars, most of this output was supplied to American Tin Plate, AS&W, and American Steel Hoop, firms with which National Steel was associated through common promoters, the so-called Moore group. National then prepared to go further, taking the offensive by building a rail mill reported to have a daily capacity of 1,200 to 1,500 tons of Bessemer rails per day, which could later be raised to 2,000 tons daily. If this latter capacity were realized, it would account for more than a quarter of the previous year's record output, ranking it roughly equal to Carnegie's Edgar Thompson works (ET) or the output of the Chicago rail mills. Only later did it become clear that National aimed to make and sell rails only when demand for bars was at low ebb.[16]

An inevitable outcome of these developments in the finishing combines and at National Steel was a reduction in purchases of semifinished steel from "primary" producers. Predictably, the latter companies reacted with plans to finish more of their own steel and in so doing invaded more of each other's product ranges as well as those of the finishing combines. In 1900 Federal Steel made 15 percent and Carnegie Steel 18 percent of the nation's crude steel. That year Federal proposed to take up the manufacture of tubes and structurals, the latter a challenge to Carnegie's long-established leadership. It threatened to build wire mills unless American Steel and Wire cancelled its plans to produce more of its own steel. AS&W gave way, but it was clear that plans to integrate backwards could be revived at any time. Carnegie Steel re-

TABLE 1.2
Capacity of steel plants of important companies, 1900–1901
(thousand gross tons)

Carnegie Steel		Pennsylvania/Maryland Steel	
Edgar Thomson	1,000	Steelton	500
Homestead	1,900	Sparrows Point	400
Duquesne	1,000	Inland Steel	
Federal Steel		Indiana Harbor	125
South Chicago	1,075	Lackawanna Steel	
Joliet	600	Scranton (to be closed)	705
Union (idle)	325	Lackawanna (building)	845
Lorain	550	Bethlehem	
Jones and Laughlin		Bethlehem Steel	335
Pittsburgh	750	Republic	
Cambria Steel		Youngstown	350
Johnstown	825	Tennessee Coal, Iron and Railroad	
National Steel		Ensley	300
Ohio plants	1,260	Colorado Fuel and Iron	
		Pueblo	250

Sources: AISA, Works Directory, 1901 and 1902.

sponded still more vigorously, initiating plans to bring it into competition with former major customers. Company officials decided to install wire-rod, wire-, and nail-making plant at Duquesne and sheet mills were contemplated for Homestead. By far Carnegie's biggest project was a wholly new, fully integrated, iron, steel, and tube works at Conneaut on Lake Erie, where costs would be at least ten dollars a ton lower than at National Tube. After many months of planning, the project was revealed to the public on 12 January 1901. At once, Edmund Converse, National Tube's president, denounced the Carnegie plans as not only losing them the 150,000 tons of plate and strip Carnegie was supplying to National Tube, but also as likely to bring back the situation before the horizontal combinations had been formed, involving "a cutthroat warfare which bankrupted the weaker ones, destroyed the profits of others and caused all manner of discrimination and uncertainty among patrons." In Gates's words, it seemed the Carnegie mill would "tear the National Tube, that Morgan had just put together, all to pieces."[17]

During 1900, tensions and stakes were raised still higher by the reconstruction of some old companies and the beginnings of new ones, though the latter were as yet insignificant in relation to the established producers. Republic Iron and Steel Company was formed in 1899 as an agglomeration of rolling-mill operations. It also controlled five blast furnaces and a few, generally ill-located, steelworks. It set in train a purposeful reconstruction, concentrating its powers in the Youngstown area. Above Pittsburgh, Union Steel Company was intending to enter the wire-rod trade and considering supporting these operations with steel capacity and blast furnaces. One of the main rail makers, Lackawanna, having decided to move operations from Scranton, was now constructing a new works on a greenfield lakeshore site just south of Buffalo. The ferment in steel also threatened the good order of the railroad business. Carnegie was actively planning new lines to release him from the Pennsylvania Railroad. At the same time he vigorously criticized it for high freight charges that made it difficult to continue to deliver structural steel into Chicago at a time when Federal had decided to spend $10 million to build its own structural steelworks.[18]

Trade conditions worsened in the latter half of 1900. As this happened, prospects of large additions to iron, steel, and rolling-mill capacity from established and new companies, through backward integration by finishing groups or through forward integration by steel makers, threatened general demoralization in an industry already marked by gross overcapacity. Uncertainty was increased by the acknowledged terrifying abilities of the Carnegie Steel Company. Two weeks after the public announcement of the Conneaut tube-mill project, Charles M. Schwab, that company's young, able, and aggressive president and Andrew Carnegie's heir apparent, sent his chief some statements outlining the financial, raw material, and capacity situations of their own operations and those of their leading rivals. The figures came with a covering note, which emphasized not only Carnegie's own present commanding position but also how it seemed likely to improve still more. "I really believe that for the next ten years the Carnegie Company will show greater earnings than all the others together. . . . I shall not feel satisfied until we are producing 500,000 tons per month *and finishing same.* And we'll do it within five years. Look at our ore and coke as compared with the others. If you continue to give me the support you have in the past will make a greater industry than even we ever dreamed of. Am anxious to get at Conneaut. Are pushing plans rapidly and will be ready for a start in the spring." Carnegie responded enthusiastically, "I like your talk—five years development at Conneaut will put C. S. Co. where she belongs."[19] Despite—indeed partly because of— Schwab's confidence, for most steel companies years of doubtful commercial prospects seemed to lie ahead.

Under such conditions it was natural that industry leaders talked of amalgamations. The amazing thing was that until a late date no one seems to have contemplated an association anywhere nearly as wide as that which eventually took shape. When this happened a positive, forward-looking emphasis could at last be added to what might otherwise have been merely a larger version of an often-repeated defensive action. Naturally, such an outcome would owe much to the disposition of the principal characters involved. They were responding to a situation for which no one of them had been responsible, though some of them had played important parts in shaping it. Some saw possibilities where most saw none, few, or even disaster, but a few proved to have the organizing power and financial acumen to transform visions or vague schemes into reality. To attain any happy outcome required a catalyst. In the

TABLE 1.3
The Carnegie, Federal, and National Steel Companies in 1900
according to C. M. Schwab

Company	Capital (millions)	Steel output (th. tons)	Earnings (millions)	Earnings per ton of steel	Earnings per $100 capital
Carnegie Steel	$320	2,970	$40	$13.47	$12.50
Federal Steel (Chicago plants)	$126	1,225	$10	$8.16	$7.94
National Steel	$63	1,400	$8	$5.71	$12.70

Source: C. M. Schwab to A. Carnegie, 24 January 1901, ACLC.

circumstances of the time, it would involve not only a persuasive idea but also a forceful proponent and much patient analysis of practicalities.

The leading actor so far in the industry was to play no subsequent part in the drama. During 1900 the readiness and vigor with which he was prepared to fight old, new, or reconstructed competitors alike had belied Andrew Carnegie's sixty-five years. Though sentimentally attached to the industry so that his attitude to major restructuring for years varied according to his mood, by this time he was generally willing to contemplate retirement from active involvement in steel. This would enable him to devote much of his time to the distribution of the riches his controlling interest in Carnegie Steel would bring him if it became part of a major public company. For some of the others who negotiated and planned for a wide amalgamation, the hope of securing Carnegie's elimination from the trade was a prime incentive to carry the work through to completion. J. P. Morgan was the next most important influence, though largely as impresario rather than performer. A main concern was the maintenance of orderly business and thereby of the conditions under which the companies in whose formation he had already been involved—Federal Steel, National Steel, and to a smaller extent American Bridge—could become firmly established and financially successful. In his recent pugnacious mood, Carnegie had proposed to build new railroads to free his company from dependence on trunk-line operators. This would strike at another key sector in which Morgan was involved. Below these two principals, future prospects for the greater number of those men already occupying prominent positions as presidents or directors of finished product or primary steel companies were tied up with the wider framework of planning and negotiation. If their concerns were merged into a larger group, they could expect big financial rewards.

As early as 1899, as the product groups were taking shape, there had been schemes for the combination of considerable sections of the industry. Elbert H. Gary, with whom Morgan was involved in the formation of Federal Steel, was said to have afterward urged the banker to purchase Carnegie Steel. There were various vague ideas for linking Federal and Carnegie. In spring that year there was a scheme, which nearly came to fruition, for cooperation between the Moore group, which had promoted a number of the product groups as well as National Steel, and certain of the Carnegie top management, notably Henry Phipps and Henry Clay Frick. Together they planned to buy out Carnegie and float a reconstructed Carnegie Steel as a major public company, with Frick taking over Carnegie's top place. One after another such projects were mooted; followed up by a flood of rumors, hopes, and fears; and then failed to materialize. The following year, Carnegie Steel was recon-

structed, but Frick was expelled from active involvement. The company retained its partnership status and prepared for keener competition ahead, in a
business that it seemed would continue to be dominated by the cut and thrust
of the survival of the fittest. At this critical time of darkening prospects, yet another way out was first tantalizingly outlined and then assiduously pursued.

The vision of better things that might be shaped for the future was presented by Carnegie Steel president Charles M. Schwab in what has come to be
seen as probably the most momentous after-dinner speech ever delivered by
an industrial executive. Unfortunately, there is no transcript of what he said at
the banquet given in his honor on the evening of Wednesday, 12 Decem-
ber1900, at the University Club in New York; within a year he had covered
some of the same ground in an address in Chicago of which a record does remain.[20] Many years afterward Schwab recalled some of what he said: "I chose
as my subject my idea of the future development of manufacturing in the
United States, especially as relating to our industry. . . . I explained at that
meeting the very great advantages that would result in manufacture from
such an organization as the United States Steel Corporation, and I gave my
reasons in detail to them." A key principle was that "one mill should run on
one product and not one mill on 50 products as was then the practice." Such
specialization would bring large cost savings. A further consideration, which
might clash with the last, was the possibility of general reductions in selling
costs, including fewer long hauls on products as plants concentrated on supplying their local or regional outlets. In any event there would be large
economies in distribution. He added, "In my proposal there was no thought of
limitation of production or the maintenance of price."[21]

There have been suggestions that Carnegie encouraged Schwab to make
such a speech, which taken along with the obvious preparations being made
by Carnegie Steel to wage a more effective economic war against its competitors might help along a profitable buy-out. Against this interpretation is
Carnegie's distaste for finance and company promoters, but there is at least
some evidence supporting such a conspiracy theory. Carnegie was reported to
have stayed at the banquet for only a short time before leaving to address the
Pennsylvania Society, though ten years later Schwab included him in the list
of those he recollected as seated at the table.[22] A discrete departure could
spare an emotional Carnegie the need to hear what he knew was inevitable.
At some point, probably a few days before the dinner, Carnegie had sent a
short message to his cousin, George Lauder. His words do not rule out the possibility that he recognized that Carnegie Steel interests were being promoted:
"Schwab's dinner here remarkable. Mr. Schwab tells me every one invited ac-

cepted and really the biggest men in New York. He is a favorite indeed and this makes him more valuable for us."[23] Whatever Carnegie's prior knowledge, the University Club speech initiated a period of activity out of which, within less than five months, the United States Steel Corporation emerged.

After his speech Schwab had half an hour of close conversation with J. P. Morgan before returning to Pittsburgh. The next vital step came when he was summoned to a meeting at Morgan's New York home. The date is unknown, but it seems to have been about three weeks after the University Club dinner, that is, during the first few days of 1900. This time Morgan asked Schwab to compile a list of companies that might be incorporated into a giant group designed to achieve the economies he had outlined on 12 December. When completed, this list also included Schwab's estimate of each company's real worth. Morgan then asked him to enquire at what price Carnegie would sell. Schwab recalled for the Stanley Committee that it was "about a week later" that he spent the day with Carnegie and passed on Morgan's message; other evidence suggests that their meeting and Carnegie's very rough draft asking $480 million for the Carnegie Company should be dated a further two weeks later, probably toward the end of January. On the other hand, there is some circumstantial evidence that Carnegie may have known what was going on much earlier than this, indeed, even before the public announcement of the Conneaut project. On 7 January 1900 from New York he sent Schwab in Pittsburgh an intriguingly worded telegram. It seemed not only to refer to the tube mill but possibly also to weightier matters. It may suggest, however, that Schwab had implied that his contacts with Morgan were about no more than the latter's wish to stop Conneaut: "Don't think it necessary for you to come on to see that gentleman. I should rather you stay at home, and return to your youthful, bright appearance. Take care of your health. Besides, the dinner of Wednesday [a dinner of eighty-nine officials of Carnegie Steel at the Schenley Hotel, Pittsburgh, on 9 January] may require you to lie up for a day or two. Think you could write a note which would meet the purpose and then announce soon, if you met him he might try to stop us and refusal would be trying, better write tonight. Am down town today on matters. Saw the important man today. Everything looks well. Anything new? Andrew Carnegie."[24] Working from the very different recollections of Judge E. H. Gary, then head of Federal Steel, his biographer, Ida Tarbell, provided a different listing and chronology of events.[25] Whatever the true timescale, Morgan accepted Carnegie's price. After that, with some help from Schwab, Gary decided on the properties they were interested in and began negotiations with their controllers.

   Federal and Carnegie Steel were the core operations of the combination. It is clear that some of the other firms or plants were brought in only as parts of a corporate package. Years later during the dissolution suit, Daniel Reid recalled that those interested in National Steel, American Tin Plate, American Sheet Steel, and American Steel Hoop would not entertain the idea of selling any one of them without selling them all. Negotiations over these properties took several days, Judge Moore "going back and forth between his office and Mr. Morgan's office until 7 or 8 o'clock at night."[26] The result of this sort of procedure was that properties that were not very attractive were willy-nilly included along with prime businesses. The new corporation would begin business not so much as an amalgamation but as an agglomeration.

   It is important to put the work of these individual actors into perspective. The "heroic" view of business history is less popular today than in a past that tended to either idolize or execrate its tycoons. The most outstanding entrepreneur, "captain of industry," or "master of capital" can only be successful if conditions are propitious. He or she—and at that time there was effectively only the former—must await opportunity. Many, however, may not see or only partially recognize the potentialities of a situation or, having realized it, may yet bungle in carrying through the plans. In the course of a concise account of the new combination, Peter Temin recognized that it solved the crisis created by vertical and horizontal integration in the 1890s, made easier the running of pools, and yielded great promotional profits. Still, he rather frowned on "the conventional story" with "all the appropriate drama."[27] Long before, Frick's biographer, George Harvey, had made what seemed a reasonable conclusion: "It was the stern reality of the actual situation, in vastly greater measure than either the eager pleading of Judge Gary recounted by his biographer [Tarbell] or the eloquent portrayal of possibilities by Mr. Schwab at a celebrated dinner party, that finally impelled Mr. Morgan to essay the greatest undertaking of his career."[28] Perhaps, after all, this much of the heroic may be allowed, that though the genesis of the new group was in part due to the stark situation and prospects of the time, largely a response to a careful assessment of dangers and possible rewards by many business leaders, it was successfully carried through only as the result of a grasp of even greater possibilities by a mere handful of men.

   In the early part of February 1901, Elbert H. Gary informed the press that moves were underway "for the acquisition of the properties of some of the largest iron and steel companies of this country." On Tuesday, 26 February, Carnegie signed a formal agreement to sell his company. A few days later, notice of intention to form what was now for the first time referred to as the

United States Steel Corporation was conveyed to the shareholders of the companies proposed to be incorporated. On 7 March Carnegie cabled Schwab, "saw Morgan today all goes well." He followed this up with a letter on the same day, ending with the postscript that he "called on Mr. Morgan this AM to get leave of absence from my Boss. He gave it."[29] What for Carnegie marked the end of one of the most successful of all industrial careers signaled the beginning of a new industrial era. On Monday, 1 April 1901, the United States Steel Corporation began business.

Early Years of Industry Leadership, 1901–1904

When asked why he had helped in the formation of US Steel, John Warne Gates replied, "To convert a lot of doubtful assets into cash."[1] He had been hurt by J. P. Morgan's refusal to include him on the Board of Directors. Even if this showed in the frank cynicism of his statement, there was at least an element of truth in Gates's words, for much of the nominal capital did not represent tangible assets. Like most combinations of the time, US Steel was grossly overcapitalized. It is true that, shortly after trading began, Schwab set the value of its properties, minerals, and cash at $1.466 billion, but a decade later the Bureau of Corporations estimated its real property at the start of operations at $682 million; by this assessment, $721 million of the capitalization was "water." In fact, it was extremely difficult to make a well-reasoned evaluation of the immense agglomeration of physical plant and even more so of the mineral resources. In summer 1902 Schwab valued the Corporation's holdings of iron ore at an extraordinary $700 million, compared with only $348 million for production plant and real estate. (That year US Steel shipped 16.1 million tons of Lake Superior ore, 58.3 percent of the total.) The "principles" guiding Schwab's ore valuation were, firstly, that the holdings could not be duplicated at any price and, secondly, that US Steel would have to pay that much to obtain this amount of ore. His estimate for their coal and coke fields was $100 million; at the same time, Thomas Lynch, president of H. C. Frick Coke Company and who knew more about this section of the business than Schwab, made what he regarded as a conservative estimate of the value of their Connellsville coking coal alone: $157 million.[2]

The creation of US Steel more or less coincided with major reconstruction at other older operations and with the establishment of some new firms that

TABLE 2.1
**Valuation of the United States Steel Corporation at its inception according to
the Bureau of Corporations, 1911 (millions)**

Manufacturing properties	$250	Transport properties	$91.5
Ore properties	$100	All other	$160.5
Coal and coke properties	$80	TOTAL	$682

Source: U.S. Commissioner of Corporations, *Report on the Steel Industry,* vol. 1, *Organization, Investment, Profits, and Position of The United States Steel Corporation* (Washington, D.C.: GPO, 1911), vii, 25.

were eventually important competitors. Lackawanna Steel was building its new Lake Erie shore plant. Pennsylvania Steel and its offshoot Maryland Steel of Sparrows Point were reorganizing. It was recognized that if these last two companies had had sufficient ore and coking-coal resources of their own, they could have saved, as compared with open market purchases, $2.9 million in 1901 and $1.37 million the following year. In a circular to stockholders in

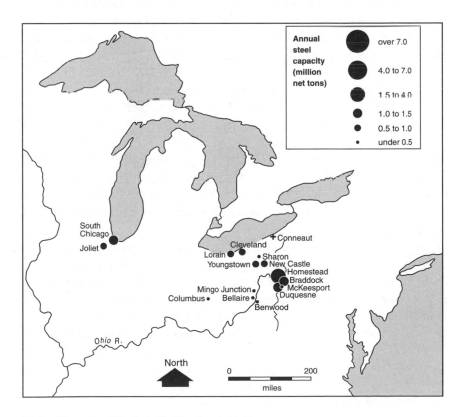

US Steel Integrated Works in the Northeast, 1901

January 1901, E. C. Felton, the Pennsylvania president, noted that the two companies' capital was too small; more was needed to keep up with the standards of the time in steel and shipbuilding and to acquire adequate mineral resources, otherwise they could not remain profitable. In April 1901 the two operations were reorganized as the Pennsylvania Steel Company of New Jersey.[3] The steel-making company Jones and Laughlin, having refused a $30 million offer for inclusion in US Steel, set out to reconstruct, expand, and keep pace as a successful competitor. Youngstown Sheet and Tube began work on its first plant east of Youngstown in the same month as US Steel was incorporated. In February the American Rolling Mill Company tapped the first open-hearth steel from its Middletown works, and in the course of the year an as yet diminutive Inland Steel bought a site at Indiana Harbor for a new open-hearth plant. Pittsburgh Steel Company was also organized that year, operated for a time at Glassport, and by 1903 was developing a new site at Monessen.

In a few cases, new or recently established companies were soon absorbed into US Steel. Outstanding examples were two newcomers on the Monongahela River well above the Pittsburgh-area mill towns that Carnegie Steel had dominated. The Union Steel Company was the brainchild of William H. Donner, who in 1898 had sold his interest in two tinplate works, one in Indiana and the other at Monessen, to American Tin Plate. By May 1900 Union Steel was in existence with Donner as president, Andrew Mellon as vice president, and Richard Mellon as treasurer; having left Carnegie Steel, Henry Clay Frick was also associated with this venture. Its capital was then $1 million, and its plant was built across river from the little settlement of Webster at the new town of Donora. Within a year operations were being greatly extended. Though starting with purchases of steel from the Carnegie Steel division of US Steel, Donner's company now planned an open-hearth plant, supported by two 600-ton-per-day blast furnaces, to meet all of its steel needs. Union bought its way into the Mesabi iron ore range and gained control over the Republic Coke Company, which owned 3,200 acres of coking coal in the Lower Connellsville district. In autumn 1902 Union acquired Sharon Steel, a company similar in size and product range, creating a new combination with $50 million capital. It was intimated that it would build a 2,000-ton-per-day rail mill at one of its works, and options were reportedly secured on land for docks at Elk Harbor, east of Conneaut, with a rail link from there to Donora. In mid-December 1902 US Steel paid $45 million to absorb Union Steel; its spiraling expansion plans thus reaping their reward. As one report put it, Donner and company "have been given large newspaper space in the past two weeks. . . . Their purpose now becomes apparent. . . . It is hardly necessary to say that the rail mill talked of for Donora will not be built."[4]

The other important acquisition by US Steel in this early period was at Clairton, a few miles north of Donora. In some respects the Clairton works represented the third attempt within twenty years by Pittsburgh's special steel interests to provide themselves with a low-cost, integrated source of semifinished materials. Earlier attempts at Homestead and Duquesne had ended with the plants being cheaply acquired and thereafter successfully operated by the Carnegie interests. In 1901 thirteen special steel producers, among whom were the Park family, established the Crucible Steel Company of America. The Clairton works, which they erected, was equipped with blast furnaces, an open-hearth melting shop, and rolling mills. As with the earlier projects, trouble was soon encountered with expenses larger than expected coupled with a recession of trade in 1903. It was agreed that US Steel should take over the works for ten years, supplying Crucible with up to 200,000 tons of steel annually at below market prices. (In 1911, as the contract neared its end, Crucible set out to build its own fully integrated works on the Ohio River at Midland, Pennsylvania.)

One acquisition of the early US Steel years was of highly questionable value. The iron- and steel-making operations around Troy, New York, on the Upper Hudson River had played a distinguished place in the early history of bulk steel making in the United States during the third quarter of the nineteenth century. Now the area seemed better suited to become an industrial museum than remain an important player in contemporary steel production. The Troy Steel Products Company was in receivership, but one of its directors was Junius Morgan and another was Henry H. Rogers, a US Steel director. Strong opposition to a proposal that US Steel should buy Troy came from Schwab, who claimed that $1.5 million would have to be spent to get the plant into shape to operate; from Edmund Converse, who put the rehabilitation cost at $300,000 higher than Schwab; and from Elbert H. Gary. The executive committee rejected the plan.[5] However, later that year the board approved the purchase of Troy by American Steel and Wire from 1 January 1903 for $1.1 million. There was speculation that it might be a source of supply for the AS&W works at Worcester and Trenton, New Jersey, but in fact, Troy was never operated. Its steel plant was dismantled in 1907, and its blast furnaces stood idle until they too were dismantled in 1917. Gary stressed that conditions had changed since they first bought it; Schwab much later suggested the purchase had been made to satisfy Rogers.[6]

Collectively, the additions to US Steel capacity during its first three years were substantial. The purchases of Union, Clairton, the Shelby Steel Tube Company, and two small eastern operations at Trenton and Troy added 16.5 percent to its blast furnace capacity, 13.3 percent in steel, and 14.3 percent to

the rolling-mill tonnages with which it began in April 1901.[7] Though expanded by US Steel, neither Donora nor Clairton were ever to become among
its major operations. An additional motivation to take over Union Steel seems
to have been the wish to ensure that Henry Clay Frick's unquestioned talents
and commitment were devoted to US Steel rather than to a rival. In a number
of other instances, absorption by the Corporation forced out leading figures
whose abilities were transferred to building up a competing concern, usually
with the aid of the money received for the sale. Some of these new ventures by
displaced capitalists or top managers were small, in a few instances they were
of considerable importance, even if many years down the road. Donner
moved on after Donora was absorbed and some years later played an important part at Cambria Steel and later at Pennsylvania before building up Donner Steel, which in 1930 became part of the new Republic Steel Corporation.
Shelby Steel Tube Company had been formed early in 1900 as a fairly small
combination that included the main makers of seamless tubing. National
Tube wanted to enter the field. When Shelby was absorbed shortly after the in-
auguration of US Steel, its former president, W. E. Miller, moved to Elyria,
Ohio, and formed Elyria Iron and Steel.[8] When US Steel was formed, Wallace
H. Rowe was forty years old. He had been active in turn at the Braddock Wire
Company, Consolidated Steel and Wire, and American Steel and Wire. He
stayed on at the AS&W subsidiary of US Steel for a time but later resigned, disposed of his holdings, and became involved in the growth of the Pittsburgh
Steel Company, of which he became president. At the beginning of the century the small firm of Kirkpatrick and Company operated both a tinplate and
a sheet mill. Almost immediately after their plant was absorbed by US Steel, its
two principals, Henry E. Sheldon and Alfred Hicks, founded the Allegheny
Steel and Iron Company, which after four decades of growth and merger became the world's largest stainless and specialty steel maker, Allegheny-Lud-
lum.[9] There were more immediate cases of stimulation of new competition.
Disappointed at US Steel, yet enriched by profits made from selling American
Steel and Wire Company to the Corporation, John Wayne Gates for a short
time acquired control of Colorado Fuel and Iron's Pueblo operation before
turning his attention to the Tennessee Coal, Iron, and Railroad Company. In
summer 1902 the president of National Tube, Frank J. Hearne, resigned, reportedly on grounds of illness. A year later, on Frick's recommendation,
Hearne was appointed president of Colorado Fuel and Iron (CFI).

There was one even more important early resignation and transfer of capital and entrepreneurial skill from US Steel. It came as a result of a fundamental difference of thought and experience among the directors and chief
officers, based on their previous business careers. Those who had worked for

Carnegie found it difficult to work in reasonable amity with rival companies, rather than competing ruthlessly with them as in the past. Frick himself found it hard to make this adjustment. James Gayley later explained to the Stanley Committee how tough he had found it: "I was brought up in the school of keen competition. Judge Gary promulgated the idea that competition was unprofitable. He was opposed to the old method of going out into the market and slashing prices in order to get business. Personally, I was opposed at the outset to the views of Judge Gary. I had been trained under the old competitive conditions. I was finally won over to the new plan."[10] In the case of Schwab, there was no eventual reconciliation with Gary's views.

In addition to marketing philosophy, strong differences in style and personality separated the two principal officers of US Steel: its president, Charles M. Schwab, and the executive committee chairman, Elbert H. Gary. Schwab was used to the exercise of the decision-making powers of an executive industrial officer; Gary sought to control from the center a complex business corporation. Significantly, the former met daily with his "cabinet" of closely associated steel executives. Within three months of the formation of US Steel, Schwab brought before the executive committee his unhappiness with the by-law that vested powers in the Board of Directors, the executive committee, and in the interval between the meetings of the latter in its chairman. The usually staid minutes of the executive committee can scarcely disguise the tension of the meeting of 2 July: "The President informed the committee that he wanted to continue his own organization and wants every man to feel that there is a head; in other words, he wants to conduct this organization as similar organizations have been conducted in the past." Gary hoped that they would not get into personalities. He thought J. P. Morgan should be consulted and wanted to "avoid getting into a discussion before the Board of Directors that might lead us into trouble. . . . The Chairman dislikes to have anything done which will get us into trouble and particularly at this time." On a motion from Percival Roberts, Schwab's aim for supreme powers was defeated.[11] This was the first of a number of reverses that marked Schwab's decline in power and the rise to primacy of Gary.

Though now based in New York, as a former mill superintendent Schwab's natural sympathies were with the directors of the constituent companies and in the centers of steel production. The latter were not completely at the mercy of 71 Broadway, the Corporation's headquarters in New York City. As John Garraty once recorded, there were some positive advantages for mill supervisors in being in charge of just one unit among so many: "Plant executives faced with practical problems sometimes found it difficult or inconvenient to comply with orders from New York; when this was the case they often kept silent and did as they pleased."[12] In contrast, for a man of Gary's background in company law and business organization, 71 Broadway was the most appropriate powerhouse and center for decision making and control. This issue like others between the two men was complicated by a profound clash of personality, by differences in their ideas about behavior appropriate to a top executive, and the rather obscure "illnesses" with which Schwab disguised his disillusionment. Given their earlier association and the natural caution of a finance house, it is not surprising that in the end the vital support of J. P. Morgan went to Gary. In summer 1903 Schwab resigned the presidency. At the same time the executive committee was disbanded. Two years before, already frustrated by what he regarded as interference with his powers of controlling and directing the Corporation in a manner he was uniquely well endowed to undertake, Schwab had bought a controlling interest in the then rather small and highly specialized Bethlehem Steel Company. He was then said to regard the investment as nothing more than a "flier," but a year after he ceased to be US Steel president, Schwab also gave up his directorship and embarked on a long-term and very purposeful process of building Bethlehem into a major steel company.

Schwab was replaced as president by his long-time former Carnegie Steel associate, William Corey. After his own disagreements with Gary, Corey resigned early in 1911. Continuing discontent with Gary policy caused the gradual withdrawal of other men trained in very different business methods at Carnegie Steel. James Gayley, only fifty-three years old, resigned from his first vice presidency in 1908. Corey was closely followed by the forty-five-year-old William Dickson. In 1915 Alva Dinkey, who had been Corey's successor both as general superintendent at Homestead and president of Carnegie Steel, also withdrew and joined Corey and Dickson in controlling the Midvale Steel and Ordnance Company.

At its incorporation US Steel comprised 213 different manufacturing plants: 100 of them in Pennsylvania, 51 in Ohio, 15 in Illinois, 12 in Indiana, and 12 in New York. It had forty-one mines and almost one thousand miles of railroad.[13] From the start it was expected to follow the practice of earlier combinations in closing those plants judged inefficient or badly located. Within a few weeks, Schwab had been to visit works in Ohio, and his tour was widely viewed as a precursor to sweeping, and for some places very painful, rationalization—in popular reports it was labeled a "dooming" tour. That summer there was a rumor, soon denied, that the Worcester operations of American Steel and Wire might be moved nearer the coal and iron regions.[14] When Schwab gave testimony to the Industrial Commission in May 1901, he assured its members that all US Steel works were in operation. Even so, some closures were inevitable. Records indicate that from 1901 through 1903, at least thirty plants were abandoned, one was partially abandoned, and two were merged with other operations. In recession conditions during the latter part of 1903, some of the least productive operations were demolished. The integrated Union Steel works in Chicago, idle when US Steel was formed, was dismantled. Other works abandoned were smaller and, by location or other circumstances, obvious candidates from the start. By 1903 American Sheet and Tin Plate had closed sixteen plants, American Steel and Wire and American Bridge each shut down four, National Tube three, and both American Steel Hoop and Shelby Steel two. (A different assessment is that AS&W alone dismantled seven works in 1901–1903.)[15]

Some of the conditions behind the closure program, and the dependence of the companies on approval from the center, may be appreciated from the consideration by the Board of the Carnegie Steel Company of one Ohio and three outlying Pennsylvanian plants it had inherited from American Steel Hoop Company. In spring 1905 Dinkey reported that the company was badly oversold for bars and had asked the US Steel president for money to build an eighteen-inch mill at their recently acquired Clairton works. If this was approved, they would be able to avoid reopening the mill at Warren, a plant in a very dilapidated state, affected by a strike, and whose best conversion cost would be $7.50 a ton as compared with $4.00 at a new Clairton mill. Costing $175,000 the new mill could be built in ten weeks, and its capacity would be double that of the Warren mill. Dinkey grew enthusiastic about the virtues of concentration: "All the arguments are in favor of building the mill; the finishing mill would be where it belongs, freights would be saved, one more isolated plant would be abandoned, and the organization thus made more compact." A few weeks later he told his colleagues that approval had been received from 71 Broadway for the removal of two mills from Girard to Duquesne, where it was expected that their output would double. Another mill would be transferred from Girard to Greenville, where it would operate as a "boom" plant—that is, working only when other mills were over stretched. Dinkey reported that Greenville "is on the line of the Bessemer road and is the only one of the isolated plants that I think we should keep up. By these changes, we shall abandon all the plants I imagine we want to abandon, except Duncansville. I think when that plant shuts down next time, it should be scrapped, as there does not seem to be any shift possible by which we could make reasonable costs." The Warren, Girard, and Duncansville mills were all written off at this time; Greenville was last worked in 1918.[16]

There is no doubt that some of the cost economies of consolidation that Schwab outlined in his speech in December 1900 were quickly realized, though others took much longer to accomplish. The February 1902 preliminary report to US Steel stockholders recognized the successes and the remaining work required toward savings. Its comments were very general "The management feels fully justified in stating that much that at the time of organization was hoped to be accomplished in the way of avoiding wasteful expenditure for unnecessary enlargement of plants by various prominent steel companies, and of establishing harmonious cooperation among them has been successfully achieved. The several companies have effected many economies which have been attended with most satisfactory results, and the outlook for further improvement in this direction is most gratifying." A "diversified management has been dispensed with as far as possible, and the several companies have endeavored to adopt similar methods as far as suited to their respective businesses." They had concentrated on the most favorable locations and tried to achieve the best costs of production and distribution. "Economies in manufacturing thus far have been quite remarkable, but the end is not nearly reached."[17] James Gayley, first vice president, stated in 1906, "Improving our own plants is the keynote of the United States Steel Corporation," and about the same time Gary claimed that in five years they had cut costs of manufacture by about 10 percent.[18] Meetings of plant and department managers made available more generally the knowledge and experience of the best practice. As Farrell told the Stanley Committee years later, a US Steel open hearth committee of twenty-four members met in New York every two or three months. A blast furnace committee met every month, and because of its work efficiency had increased so that now forty-seven Carnegie Steel blast furnaces produced more than fifty-one had done previously.[19] Far more important than these relatively small gains at Carnegie plants was the success of the technical committees in spreading the established excellence of that company's practice to other US Steel works, for as Corey disclosed to the same committee, there had been relatively small improvement at Carnegie Steel since 1901 because its practice was already so good. Paradoxically, this wider spread of high standards, combined with the reduction of cross hauling within US Steel, helped reduce the area within which the Pittsburgh plants were competitive. Savings could also be gained from organizational changes. In 1903 two important formerly separate subsidiaries were merged as the American Sheet and Tin Plate Company. The following year a director of US Steel claimed that this move alone cut annual operating costs by $250,000.[20]

Despite popular expectations that the mammoth new trust would crush its much smaller competitors, it soon became clear that US Steel would have to fight hard to hold its own. In May 1901 before the Industrial Commission, Schwab reckoned that in broad terms the Corporation controlled 65–75 percent of the national steel business. He forecast that in very prosperous times their percentage would be smaller than this and in less buoyant periods very much larger.[21] That year, far and away the best ever to date for the industry, US

Steel's proportion of raw steel production was 65.7 percent. Afterward, national output fell 2.8 percent in 1903 and by a further 4.7 percent in 1904. US Steel's share of this falling total declined from 65.2 percent in 1902 to 63.1 percent and then to 60.7 percent; after this in no year was it ever again over 60 percent. In short, Schwab's anticipations of better performances in times of difficulty, projected from years of experience at Carnegie Steel, were quickly disproved. In other respects too early indicators were disappointing. For instance, earnings in relation to capital or per ton of steel were less good than in the predecessor companies as a whole, and in the former respect these were particularly behind what had recently happened at Carnegie Steel. Those who directed the fortunes of US Steel were hard pressed to do all they could to hold the line; the Corporation was on the defensive, and its rivals were attacking. There had been an interesting sentence in the circular that J. P. Morgan had distributed on 2 March 1901 concerning the forthcoming amalgamation. After writing about 1900 net earnings and financial arrangements in the constituent companies, the circular went on to mention, "It is expected that by the consummation of the proposed arrangement, the necessity of large deductions heretofore made on account of expenditures for improvements will be avoided."[22] In fact, US Steel did make large new investments, but in an industry in which technical change and increase in production were both moving ahead rapidly, the hope of saving on outlays should not have been entertained even for a moment.

TABLE 2.2
The performance of US Steel, 1901–1904, compared with its predecessor companies in 1900

	Capital (millions)	Steel output (th. tons)	Earnings (millions)	Earnings per ton of crude steel	Earnings per $100 capital
1900					
Eight companies	$809	6,485	$88	$13.57	$10.88
(of which Carnegie)	($320)	(2,970)	($40)	($13.47)	($12.50)
United States Steel Corporation					
1901[a]	$1,403	9,917	$104	$10.49	$7.41
1902	n.a.	10,902	$90	$8.28	$6.44
1903	n.a.	10,275	$55	$5.39	$3.95
1904	n.a.	9,422	$30	$3.20	$2.15

Sources: C. M. Schwab testimony, U.S. Industrial Commission, *Reports,* vol. 13 (Washington, D.C.: GPO, 1901); US Steel annual reports and U.S. Commissioner of Corporations, *Report on the Steel Industry,* vol. 1, *Organization, Investment, Profits, and Position of The United States Steel Corporation* (Washington, D.C.: GPO, 1911).
[a] Bureau of Corporations calculations for whole year.

Judge Gary's "Umbrella"

The Advantages and Disadvantages of a Managed Industry

The steel industry, during the quarter-century beginning with the establishment of US Steel, had as its unquestioned leader the imposing figure of Elbert H. Gary. He epitomized not only the might of the preeminent operator in what was still the nation's leading manufacturing sector, but, as even those who disagreed with many of his policies accepted, he also presented an impressive public face for the company and industry, one of dignity, responsibility, and rectitude. As W. B. Dickson, who had previously worked for Carnegie Steel and was often in disagreement with Gary, recognized, in its early years, US Steel "was menaced by political action from the White House. . . . The Judge, by his conciliatory attitude, steered the Corporation through these troubled waters in a masterly fashion."[1] Gary had stressed that US Steel was not to be managed in order to manipulate the stock market. Under his guidance the steadier routines of a mature manufacturing economy replaced cutthroat competition. Stable, or relatively steady, prices prevailed over more erratic tendencies that had previously reflected the changing opportunities presented by the state of demand and the inability of understandings, pools, and other arrangements to restrain unfettered competition. For most products there was now a well-established geographical pricing system under which all producers knew the prices in each area or location of the country. Together, collaborative action and these stable, open price structures helped preserve existing centers of production yet at the same time provided shelter under which new producers could not only compete but also earn higher than average profits.

At first sight some of these outcomes were paradoxical. The Gary system did not positively discourage change, but it helped insulate the industry from the immediate effects of such changes on established producers. With the benefit of hindsight, this obviously created problems for the future of the industry's leading firm. Cooperation, steady prices, and their known areal pattern were essential parts of a paternalistic era, or of what with less respect was referred to in some quarters as "Judge Gary's umbrella." It was easy to forget that, though providing useful protection from the elements, umbrellas are of little use in real storms. Even during Gary's lifetime this was occasionally demonstrated; shortly after his death the whole vast system was swept away by weather of unprecedented severity in the greatest crisis in the history of American capitalism. But that remained far ahead, and for over twenty-five years of US Steel's first century, Elbert H. Gary led, spoke for, and dominated both the Corporation and the industry.

In the early 1920s the economist John R. Commons reckoned that possession of 40–60 percent of the business in a particular industry was sufficient to induce rivals to adopt that company's pricing policy.[2] Such a share might be the outcome of competition, but in the case of US Steel it seems to have resulted in part at least from establishing a target. Gary accepted William Jennings Bryan's 1906 dictum that control of more than 50 percent of an industry's capacity was a rough guide to the possibility that a company might be infringing on at least the spirit of antitrust controls.[3] Starting with well in excess of that share in most of its main activities, US Steel was under suspicion from the start. It was the natural leader in prices, and the levels that ruled during its early years were steadier but generally higher than in the last decade of relatively unfettered competition. In addition to an understandable wish for greater profits, there was another important reason for this. According to estimates made a decade later by the Bureau of Corporations, "water" accounted for about half the original nominal capital of the Corporation, or roughly as much as the whole common stock. Clearly US Steel had every incentive to set prices at a level that would enable net earnings to cover this excess. What was desirable for it could be beneficial to other companies, though not to consumers. For many rivals this was one factor encouraging rapid growth.

At the start, not only the rest of the industry but also some of Gary's fellow directors had to be persuaded of the advantages of price stability as well as his other new ways of conducting business. Schwab, Frick, Corey, and others at a slightly lower level, whose business methods had been learned at Carnegie Steel but who now occupied key or important positions in the new concern, needed time to adjust. As president, Schwab found himself generally at odds with his chairman. Recollecting the situation long afterward, he was perhaps unduly harsh in his judgment: "Judge Gary, who had no real knowledge of the steel business, forever opposed me on some of the methods and principles that I had seen worked out at Carnegie Steel—methods that had made the Carnegie Company the most successful in the world."[4] The preliminary report of US Steel, submitted on 17 February 1902, ended with a section entitled "Policy as to Prices." It was a statement that would not have been issued by its predecessor firms and was presented in the tones used by a highly responsible, magisterial, industrial leader. "The demand for the products of the several companies has been so great that prices could easily have been advanced. Indeed, higher prices have been voluntarily offered by consumers who were anxious for immediate execution of orders, but the companies have firmly maintained the position of not advancing prices, believing that the existing prices were sufficient to yield a fair return on capital and maintain the properties in satisfactory physical condition, and that the many collateral advantages to be gained in the long run by refusing to advance prices would be of substantial and lasting value, not only to the companies, but also to the general business interests of the country."[5] Not content to gradually convert colleagues or his stockholders, Gary publicized his wish for stability, though the implicit assumption that cost conditions were the same for US Steel and its competitors had little justification. Not only did he automatically assume the role of price leader but also the accompanying responsibility to police the system. Late in 1902, asked whether US Steel contemplated making reductions, he solemnly, indeed rather pompously, stated: "The policy of our corporation is to maintain reasonable and steady prices. We use our influence against an increase in prices and also against a decrease in prices if those which obtain are fair and reasonable. If competitors in any locality and concerning any commodity cut prices in order to secure an advantage, our people immediately

TABLE 3.1
Selected average prices for steel products, 1898–1901 (price per gross ton)

Product	1898	1899	1900		1901
			March	October	May
Billets	$15.18	$29.81	$33.00	$16.50	$24.00
Plates	$24.23	$49.36	$45.25	$24.19	$35.39
Rails	$17.63	$28.13	$35.00	$26.00	$28.00
Structurals	$26.25	$40.49	$50.40	$33.60	$35.84
Bars	$21.32	$41.36	$50.40	$24.42	$31.58
Black sheets	$42.28	$60.24	$67.20	$62.72	$71.68

Source: E. Jones, *The Trust Problem in the United States* (New York: Macmillan, 1922), 197, 203.

inquire carefully into all the facts. If there is justification for a decrease in prices our people promptly meet the conditions and all our customers are treated alike. We desire and seek the friendship of our competitors, but look first to the protection of our customers."[6]

For most products, price oscillations through the early years of the century were smaller than in the 1890s. Plate prices had varied quite sharply over the short period preceding the formation of the Corporation. In February 1899 its price in Pittsburgh was $29.12 per gross ton. Increased demand caused the plate pool to raise the price to as much as $67.20 by August. This high level encouraged entry to the trade by several new producers. In 1900 consumption fell away and the fight for orders became keener. By July the price was $23.52, a level at which the business was sufficiently unprofitable that several mills withdrew from production. Revived demand in 1901 and 1902, particularly from shipyards and steel-car manufacturers, enabled US Steel to increase the price to $35.84. After that, variations were kept within a narrower range. With other products too the changes over time were less drastic than before, but average prices were higher.[7]

For no product was the contrast more remarkable than with rails. During the ten years after 1891, the price of Bessemer rails at works in Pennsylvania had ranged from a high of $32.29 a gross ton to a low of $17.62; each year was marked by considerable, often dramatic, variations from the last. This volatility continued to the eve of the creation of US Steel. In July 1900 the price was $35 a gross ton; by year's end $26. Rail prices were considered at the first meeting of the US Steel Executive Committee on 9 April 1900. The general opinion of those present was that there should be an increase, though one director, Percival Roberts, expressed doubts about a rise "at this early stage of the organization of the company." The minutes recorded that "On this question the President [Schwab] believes it to be good policy to advance the price now and that, in his judgement, it would attract but little attention."[8] Next

TABLE 3.2
Steel billet prices, 1891–1900 and 1901–1910 (price per ton)

	1891–1900	*1901–1910*
Average annual price	$20.98	$26.17
Highest annual price	$31.12	$30.57
Lowest annual price	$15.08	$22.10

Source: P. Temin, *Iron and Steel in Nineteenth Century America: An Economic Enquiry* (Cambridge, Mass.: MIT Press, 1964), 284–85.

month there was a rise to $28.00. In the previous sixteen months the price for rails had changed five times; this new figure would remain unchanged for fifteen years. In the middle of the first decade of the new century, production of rails from open-hearth steel began to increase rapidly. For this new product $2.00 was added to the Bessemer price. Unquestionably, these were managed prices that in no way closely reflected changes in costs in time, space, or at varying levels of production. At the end of the decade, an enquiry by the Bureau of Corporations exposed something of the wide range of production situations that lay behind this simple, stable price structure. In the years 1902–1906, existing plant to produce Bessemer rails represented an investment ranging from $55 to $80 per ton of product. The average cost of production—excluding transfer margins from one department to another of the same works—in the operations studied was $18.80 a ton, giving at the ruling price a profit per ton of $9.20, or roughly 11–17 percent on the capital involved.[9] Despite this, the steady and fairly high price failed to encourage many new entrants to the trade, for which there was a fairly simple explanation. Although the tonnage of rails produced reached its highest level ever in 1906, this was no longer seen as a sector that promised further growth in demand on a scale that might justify the large outlay of capital—at the new Gary works, the rail mill represented an investment of about $25 million. In testimony to the Stanley Committee on 2 June 1911, Gary himself made clear that the rail price was "managed." During the course of a few minutes, he said both, "The cost of rails has increased, I should say, at least $3 a ton during the last ten years," and "The prices of rails have not changed for about ten years."[10] In the simple words of Frank Taussig, "No sensible person will believe that the price could be held at this precise figure, without a variation from month to month or year to year, except through the abrogation of competitive bidding."[11]

TABLE 3.3
Production of finished rolled iron and steel, 1905–1925
(thousand gross tons)

Year	Rails	Structurals	Plates and sheets	Wire rods	All products
1905	3,376	1,660	3,532	1,809	16,840
1910	3,636	2,267	4,955	2,242	21,621
1915	2,204	2,437	6,078	3,096	24,393
1920	2,604	3,307	9,338	3,137	32,348
1925	2,785	3,604	9,808	2,845	33,387

Sources: AISI annual statistical reports.

For some products, fairly high and, above all, stable prices combined with prospects of considerable increases in consumption did call forth large extensions in capacity. Many of these investments were made by the "independents." Even for heavy products, so long as demand was growing—and this was expected to continue—the new price regime encouraged increases of capacity and production. This was the case in the production of heavy structurals for building construction. The price was fixed by US Steel soon after its formation at $1.60 per 100 pounds, and despite market fluctuations this price was long maintained. With flat-rolled steels, particularly thinner grades such as sheet and tinplate, market growth was rapid, and even more of the increased production was by companies other than US Steel. These changes would eventually result in the latter suffering from a disproportionately large share of slow growth products.

It was in times of trade recession that Gary's policy of maintaining prices could be particularly frustrating for those of his colleagues who had been taught to go after business even if this meant cutting prices. The first crisis built up over 1903–1904 as the industry's levels of activity fell away from the high of 1902. This led Frick to sell most of his stock and Morgan to call on him for advice.[12] Some months later the new president, William Corey, received a letter from his first vice president, William Dickson, previously like him a senior official at Carnegie Steel. Dickson recognized that, in terms of improved efficiency and cost reduction, they had accomplished even more than expected three years before, but their price policy meant that some of their constituent companies were "less [well] regarded by their competitors and by large buyers than before the organization of the Corporation." As a result of holding prices steady, they were operating at not much more than 50 percent of capacity, and "as an inevitable result, costs have increased, works standing idle have deteriorated more than if they had been in operation; the men are disheartened and a certain amount of apathy exists." He wanted a return to "open competition." "We are even now looked upon by vigorous competitors as an 'easy mark.' If the Corporation can only make $65,000,000, or less, it is better by all odds to make this profit on a full output at competitive prices than by half output at artificial prices. The result would be a more efficient organization, lowest possible costs, and the prevention of the building up of additional competition."[13] In fall 1907, trade began to fall away much more sharply, and in 1908 national output of rolled iron and steel was down by eight million tons, amounting to 59.5 percent of the 1907 tonnage. Inviting competitors to confer with them, US Steel led a campaign to prevent price-cutting. Gary called the principal officers of the subsidiary companies to a meeting in his office at 71 Broadway. He opened it with an address in which he emphasized the desirability of maintaining the price structure despite the diminishing demand, and he then called on each officer for his views. As one who was present recalled, "Being mere human beings with a natural instinct for self-preservation, the resultings were strikingly in accordance with his own." Dickson, who did dare to disagree, again recommended a very different reaction—that they should drastically cut prices so as to keep the mills at work and instruct their salesmen to sell for immediate delivery. Having conclusively won his case against this "Carnegie" policy, as soon as the meeting was over Gary sent out telegrams that resulted in the closure of eight steel plants in Pennsylvania, Ohio, and West Virginia. They remained inactive for an average of almost thirteen months. The decline in US Steel shipments in 1908 was only marginally worse than the national average, at 59.2 percent of the previous year. But its utilization rate, which had reached over 100 percent of nominal capacity in 1906, was down to 88.6 percent in 1907 and only 50.3 percent next year. By early 1909 price-cutting was so common as to "practically isolate our companies in their endeavor to maintain their official prices, resulting not only in the loss of business but of their prestige in the trade and the demoralization of their selling organisations."[14]

Whereas Gary held out strongly against price-cutting, the rivals of US Steel were not so firmly wedded to this policy. Outwardly, a new era of accord had arrived with the leadership of US Steel; as Herbert Casson put it, "Big and little, all the iron and steel companies seem to live as happily together as a basketful of kittens."[15] The appearance of universal harmony was deceptive. Whether in the pricing of finished steel or the associated geographical pricing system, it was not always easy to keep a united front. At times of recession, the arts and mechanisms of persuasion might need to be deployed to induce rival companies to maintain prices when they were strongly tempted to cut, or "shade," them in order to lower their costs by "running full," thus covering the heavy overheads on highly capitalized plant. To administer a personal appeal for industry-wide conformity in these dangerous conditions, Gary instituted the series of dinners that bore his name. On these combined social and business occasions, he would exhort the chairmen, presidents, or major owners of other steel companies to maintain rank. One later commentator reckoned that Gary addressed them like a revivalist preacher, but it seems that the advocacy of quiet reasonableness would be a more appropriate description. The veteran Valleys ironmaster Joseph G. Butler, who once eulogized Gary as "one of the great men of the present century," regarded these occasions as a beneficial part of a new order replacing the old times in which "the manufacture of steel was a merciless game."[16]

Although at the time he too praised Gary, Schwab was later and in private conversation much more skeptical. The apparent amity and unanimity of these occasions were, he believed, followed immediately by price-cutting and therefore "did absolutely nothing for the steel business except to give the industry a bad name."[17] This may be the harsh judgment of a disappointed rival—the two men being temperamentally so incompatible—but the evidence reveals that there was much less concord and brotherhood than was claimed by Gary. On a number of occasions when depression conditions arrived pursuit of overt self-interest took over. For instance, in autumn 1907, from the beginning of October to the close of the year, the number of active blast furnaces fell from 329 to 139. Early in November, US Steel was involved in the "rescue" of the Tennessee Coal, Iron, and Railroad Company (TCI), adding that capacity to its existing facilities. By the middle of that month, the situation appears to have been monitored closely, for the report that Corey then submitted to the US Steel Finance Committee included "percentages of orders recently booked by the subsidiary companies and other companies in similar lines."[18] Two weeks after he and Frick had cleared the TCI takeover with Pres. Theodore Roosevelt, Gary called top executives and directors from all the main steel companies to the Waldorf Astoria Hotel for the first "Gary Dinner." Naturally, he delivered the main address. In the course of the evening, a committee of industry leaders was appointed to consider prices. Ten weeks later, at a second dinner, this committee reported, advising firms not to cut them. Exceptionally, on this occasion J. P. Morgan spoke, stressing that one of his motivations for the formation of US Steel had been to obtain stability throughout the industry. When they gathered again late in May 1908, Gary congratulated fellow industry leaders for their part in preserving this stability.[19] In fact, it seems to have been a near thing, with US Steel itself on the brink of price-cutting in order to preserve its position: on the morning of the day of the third dinner after a full discussion at a special meeting of the finance committee, "it was voted to be the sense of the Committee that we should recommend to subsidiary companies *No reductions in the prices of steel to be made at the present time.*"[20] In 1908, production of rolled products was lower than in any year since 1900, but billet prices fell by only 10 percent. Rail output declined by over 47 percent, but price quotations remained unchanged at $28.00 a ton.

Although 1909 overall was a good year for steel, during the early months fears remained that the administered price system might break down. Industry leaders had dined together again early in December, and the fifth Gary Dinner was held on 18 February 1909. By this time the independents were beginning to desert price maintenance. Only ten days before, Pres. E. W. Pargny of American Sheet and Tin Plate had informed Henry Bope of

Carnegie Steel that his company would have to reduce its purchase of sheet and tinplate bars, for prices were never as deflated as at present and "there is no question that the independents are not maintaining them."[21] On the very day of the February dinner, Gary cabled his fellow senior director, Henry Clay Frick, then in Paris: "Privately yourself; Careful inquiry demonstrates that all competitors materially cutting prices much to our detriment. Our presidents all here and the only way of protecting our customers favor thoroughly meeting competitive prices. After long deliberation our Finance Committee concur. Have you any suggestion to make? Immediate action seems necessary."[22] In May, Gary sent a memo to one of his vice presidents that showed how aware they were of the unreliability of their rivals, though the means whereby such commercially sensitive information had been obtained was not revealed: "The Lackawanna Steel Company sold steel for the 3,000 cars lately ordered by the New York Central at $1.07, although the price named in the contract is $1.20." Generally, Corporation threats once more brought the price-cutters to heel.[23] The sixth dinner was held at the Waldorf Hotel in mid-October 1909. Its purpose was, as Butler put it, "a purely social affair with no business . . . when every guest was overflowing with gratitude and appreciation of the services rendered by Judge Gary." It would have been much more interesting to hear the private opinions of the more than one hundred men from the independent companies and practically all the directors and senior officers of US Steel who were present.[24] Early in 1912, in its answer to the government dissolution petition, US Steel denied that at any of the Gary Dinners "was there any attempt to reach any agreement or understanding with respect either to output or prices."[25]

Behind the facade of goodwill during these dinners and the chairman's general wish for industry-wide accord, there lurked a strong propensity to wield the big stick if other companies did not conform. On 27 January 1910 there was a lunchroom discussion in the US Steel head offices about cooperation versus competition that was so striking that one of those present made a note of it the next day. In the course of the impromptu meeting, Gary stated that in the event of a steel company refusing to cooperate, those who were willing to do so should get together and take steps to discipline the outsider. When challenged as to whether this would be a conspiracy, he claimed that such an action, if necessary, could be defended before the Federal authorities. In late November that year, there was an even more extreme example of Gary's aggression. Leading US Steel executives met in his office to consider their attitude to a conference to be held at the Railroad Club that afternoon. Gary mentioned that some small sheet firms were refusing to cooperate and cutting prices. There were "certain mills not entitled to live," and those cooperating should get together to "train their guns" on these mills. When once again challenged on the legality of such action, he claimed to have assurance from President Taft and from the U.S. attorney general, George W. Wickersham, that it would be proper and legal.[26]

While maintaining steady prices, US Steel possessed advantages over its rivals in costs of production. In 1911–1912 this was made clear in relation to ore freights from Lake Erie to Pittsburgh. About half of the iron ore came over the Corporation's Bessemer and Lake Erie Railroad, and of the rest, about half by other roads to US Steel iron plants and half to other operators in the area. In 1912 the Pittsburgh Steel Company of Monessen complained to the Interstate Commerce Commission about discriminatory charging, alleging that "the only motive for this violation of law is to benefit at the expense of all other consumers of iron ore, one particular interest—the United States Steel Corporation." The freight charge was 96 cents a ton, but US Steel—and Jones and Laughlin (J&L)—received a rebate of 26 cents for the final stages of the ore delivery, which took place over their own local lines. Although he explained he was not an expert on this aspect of things, Julian Kennedy reckoned a fair rate for ore carriage on the Bessemer and Lake Erie line would be about 40 cents a ton; the traffic manager of Pittsburgh Steel claimed the cost on this specially designed and graded line was only 28.25 cents. Even on Kennedy's figure, using ore of 50 percent Fe content, local furnaces of US Steel would have a cost advantage on the ore carriage account alone of some $1.12 a ton over other local producers who received no concession.[27] At this time Kennedy reckoned that the Corporation could make and sell steel for $1 a ton less than its rivals and that both Pennsylvania Steel and Cambria Steel were dependent on support from the Pennsylvania Railroad to survive.[28] Reviewing 1911 steel profits, the *Wall Street Journal* pointed out that for US Steel, they were close to $11 a ton, whereas for independents they averaged $4 and even for the most

TABLE 3.4
US Steel production of certain products as proportion of U.S. total

Year	Rails (%)	Structurals (%)	Plates and sheets (%)
1901	59.9	62.2	64.6
1910	58.9	51.3	48.0
1920	45.1	27.9[a]	plate 35.8
			sheet 33.6

Source: U.S. Commissioner of Corporations, *Report on the Steel Industry*, vol. 1, *Organization, Investment, Profits, and Position of The United States Steel Corporation* (Washington, D.C.: GPO, 1911); and *Iron Age*, 19 February 1925, 540–46.

[a] Of this, only heavy structural shapes apply for US Steel in proportion to all structural shapes for U.S. total.

profitable of them $7 a ton. In other words, in unrestrained competition, US Steel could crush all its rivals. However, such a policy would have been wholly unacceptable to the public—the Corporation was under investigation and during 1911 the Supreme Court already had ordered the dissolution of Standard Oil of New Jersey. Moreover, though their costs may have been higher than those of US Steel, the independents benefited from the cooperation that gave them, like the larger company, stable and fairly high prices. Some gained much from the general prevalence of the single basing point, or "Pittsburgh Plus," system of pricing, which US Steel brought to perfection.

In addition to initiating and winning general acceptance for price stability, US Steel also fostered a system of spatial variations in selling prices, which permitted older centers to compete at a distance or, and much more completely, the adoption of a simple, industry-wide geographical pricing system that had the same effect. An example of the first was the billet pool arranged in July 1903 between leading producers US Steel, Jones and Laughlin, Wheeling, Cambria, Pennsylvania, Lackawanna, and Maryland. The price would be $27.00 in Pittsburgh, $28.00 in Chicago, $28.75 in New York, $39.25 in San Francisco, and $40.25 in Pueblo (Colorado); this agreement failed after a short time.[29] The fully fledged version of the geographical pricing system was "Pittsburgh Plus," which had a major impact on the location of the steel industry and inevitably affected the distribution of the steel-consuming sectors of manufacturing throughout the country.

A geographical pricing system reflects the pattern of production insofar as the leading district either dominates output or/and has capacity that is surplus to the needs of its local consuming industries, thus sales by its mills at a distance constitute the marginal supply and thereby have a disproportionate influence on prices. In the middle decades of the nineteenth century, Philadelphia, as focal point for eastern iron and steel production, had been the "basing point" for pricing much of the American output. By the last quarter of the century, as areas west of the Appalachians came to dominate, Pittsburgh took over this role. It was a dubious achievement of US Steel in its first quarter-century to preserve this nineteenth-century pattern, to formalize it, and thereby to prevent further "natural" processes of change in pricing. One outcome was that Pittsburgh mills could still compete at a distance; another was that the system provided important premiums to producers situated well away from this basing point. The latter condition encouraged more expansion at centers distant from Pittsburgh, particularly since some well beyond Pennsylvania had also become lower-cost producers. In fostering both continuity and certain changes, US Steel pricing policies helped further the already entrenched tendency to overcapacity.

"Pittsburgh Plus" pricing meant that consumers paid the mill price at Pittsburgh plus the cost of transport from there to the destination of the steel. They paid the same amount for the product whether they used Pittsburgh steel or material bought from much nearer mills. Rails were a major exception to this system, railroads collecting their own supplies or those destined for other companies at f.o.b. (free on board) mill prices. In other products, existing local pricing arrangements were sometimes given up and replaced by the new system. Until 1901, light sheets were sold on a f.o.b. mill basis. In the early years of US Steel, its American Sheet and Tin Plate Company announced it would now sell on a Pittsburgh base, though none of its mills producing sheet or tinplate were actually in that city.[30] The system became so formalized and fully identified with US Steel that the latter eventually went so far as to supply its rivals with trade literature listing price and transport cost data so as to make their calculations of quotations easier as well as more conformable to the overall industry scheme. The most obvious victims of this discriminatory structure were consumers, who had to pay more for steel with increasing length of carriage from Pittsburgh; conversely, producers distant from Pittsburgh often received a price well in excess of their combined costs of production and delivery to the buyer. Under normal trading conditions this was eminently satisfactory both for independent steel firms and for US Steel plants far from Pittsburgh. They were content to receive "phantom freight," the name given charges for carriage not actually incurred, for it increased their profits.

Doubts about price maintenance and the generally predominant position of US Steel were prime considerations in a long-term move for the Corporation's dissolution, which formed the backdrop to its operations over many of the Gary years. In 1906 President Roosevelt ordered an investigation by the U.S. Bureau of Corporations, which he had established three years before. At times over the next four years, as many as thirty auditors were employed by US Steel to extract statistics for the use of the Bureau. The final report of the investigation was submitted to President Taft in 1911. In May that year the House of Representatives authorized the appointment of a special committee of nine members, chaired by Rep. A. O. Stanley of Kentucky. After taking testimony from various witnesses, the so-called Stanley Committee adjourned on 13 April 1912 and reported its findings to Congress on 2 August, suggesting legislation to control combinations. Already, on 26 October 1911, the Justice Department had filed suit in the District Court of New Jersey requesting that US Steel and eight subsidiaries be held to be unlawful monopolies and should therefore be dissolved. Testimony began in early May 1912, some three weeks after the Stanley Committee's proceedings ended. After considering the evidence of 402 witnesses recorded in 12,151 pages of testimony and 4,000 more pages of exhibits, the Circuit Court of Appeal handed down a decision in favor of US Steel in 1915, when the four judges ruled the Corporation had not been founded as and was not at that time a monopoly. Government attorneys appealed the ruling, and not until May 1920, after protracted consideration and further delays due to World War I, did the United States Supreme Court dismiss the government suit by the vote of three justices to one.

Both the Stanley and the dissolution hearings provide a wealth of insight into the antecedents, foundations, and early business operations of US Steel. To the Stanley Committee, Gary made a statement about prices and the Gary Dinners that epitomized his character, policy, and aspirations, hinting also at what was soon to become more obvious, their limitations.

> Prices were not attempted to be fixed, were not fixed, could not be fixed, and there was no possible way of fixing them or maintaining them unless you have some way of having them fixed under Government control, or you are allowed to do it by positive agreements. It has never been possible. It never could be possible. We have never succeeded in doing so. But we have, by this friendly intercourse, prevented demoralization—sudden, wild, extreme fluctuations—destructive competition that would drive large numbers of them entirely out of business, and that would be ruinous to the customers of the steel people who had large stocks of goods on hand from time to time, and which would spread to other lines of industry. We have made no secret about it, and the public has known exactly what we have done: and if the Department of Justice, for instance, or the President, or Congress, should say, "this is not the wise thing to do or the right thing to do," you may be certain it would not be continued for one moment.[31]

Well before World War I, it was clear that, despite its low costs, high profits, and the fact that it still far exceeded all other companies, US Steel was growing less rapidly than a number of its main rivals as indicated by capacity and even by trends in earnings. Although over the twelve years since its formation it had increased its share of national pig iron output, its proportion of steel production had fallen from 65.7 percent to 53.2 percent and of finished rolled products from 50.1 percent to 47.8 percent. In 1901 American Tin Plate had contained about 75 percent of U.S. tinplate capacity, but by 1911–1912 the amount controlled by American Steel and Tin Plate, its successor company, was about 60 percent.[32] Between 1902 and 1913 US Steel shipments of steel products increased from 8.9 to 13.4 million tons. This far outpaced any other company in absolute terms, but relative growth was more rapid elsewhere. J&L turned out 486,000 tons of finished products in 1901 and 1.49 million tons in 1911. Between 1901 and 1913 Cambria's output of finished products went from 467,000 to 1.19 million tons, Bethlehem's from 18,000 to 700,000 tons, and Youngstown Sheet and Tube, where production only began in 1902, was shipping 465,000 tons by 1913.[33] Over the few years before the First World War began in Europe, it was striking how US Steel seemed to be reaching a plateau in profitability, whereas other companies, still far behind, were making greater headway.

There was another factor in the success of some of the independent companies. As suggested above, the "Pittsburgh Plus" single-basing-point system, part of the pricing structure apparently perfected under Judge Gary, when placed under pressure was by no means watertight. Burns estimated that between 1904 and 1921 about 90 percent of all sales of plates, shapes, bars, wire products, and sheets were made on a "Pittsburgh Plus" basis.[34] Though this was a high percentage of the total, there was clearly no unanimity within the industry. Colorado Fuel and Iron, operating far from Pittsburgh, understandably chose not to follow the common practice. Even some much closer firms acted independently. The Cleveland iron-making firm of Corrigan-McKinney built their own steel works in Cleveland between 1913 and 1916. Its leaders scorned Judge Gary's "umbrella," never attended any of "his" dinners,

TABLE 3.5

Financial performances of US Steel and some other leading steel companies, 1909–1914 (thousands of dollars)

Company	1909	1910	1911	1912	1913	1914
US Steel (net earnings)	131,491	141,055	104,305	108,175	137,181	71,664
Bethlehem Steel						
(net manuf. profits)	2,654	4,216	4,392	4,847	8,531	9,378[b]
Republic Iron and Steel						
(net profits)[a]	3,325	3,009	2,002	3,065	3,963	1,869
Cambria Steel						
(net earnings)	2,538	4,553	2,777	3,901	6,688	2,478
Lackawanna Steel						
(net earnings)	4,468	5,949	3,035	3,729	5,419	−556

Source: Iron Age, various issues.

[a] The Republic financial year ran from June to June.

[b] Bethlehem profits in this generally depressed year were boosted by the first large orders of war material from the Allies.

and sold the semifinished steel they then made at whatever price they could obtain.[35] In the difficult trading conditions of 1909 and 1911, Chicago-area operators showed an inclination to assert their independence as well. They attempted to get work for idle mills by quoting steel on a local basing point, though when the crisis was over they reverted to the system that yielded them higher profits. Giving evidence to the dissolution suit in spring 1913, James Farrell admitted that "Pittsburgh Plus" did not hold as closely as before because of the growth of distant centers of production, each trying, as he put it, to make the best price possible for local consumers.[36] In fact, three main factors acted against the long-term maintenance of this pricing system: growth of demand far from Pittsburgh, rising costs of access to these markets due to freight-rate increases, and new conditions under which some distant steel centers became lower-cost producers.

Having secured a dominant position in the South through the purchase of TCI, US Steel introduced a variant of its pricing system there. This had to be done because, when asked to pay "Pittsburgh Plus" prices, buyers chose instead to switch their purchases from TCI to northern steel producers. Faced with this challenge, in 1908–1909 TCI managers complained to the controllers of US Steel that they were being penalized by the existing system. On the other hand, freight rates on plates and bars were shown to be at least as favorable from Birmingham as from Pittsburgh and Chicago to 46 percent of the United States and for rails to 35 percent of the nation. US Steel reacted in a way that reduced the cost burden on consumers but at the same time diminished the area of competitiveness of its own newest center of production. This was a "Birmingham Differential," under which the base price for products was generally three dollars a ton higher than in Pittsburgh. By August 1920, when freight charges were increased, the "Birmingham Differential" had been raised to five dollars.

Despite the support provided by the single-basing-point system, it gradually became clear that the competitive position of the Pittsburgh district was worsening. In January 1911 Henry P. Bope, who had worked for the preamalgamation Carnegie Steel Company when it ruled the roost and was now vice president of the Carnegie Steel division, addressed the Pittsburgh Realty Club. Gary works was now in production, other US Steel subsidiaries were beginning to install plant there, and Bope recognized a mood of pessimism in his local area. He thought it unjustified but admitted circumstances had changed in ways that were unfavorable.

> We have the supply here, never better than today; the genius of
> our workmen, and their skill and their ability to produce a most sat-

isfactory quality in the most satisfactory way were never greater than today; but by the location of large plants in other centers, by the fact that a great demand has arisen in localities where it did not exist before—localities which can be served better by plants located near to the source of the demand—our demand has fallen off. I can make that a little plainer by stating, using Chicago as an illustration, that the plant in Chicago, in the neutral freight zone of course, can do as well as ourselves; but our neutral freight zone only goes to the western Ohio line. Everything west of that is in the domain of Chicago and Birmingham. Therefore it becomes necessary for us to look in new directions; first, to find new fields; second, to find new uses for steel.[37]

This was to become a leitmotif of the next half-century. Already the increases in rail freight rates would have injured Pittsburgh if it had not been for the basing-point pricing system; later, much sharper rate increases would put heavier strains on the ability of local mills to command distant markets. At the same time the drift of changes in production costs favored Chicago. By 1919 Judge Gary admitted that Gary works made steel over 13 percent more cheaply than Pittsburgh. A year later US Steel revealed this works had lower production costs for plate of sixty cents a ton, for bars of $4.20 a ton, and that black plate and structurals were respectively $7.20 and $8.40 less—in the last case a saving greater than the freight charge from Pittsburgh.[38]

Under such conditions it was understandable that when demand fell to low levels producers distant from Pittsburgh were unhappy with the restrictions imposed upon them by the single-basing-point system. Consumers in the same districts chafed at the high costs they were paying for steel. World War I brought these discontents to a head. In September 1917 the War Industries Board made Chicago a basing point. Ten months later Judge Gary suggested a return to "Pittsburgh Plus," which was done, resulting in an increase of $5.40 a ton to the cost of steel, which war-stimulated industries in the Chicago district were consuming in ever-larger tonnages. From 1917 to 1919

TABLE 3.6

Rail-freight charges on steel, Pittsburgh to Chicago, 1903–1922 (price per ton)

6 Nov 1903 to 1 June 1907	$3.30	20 Sept. 1917 to 25 June 1918	$4.30
1 June 1907 to 26 Oct. 1914	$3.60	25 June 1918 to 26 Aug. 1920	$5.40
26 Oct 1914 to 20 Sept 1917	$3.78	26 Aug 1920 to 1 July 1922	$7.60

Sources: U.S. Federal Trade Commission, *Report* (Washington, D.C.: GPO, 1924), 1245; *Iron Age,* 1 January 1925, 56.

the number of industrial operations in Calumet alone increased by 53 percent.[39] Already, steel output in the area was growing more than in older eastern centers. Basing-point pricing meant these changes became of prime financial concern to local steel users.

In January 1919 the Western Association of Rolled Steel Consumers was formed to campaign against "Pittsburgh Plus." It organized meetings about the issue not only in and around the Chicago area but also in cities from Minnesota to the Gulf of Mexico. Other business associations and thirty-two state

TABLE 3.7

Selected areas' output of rolled iron and steel, 1917, as a proportion of the United States, and percentage increase in production, 1914–1917

Region	1917 share of national production	1914–1917 increase in output
Pennsylvania	45 percent	65 percent
Allegheny Co.	21 percent	65 percent
Ohio	18 percent	75 percent
Illinois ⎫	–	⎰ 90 percent
Indiana ⎭	20 percent[a]	⎱ 100 percent
All other	17 percent	n.a

Source: J. E. MacMurray, president of the Western Association of Rolled Steel Consumers, quoted in *Iron Age*, 6 March 1919, 611.

[a] Figure reflects 1917 share of Illinois and Indiana combined.

TABLE 3.8

Margins between costs of production and selling prices under "Pittsburgh Plus" and the "Birmingham Differential," c.1920 (cost per ton)

Location	Bars	Plates	Structurals	Black sheets
Pittsburgh	$2.00[a]	$10.00	$2.00[a]	$10.00[a]
Chicago	$14.00[b]	$18.00[b]	$18.00[a]	$25.00[a]
Birmingham[c]	$8.00	$7.00	$7.00	n.a.
Birmingham[d]	$18.00	$18.00[b]	$18.00[b]	n.a.

Source: U.S. Temporary National Economic Committee, *The Basing Point System*, monograph 42 (Washington, D.C.: GPO, 1939–40), 119.

[a] Just over.

[b] Almost.

[c] With "Birmingham Differential."

[d] Rate if "Pittsburgh Plus" had applied.

governments in the West and South gave their support. At this time the Western Association would have been content with one new basing point, Chicago, but, as US Steel pointed out, other areas would not find this acceptable.[40] From September 1920 a 40 percent increase in freight charges further aggravated matters. After wartime capacity extensions, there was an increased risk that when an inevitable downturn came, steel works distant from Pittsburgh would strive to hold on to their increased trade by quoting on a local-price basis. The Federal Trade Commission enquiry amply proved that the Chicago area had been seriously disadvantaged by "Pittsburgh Plus." For example, American Bridge, bidding in competition with local independent fabricators, could earn three times their profit on an ordinary structural job since its materials were openly priced as from Pittsburgh but actually bought from Illinois Steel. As a result of this boost, the capacity of its Chicago-area plants was thirty times that of its nearest regional rival.[41] On the other hand, a striking if extreme instance of the damage done to an independent consumer involved the manufacturer Heil of Milwaukee. This company often bought bars from the mill that Illinois Steel then operated in nearby Bay View. The steel was collected by the purchaser's own trucks but Heil was charged as if the material had been made in Pittsburgh and carried by train to Milwaukee. Altogether, fabrication of steel in that city alone was reckoned to provide US Steel with annual unearned payments for freight amounting to over one million dollars.[42] The agricultural machinery maker Deere and Company of Moline consumed 100,000 tons of steel a year on which they paid $488,000 to cover "Pittsburgh Plus" pricing—an extra charge that, as their witness made clear, the company passed on to the farmer. Indeed, one estimate was that, all in all, 825,545 farms in the states of Illinois, Iowa, Wisconsin, and Minnesota were each being forced to pay an average of ten dollars a year as a "toll" for "Pittsburgh Plus."[43] In some markets distortion of prices caused by this pricing system was greater than the average net profit steel consumers were likely to make in fabricating operations, and as a result the latter had to give up certain lines of business.[44]

Although independent Chicago steel producers had found that the "Pittsburgh Plus" system had boosted their income, it was becoming an increasingly intolerable burden on the district's economic growth.[45] In April 1921 the Federal Trade Commission filed a complaint against US Steel. Three months later, as demand underwent a very sharp fall, the Corporation found it was in its own interest to quote plates, shapes, and bars on a Chicago base. Even then it chose to soften the impact on Pittsburgh. Chicago had lower costs of production, but the price quoted for structural steel there after summer

1921 was fixed some $2 a ton above that in Pittsburgh, even at which level it was some $4.80 lower than if "Pittsburgh Plus" was applied.[46] Chicago-area sales for sheet, tinplate, pipe, and wire products continued on a "Pittsburgh Plus" basis.

In July 1924, after years of gathering and evaluating evidence, the Federal Trade Commission issued a cease-and-desist order against the "Pittsburgh Plus" pricing system. To the surprise of some, US Steel did not challenge this ruling. Two months later it made Chicago and several other places into basing points. Already a considerable impetus had been given to extension of capacity distant from Pittsburgh. This affected both the Corporation and independent firms, adding to the effect of the westward drift of steel demand that had accompanied the gradual rounding out of the national economy geographically. On the other hand, US Steel and the industry as a whole again adopted pricing systems that represented only a gradual move in the direction of each mill having its own base price. For instance, by the beginning of 1925 wire was priced from at least nine separate basing points. But whereas the price was identical at three of them—Pittsburgh, Cleveland, and Ironton, Ohio—the mill price at Anderson, Indiana, was one dollar; at Joliet and DeKalb, Illinois, and Duluth, Minnesota, two dollars; and at Worcester, Massachusetts, and Birmingham, Alabama, three dollars higher.[47] In other words, Pittsburgh-area mills such as the AS&W operations at Donora or Rankin were able to adjust more easily to the new regime. "Pittsburgh Plus" was dead, but there was as yet no straightforward f.o.b. mill system. Moreover, the whole furor over the pricing system had shown once more how important steel was in the national economy, and therefore the extraordinary power of that industry's leading company. Indeed, in summer 1919, when Gary had suggested that the Federal Trade Commission ought to look into the complaint from the Western Association, Victor Murdock, one of the commissioners, responded, "Judge Gary, the question is a pretty big one and not only involves this proposition, but involves the whole economic structure of society."[48]

The Changing Balance of Locational Advantage and Expansion

The Rail Trade and the Gary Project

At the end of the nineteenth century, it was not at all easy to foresee how the distribution of the American steel industry would change over the next few years. Many observers, extrapolating from the production patterns of the 1890s, were convinced that the East was played out, a conclusion soon to be dramatically disproved. A few, deeply impressed by its close juxtaposition of iron ores, coal, and limestone in contrast to the long distances separating Great Lakes ore from Appalachian coal, felt the future lay in the Birmingham area. The British expert J. S. Jeans, on a visit to study American competitiveness, realized that the time he had available would not allow a visit to both Chicago and Alabama. He decided that their relative future importance meant he should go to the latter. Time proved he made the wrong choice. Yet others, and notably Andrew Carnegie, had been so completely conditioned by their past experiences that they had no time for southern prospects. There was wide recognition that the Lake Erie shore, as the break-of-bulk point between the flows of iron ore and coal/coke, had become an advantageous location for iron smelting and steel making. Late in 1895, speaking in Cleveland, Carnegie had spelled out its attractions: "Pittsburgh is no longer the best point to manufacture iron and steel in the United States. The reason is easily given and will be obvious to the dullest comprehension. It takes two tons of iron ore to make a ton of finished steel. It takes only one ton of coke to make one ton of steel or pig iron. Therefore a manufacturing plant situated upon Lake Erie, north of Pittsburgh, has not to transport by rail the two tons at all; it comes to the manufacturer by water, and he only has to transport the ton of coke from Pittsburgh to the lake. This gives to the Lake Erie manufacturer so decided an advantage that if one were about to locate iron and steel works, he would go there."[1] In fact, as always, plant efficiency and company organization proved more important than relative location. As a result, throughout the 1890s the Pittsburgh operations of Carnegie Steel—and to a lesser extent of Jones and Laughlin—were more successful than those of, for instance, the Cleveland Rolling Mill Company. Even so the benefits of a lake shore location were indicated by growth of merchant iron furnaces at its ports, in the new integrated works at Lorain, Ohio, and even more spectacularly by the 1899 decision of Lackawanna Steel to move operations from Scranton to Buffalo.

About a year before he resigned from Carnegie Steel, H. C. Frick set up a committee under the chairmanship of Daniel Clemson to study the possibilities of an iron and steel plant at Conneaut, the Lake Erie terminus of the company's Pittsburgh, Bessemer, and Lake Erie Railroad. In 1898, 67.1 percent of the railroad's traffic had been iron ore, and so it was badly in need of a return bulk cargo such as might be provided by coke destined for lakeside furnaces.[2] In 1900 the expansion of the National Tube Company and its decreasing purchases of semifinished steel from Carnegie caused the latter to push ahead with a planned Conneaut tube works. By the second week of January 1901, the scheme was sufficiently advanced for a public announcement of an integrated works of two, and possibly four, blast furnaces, up to twenty fifty-ton open-hearth furnaces, and tube mills on part of a site of about 5,000 acres east of Conneaut Harbor. After only minor site work had been done, Carnegie Steel passed into the United States Steel Corporation and work stopped.[3]

When those in charge of US Steel began to consider future patterns of expansion, it was obvious that the attractions of Conneaut were still considerable. The imbalance in traffic on the Conneaut-Pittsburgh line was more or less as great as ever—in 1901, 66.8 percent of its total traffic was iron ore, as much as 70.1 percent the following year. It was also obvious that there was need for much improvement at the National Tube works in McKeesport, Pennsylvania. Its blast furnaces were new, but it had only Bessemer converters and had to bring in any open-hearth steel it used. Exceptionally, in this case a knowledgeable foreign visitor was not impressed by US Steel's practice: "Generally speaking, the works strike me as being unduly crowded, and there is a lack of that method and order about them which one expects to find in a plant of the most modern type."[4] Yet early in 1903, any surviving hope that such deficiencies might mean that the Corporation would take up the plans for Conneaut were dashed. The former Federal Steel plant at Lorain also had an ample site and was producing 1,200–1,400 tons more steel each day than was used in its girder-rail mill. Instead of this surplus being sold as billets, it was decided that the works should be extended with two new blast furnaces, additional rolling capacity, and pipe and tube mills. As compared with the planned outlay of $12 million at Conneaut, extensions to the existing Lorain operations would cost only $8 million. The new plant was built between summer 1903 and 1905.[5] US Steel retained the Conneaut site as an undeveloped tract.

In the early years of the new century, national steel expansion was going ahead at an exceptional rate: the estimated increase between 1898 and 1904 alone was 72.1 percent. As the dominant force in the industry, US Steel was involved in extensions despite its already huge size. Between 1902 and 1904, its rated-steel capacity went up from 11.2 to 12.9 million tons and over the next two years reached 15.1 million tons.[6] These figures represented the net increase resulting from the acquisition of a few more operations, notably Union and Clairton, closure of a few small steel plants, and (on the plus side) a large number of "rounding out" additions to plants, a few of which seem to have been a case of making the best of a bad job (for example, the investment of $150,000 at the integrated but badly located Columbus, Ohio, works to enable it to make sheet bars).[7] No important completely new ventures on virgin sites were set in motion, and consequently it was difficult to make out the drift of thinking about a long-term strategy for the distribution of Corporation capacity. On the other hand, it was clear that the two great contenders for further large-scale investment were its existing poles of capacity in Pittsburgh and Chicago.

At its formation, the heartland of US Steel productive power was unquestionably in Pittsburgh; in terms of capacity, Chicago was then only a little way ahead of the Valleys district. By 1901 the 325,000-ton Union works was idle, and the other two main Chicago plants brought into the Corporation by Federal Steel—one in South Chicago and the other at Joliet—had a combined ingot capacity of 1.67 million tons. The three Carnegie plants of Homestead, Edgar Thomson, and Duquesne, together with National Tube at McKeesport, aggregated 4.2 million tons of steel production. Some early commentators expected still more concentration of operations around Pittsburgh.[8] Expansion over the first three years gave no clear indications one way or another of shifts in assessment of locational advantage; if anything, Pittsburgh seemed to have the edge over Chicago. Union works was demolished. Capacity at South Chicago and Joliet increased by 145,000 tons; Pittsburgh works were extended by 185,000 tons. Additionally, the Pittsburgh area gained from the purchase of Donora, with 300,000 tons capacity, and the 400,000 tons at Clairton. Late in 1902 it was decided that a new forty-inch plate mill should be installed at Homestead.[9] The following February, James Gayley submitted a list of required major improvements. With schemes already approved, his list brought required spending to $36.5 million. The program was expected to increase US Steel's annual capacity by 2.7 million tons of products. More important, it would increase annual earnings capacity by $7 million and save $5 million in costs. Over three years the increase in overall profits would cover the capital outlay. Two weeks later Gary announced these plans, "which have been under careful consideration and preparation for over a year for harmonizing, extending, and rounding out the various plants." Again, the signals as to locational choice were ambiguous, National Tube plants in Pittsburgh and at Lorain receiving the lion's share of spending. Another important factor in interpreting these expenditure schedules was that the superb quality plant and efficiency of practice in the former Carnegie Steel operations indicated that the properties bought from other groups were those requiring immediate attention. In other words, the expenditures for operations beyond Pittsburgh did not necessarily indicate in what area executives believed the future of US Steel would be.

Yet it was undeniable that various factors were moving against Pittsburgh. Fuel economy and byproduct coking together were strengthening the "pull" of ore-oriented, or more specifically ore-port, locations as against those nearer the coking-coal fields. Pittsburgh was turning increasingly to open-hearth steel—Duquesne made its first in 1900, and Homestead had long been the most important open-hearth plant in the nation—but the rapid advance of this process made attractive those locations where scrap was available from metal processing industries on a larger scale relative to the size of steel capacity than was the case in western Pennsylvania. Markets for steel were bigger

TABLE 4.1
US Steel plant expenditures announced in March 1903
(thousands of dollars)

Illinois Steel Company	
South Works	5,075
Joliet	1,470
National Tube Company	
Lorain	8,646
McKeesport	9,256
Carnegie Steel Company	
Homestead	1,135
Edgar Thomson	275
Duquesne	330
All other plants	8,207

Source: Iron Age, 5 March 1903, 29.

or closer at hand for a number of other centers. A further increasing liability for operations around Pittsburgh was the huge growth in demands on and consequent bottlenecks in its facilities for bulk transport. Between 1897 and 1905, pig iron production in Allegheny County rose from 2.7 to 5.4 million tons and crude steel production from 2.8 to 5.7 million. In 1897 the area's combined inward and outward freight movements had been 46 million tons; by 1905 the total reached 103 million tons, of which the railroads handled 92 million tons.[10] The whole industrial district seemed to be choking because of its very success. For a time it was even suggested that Youngstown might gain because it offered more room for development than Pittsburgh, an idea that received a further fillip when, at the end of 1906, it was announced that Carnegie Steel was to build two new blast furnaces, twelve open-hearths, and blooming and bar mills at the Ohio works it had inherited from National Steel.[11] In fact, by that time it had already become clear that the real growth area lay much farther west.

No one disputed that Chicago had impressive advantages in steel making, though the experience of the 1890s had taught the painful lesson that superior management elsewhere might cancel out the intrinsic advantages of lakeside location and command of huge tracts of the Midwestern and western markets. Now that the bulk of Chicago steel making, like that of its Pittsburgh rivals, was included within US Steel, competition between them for current business and for further major extensions could be conducted on a more objective basis; the disruptive influence of Carnegie and his exceptional team of managers had been cancelled out. Within four years of the formation of the Corporation, this new situation led to the announcement of the only completely new integrated works that US Steel freely chose to build in its first half-century. Things came to a head in the product in which Carnegie Steel had scored its most spectacular victories, steel rails.

In every year from 1897 to 1901, Edgar Thomson had produced more rails than the entire Chicago area. This achievement was a reflection not of natural advantage but of superior efficiency. Even in his days as head of Carnegie Steel, Schwab had recognized that for at least a limited range of products, notably rails, Chicago had advantages over Pittsburgh. For many years the biggest outlets for rails lay west of Chicago. Within the new US Steel, whose existence promised at least the beginnings of an overall strategy for production and distribution and in which efforts were made to bring all plants up toward the standards of the best, Carnegie Steel soon began to feel the erosion of its hard-won special position in the trade. Edgar Thomson's highest ever share of national rail production was reached in the highly competitive conditions of 1897: 29 percent of a total output of 1.6 million gross tons. In

1901 its share was 24.9 percent and by 1907, 20.8 percent. National production of steel rails of all types in 1910 was almost exactly the same as in 1907; that year, although Carnegie Steel's total finished-steel production was 0.8 percent down, its rail tonnage was only a little over two-thirds the level of three years previous.[12] The reasons for decline were far from simple. One confusing factor was the wholesale switch from Bessemer to open-hearth rails. By 1910, 48 percent of national and of US Steel rail output was in open-hearth steel. Edgar Thomson produced small tonnages of these, but from shipped-in ingots. More striking was the fact that in 1910, this works made only 34 percent of the Corporation's Bessemer rails; its capacity was about one million tons, but from the start of the recession in late 1907 to spring 1908, it ran at less than half this rate.[13]

The board minutes of the Carnegie Steel division of US Steel reveal the worsening situation and its main causes. In 1903 Edgar Thomson rolled 735,000 gross tons of rails as compared with 717,000 tons from South Chicago, but the setting of a stable price for rails—$28 a ton f.o.b. works—had already removed the Pittsburgh work's traditional resort to price-cutting in order to go for business, and more and more of this business was in the West. Sometimes things worked out well. For instance, in March 1905 the banker and US Steel director George Baker wired Henry Bope at Carnegie Steel that the Santa Fe Railroad was in the market for 25,000 tons of rails for prompt delivery, which Illinois Steel could not possibly supply. As a result, it was planned that Carnegie would take over an order for the Pennsylvania Railroad lines east of Chicago and Illinois would make the deliveries to the Santa Fe, saving on overall delivery costs; a good instance of the distribution economies anticipated for the combination.[14] Unfortunately, a gradual loss of business from Pittsburgh was more usual than a mutually beneficial quid pro quo.

TABLE 4.2
Production of heavy standard rails by works of US Steel, 1910
(thousand gross tons)

	Bessemer rails	Open-hearth rails	
Edgar Thomson	314	Gary	435
South Chicago	610	ET and two other "northern" works	96
		Ensley	323
Total	924		854

Source: U.S. Commissioner of Corporations, *Report on the Steel Industry,* 2 vols. in 3 pts., pt. 3, *Costs of Production* (Washington, D.C.: GPO, 1913), 461–65.

Edgar Thomson iron and steel works, Pennsylvania, 1912. *Courtesy of USX Corporation*

In July 1905 a new rail mill began work at Edgar Thomson, but by early autumn that year Carnegie Steel had only taken about two month's work of rail orders for the following year whereas Illinois Steel was fully booked to November. Now Bope was apprehensive about taking over work from Illinois: "so far as possible we want to avoid taking very much tonnage at $28.00 Chicago, but we shall probably have to take care of some of the trunk lines on this basis." Before the year's end, they had approval from US Steel to spend $2.2 million on improvements at Edgar Thomson in ore handling, the Bessemer plant, and rail mill.[15] But the situation worsened. By spring 1909 there seemed a likelihood that Edgar Thomson's two large rail mills would have to close. Both Dinkey and Bope felt the urgency of the situation. It was all so strikingly different from the situation of an independent Carnegie Steel Company less than a decade earlier. Illinois Steel had a good deal of rail tonnage on order, "but we do not want to have to ask to have any of it transferred to us because there would be a loss of about $2.50 in freight. . . . Nearly everything that goes to Gary formerly came to us. The only field we have to draw from to replace that is export." (Soon afterward, Edgar Thomson accounted for the whole of US Steel's export rail tonnage, though in summer 1909 TCI was allocated an Argentine order.) The following year already looked like being a banner year for rails—it was in fact second only to 1906—but Bope and

Dinkey reported: "[We] cannot see where . . . Edgar Thomson, pocketed as it is, is going to have anything like the tonnage of rails for which it has capacity. With Gary in shape to be shoved up to practically 5,000 tons per day, and with South Chicago in addition, it looks to me as if Edgar Thomson will get only the leavings." If they had to make open hearth rather than Bessemer rails, Bope thought US Steel would have to do this at their Youngstown works, where open-hearth steel was first produced in January 1909, for they needed all their melting shop capacity in Pittsburgh for other products. He drew a dismal conclusion: with rail demand now mainly in the west and southwest; "I cannot see any great future for ET." They might turn over production to sheet bars, though most American Sheet and Tin Plate works used open-hearth bars. In 1907 Edgar Thomson began to roll billets, in late 1909 tie plates, and in early 1910 sheet bar. In 1910 Carnegie Steel, very carefully assessing the national rail situation, put capacity to produce at 5 million tons; under the best circumstances it was unlikely that there would be demand for more than two-thirds of it. Two years later, though they had hopes from a new rail mill of high capacity and quality, orders for 1912 were already to the extent of three-quarters for open-hearth rails.[16] In 1913 Edgar Thomson was equipped with a melting shop, after which open-hearth rails became important. But the heavy rail trade continued to slip away so that by 1919 Edgar Thomson was a smaller producer than TCI and only a little over one-third as big as Illinois Steel. In 1905 rails had represented 14.7 percent of Allegheny County's rolled iron and steel output; by 1919 they accounted for only 5.2 percent.[17] Strangely, what for years seemed a calamity promising at best an uncertain future was by the 1930s to open up for this works a completely new, expanding, and much more secure place in a completely different section of the finished-steel industry. In the meantime not only the rail trade but also other sections of production were moving strongly westward.

As president of US Steel, Schwab seems to have blocked attempts by Chicago interests to secure major new mill projects there. On 3 February 1903, the spending of $1.1 million on a structural mill at South works was approved but then sometime later postponed. Not until October 1904 was it again given the go-ahead, by which time the estimated cost approached $1.3 million.[18] Much later, Eugene J. Buffington, the first president of Illinois Steel after its acquisition by US Steel, stated that during the early years of the Corporation, his company had acted as selling agent in the Chicago market area for structurals rolled in Pittsburgh. Not until 1906 did South works get its structural mill. By this time, not only changes in products but also in the balance between Bessemer and open-hearth steel favored Midwestern expansion as compared with that at the head of the Ohio River. Open, flat, extensive, and cheaply bought sites at the foot of Lake Michigan contrasted with the few, small, awkwardly shaped, and very costly tracts now available along the rivers around Pittsburgh. Most decisive of all, the centers of demand had long been shifting away from western Pennsylvania, whereas consumption in the huge area tributary to Chicago had grown. Chicago metalworking far exceeded its metal-making industry; around Pittsburgh the reverse was true. In 1905 a US Steel committee headed by Buffington computed that Chicago district output of steel was 2 million tons less than demand. It reckoned rail production in that area should be increased by 450,000 tons.[19]

It seems that not until mid-1905 did US Steel executives consider a first broad overall plant-development strategy. On 13 June all members attending the Finance Committee agreed: "the subject matter of formulating a comprehensive plan for the future development of all the properties in which the United States Steel Corporation is interested, *considered as a unit*, was referred to a special committee to be appointed by the chairman of the Board for consideration and report." In short, schemes were no longer to be considered, approved, or rejected piecemeal but rather considered as a whole and comparatively. Such an approach soon made clear that expansion planning was moving in favor of Chicago. Some ten weeks after the announcement of the overall review decision, officials approved proposals to build a new blast furnace and ten open-hearth furnaces at Illinois Steel. The question of acquiring more land near Chicago came up, for which there were various possibilities. Some space could be found southwest of Joliet, but this was ruled out because it was undesirable to build large new plant away from the lakeshore. There

TABLE 4.3

Geographical distribution of US Steel blast furnaces and open-hearth furnaces, 1906

Location	Blast furnaces	Open-hearth furnaces
Within 50 miles of Pittsburgh	48	111[a]
Within 50 miles of Chicago	21	24
Valleys district	9	18
Lake Erie shore	10	6
East of Alleghenies	2	19
Other	7	2
Total	97	180

Source: Iron Age, 16 August 1906, 406–8.

[a] Sixty of these were at Homestead.

were sites near Whiting, Indiana, including a substantial stretch toward the canal that Frick had owned for some years. Gradually, however, attention focused on the so-called Holt tract and the purchase of neighboring parcels of land.[20] Inevitably, news that the Corporation had purchased a very large area of lakeshore land in Indiana leaked to the press, and by November Gary felt it necessary to make a statement. Its wording seemed to confirm Brody's claim that if conditions, and particularly Illinois state politics, had been different, US Steel might have extended its South Chicago works instead of building a new one.

> A large body of land, approximately 2,500 acres, on the shore of Lake Michigan, in Indiana has recently been acquired in the interests of the United States Steel Corporation or its subsidiary companies. The Illinois Steel Company in the near future will probably make important improvements on this property for manufacturing purposes. The extent of the improvements will depend upon the attitude of the public authorities in Indiana and Illinois. Formerly the concentration of plants of this company has been carried out at its location at South Chicago. During the last ten years its works, known as the North Works and Bridgeport Works, within the city limits of Chicago, have been partially removed to the South Chicago location. This location is well adapted to its business in many ways, and the same would be further developed if there were sufficient room at that point. The disposition of the Indiana officials seems to be to encourage the location of industries in Indiana in every reasonable way, and therefore the recent purchases have been decided upon. In any event there will probably be erected on this site blast furnaces, open-hearth furnaces, by-product coke ovens, and various mills for a diversity of steel products.[21]

On Thursday, 28 December 1905, a report was submitted by a special committee given the task of investigating and making recommendations about the possible new plant. Pittsburgh and Chicago ancestries of US Steel were equally represented in the committee membership. The following day their report was considered by the Finance Committee, which decided that "provided the cooperation of the public officials in Indiana and representatives in Congress be secured as promised," US Steel should build a complete works with docks, coke ovens, blast furnaces, open-hearth steel furnaces, blooming and slabbing mills, rolling mills to produce rails, structurals, plate and bars, a foundry, and axle and machine works. Within less than three weeks, a planning committee had elaborated the plan, recommending a start be made at once on four blast furnaces, twenty-four open-hearth furnaces, a blooming mill, four bar mills, and a rail mill.[22]

The site chosen for the new works, soon to be given the name Gary, was part of a nine-thousand-acre tract with a ten-mile frontage along the shore of Lake Michigan. It contained sandy hillocks, sloughs in the depressions, and a scatter of thickets of scrub oak. Sections of three railroads (in all fifty-one miles of track) had to be relocated, and two miles of the Grand Calumet River were straightened. Construction began in summer 1906 but was slowed by depressed trade in late 1907 and throughout 1908. Even so, in some respects the size of the project increased. In location, site, and layout, the works represented an ideal never previously equaled and for another half-century scarcely ever bettered. As a very large, flat site had been secured and planned from the start for a major works, the various stages of production were placed so as to secure the most efficient materials flow while allowing each section room for future extension of capacity. The result was a world away from the cramped sites, complicated layouts, and confused interdepartmental transfers so common where operations had grown up and expanded bit by bit, especially if limited by difficult conditions of local relief. Unflattering comparisons

2. Site of Gary works, Indiana, showing the excavation of the dunes for the foundations of the open-hearth plant, 3 August 1905. *Courtesy of USX Corporation*

with the Corporation's Pittsburgh-area sites and plant layouts were inevitable. Such differences were by no means immaterial—heat loss and extra movements inside a plant may add to overall costs of production as effectively as long hauls for raw materials or products.

The first Gary blast furnace was blown in during December 1908, and over the next twelve months five more furnaces were commissioned. Open-hearth steel and rails were first produced in February 1909. By now growth in steel demand in the West and down the Mississippi Valley was so rapid that even this huge addition to Chicago-area plant was recognized as insufficient. By 1910 western railroads reportedly were disinclined to pay for the haulage of rails from Pittsburgh so long as they could place orders in Chicago. As a result, whereas mills in the latter had order books filled for months ahead, Pittsburgh and eastern works had so little business that they were forced to roll billets and other specialties. Rails sold for $28 a ton; billets at Pittsburgh mills averaged only $25.38 that year.[23] Meanwhile, Gary's huge lakeside tract provided an attractive location and ample room for other subsidiaries that needed to buy semifinished steel. American Bridge was developing a neighboring site before the end of 1909, and the next year American Sheet and Tin Plate began to build sheet mills. By the early 1920s, there was room enough to plan a major new tubular products plant, though in fact a fully integrated Gary tube works was never built.[24] By the time of its tenth anniversary, the iron and steel producing capacity of US Steel was roughly double that at the time of its incorporation, and much more of it was already west of Pennsylvania.

Additional Gary expansion was favored by the triumph of byproduct coking and the open-hearth process along with further changes in the geography of national economic growth. Between 1899 and 1909 the value added by all manufacturing industries in Illinois and Indiana had risen from 84 percent to

TABLE 4.4
The expansion of US Steel capacity, 1901–1909, 1911 (thousand gross tons)

	Capacity 1 April 1901	Added by purchase of		Additions and improvements		Capacity 1 Jan 1911
		Union & Clairton	TCI	To 1/1/1909	1/1/1909– 1/1/1911	
Blast furnaces	7,440	1,228	1,000	5,322	1,250	16,240
Steel ingots	9,425	1,258	500	5,887	968	18,039
Rolled products	7,719	1,103	400	3,678	1,645	14,547

Sources: US Steel, "Annual Report for 1908"; U.S. Commissioner of Corporations, *Report on the Steel Industry*, 2 vols. in 3 pts. (Washington, D.C., GPO, 1911, 1913), pt. 3, p. 269.

96.1 percent of that in Pennsylvania. The requirements of World War I possibly account for the reduction by 1919 to 85.7 percent, but ten years later these Midwestern states were 18.5 percent ahead of Pennsylvania, and the gap was to widen further. Here and farther west was the natural market area for Gary. In 1904 the Chicago district plants of US Steel had a combined capacity of 2.04 million net tons as compared with 5.69 million in the Pittsburgh area. Sixteen years later the Pittsburgh-district works aggregated 7.79 million tons, while Chicago plants were now capable of 6.86 million tons. Within a decade of production beginning, Gary had already become not only the largest US Steel plant (with South works second) but also the largest in the United States and a model for the industry worldwide.

Government, Business, and Industrial Development

The Cases of Duluth, Birmingham, and a Canadian Plant

Despite a firm, frequently reiterated, national commitment to free enterprise, government had always played a most important role in the growth of the iron and steel industry. The most sustained form of involvement was in tariff policy. Types and levels of duties changed, but the fact of protection, in early years from larger and better endowed national industries and later from those that had much lower labor costs, was generation after generation an essential backdrop to the industry. In some instances, as with the iron-rail business in the second quarter of the nineteenth century, Bessemer rails in the 1870s, and twenty years later with tinplate, duties were nicely adjusted with the express intention of building up a major home industry. In other cases sectional interests diverged and the federal government responded first to one lobby and then to another. A notable example of this was advocacy by eastern steel producers of reduced duties on imports of foreign iron ore and opposition to all such proposals from firms and congressmen from interior centers of production. From an early stage, individual states also played a part, as for instance with attempts to stimulate the introduction of methods of production designed to use Pennsylvania anthracite or coke in smelting. The creation of the United States Steel Corporation produced very different forms of government involvement. As an operating giant, it was exposed to scrutiny by federal administrations beginning to respond to what were increasingly seen as the dangers of trusts. But it was also vulnerable to leverage from those few local areas that controlled a large part of the raw materials on which its operations depended. An important example of the first was associated with the absorption of the Tennessee Coal, Iron, and Railroad Company in 1907. Over a longer period, the state government of Minnesota used the immense power given to it by its possession of the largest domestic iron-ore deposits to stimulate its own wider economic and social development. US Steel also became involved in international ventures and, therefore, with foreign governments.

Between the world wars, one graphic account of the American industrial economy likened it to a massive inverted pyramid. The larger part of this figure's broad, up-ended base consisted of the myriad of metal using industries, interrelated with each other and employing billions of dollars in capital and millions of workers, drawing on the primary metal industries below, of which the outstanding group was iron and steel. The apex of the inverted pyramid rested on the small area of the nation from which the iron ore for the furnaces was drawn, above all the Mesabi Range in northern Minnesota. Only in the last two decades of the nineteenth century did the Upper Great Lakes ranges become the main source of ore for the steel industry. In 1880 they had provided 23.6 percent of national ore production, but by 1900 their share was 74.4 percent. Michigan was the leading producer until the second half of the 1890s, but Minnesota was already growing rapidly in importance. The Vermilion Range first shipped ore in 1884, and the Cuyuna Range was not fully opened until 1911. Mesabi, opened in 1892, soon became far and away the main supply source. It shipped 0.6 million tons of ore in 1893, 7.8 million by 1900, and ten years later 29.2 million tons, almost one-quarter of the world's total. As late as 1890, the Mesabi Range had been an archetypal "trackless waste." Two years later it received its first railroad connection to the lake head, the Duluth, Missabe, and Northern. A few years later the main part of the range was also reached by the Great Northern Railroad. Mining towns were established along the range. Population in St. Louis County, which included both Mesabi and Vermilion, increased from under 45,000 in 1890 to 163,000 by 1910.

Its main predecessor companies and US Steel itself played a key part in this growth; in turn, Minnesota ore and that from Mesabi in particular became of overwhelming importance to the Corporation. In 1901, 60.9 percent of all the ore mined by US Steel came from operations in Minnesota; by 1910, this had grown to 76.2 percent. As early as 1902–1904, about half of the ore was worked by the open-pit method. By 1920 the proportion was nearly 76 percent. The change in method of extraction was the occasion for an important test of strength between US Steel and one of the leading mining communities its activities had created. Hibbing, Minnesota, had been settled in 1892; it was laid out as a village during the following year. It contained 2,481 people in 1900, by 1910 8,832, and in 1920 15,089. During World War I, US Steel

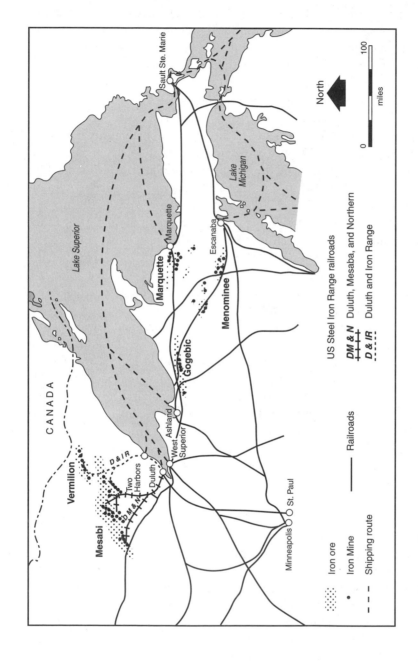

US Steel in the Upper Lakes Iron Range Country, 1908

merged three formerly separate mining operations to form the great Hull-Rust-Mahoning open pit, which lay on the edge of Hibbing, and its further growth threatened to swallow much of the town. To tackle this problem, by 1917 US Steel's Oliver Mining Company had bought a forty-acre tract of the town, containing about three-quarters of the whole of the settlement, and began to move it building-by-building to a new site 1.5 miles away. By the mid-1920s the open pit covered an area of 1.75 by 0.5–1 miles. In 1926 this mine produced 4.5 million tons of ore. Already some of the local community had tried to press claims against the Corporation. The following, conveying the spirit in which their efforts were received, was recorded in US Steel minutes on 16 January 1923: "The Finance Committee listened to a very long oral statement by Mayor John M. Gannon of Hibbing, Minn., in regard to the lawsuit which has been initiated for the purpose of having vacated the streets and alleys upon the 40 acres of ground in which a part of the village of Hibbing was formerly located, all of which, including buildings and improvements, have been purchased by the Corporation. With the statement that the Committee was in doubt whether it had legal right to adopting of the suggestions made by the Mayor, it took the whole matter under advisement." Two weeks later the finance committee took a more definite line: "Having listened to the able and vigorous presentation by Mayor Gannon and his associates, [the committee] has reached the unanimous conclusion that there is no legal, equitable, or moral liability against the Corporation or any subsidiary and it is therefore resolved that said claims must be disallowed." However, by May Elbert H. Gary suggested to his colleagues that as the courts would probably award about $300,000 to the people of Hibbing, US Steel should make provision to pay them.[1] As well as monetary compensation, Hibbing, like other mining communities of the iron-range country, was provided with splendid public facilities; in its case, a fine high school and junior college were built between 1920 and 1924.

The State of Minnesota had far more leverage than individual dependent mining communities. For many years it used it so effectively as to force an important new direction in US Steel plant-development policy. Time and again in the past, hopes that the growing importance of the Upper Great Lakes as a source of iron ore would be accompanied by iron and steel making facilities had been disappointed. In the older centers of the trade there was skepticism about the area's prospects. Years after the event, it remained a continuing occasion for expressions of regret by older members of Andrew Carnegie's business group that they had once been persuaded to invest in an iron furnace at Escanaba, Michigan (on the Upper Peninsula). A number of projects or actual attempts to make steel and finished products in the region had failed.[2] In

1892, in a mood of strange euphoria, A. D. Thompson of Duluth went so far as to predict that within twenty years the largest steel-making center of the United States would be at the head of the Great Lakes.[3] Events proved such rosy anticipations wildly awry. Yet in the latter part of Thompson's forecast period there seemed at last good reason for optimism.

Lake ore now fed most of the nation's blast furnaces, but the ore carriers that left Upper Lakes docks for Lower Lakes ports commonly made the return journey in ballast. It might be sensible to replace this with "back haul" coal—at freight charges eventually about half those on ore carried down the Lakes. Despite the reported opposition of Pittsburgh interests to the scattering of works, these considerations as well as byproduct coking as part of an inte-grated iron and steel operation seemed to promise a break with the tradition that ore was moved to the coalfields rather than vice-versa. Moreover, eco-nomic activity was increasing rapidly in the northern Great Plains, the popu-lation of seven states in that region increasing from 8.9 million in 1890 to 11.6 million twenty years later. Through the main lines and networks of the Great Northern and Northern Pacific Railroads, the head of Lake Superior seemed to be well placed to supply steel demand in the region. The lakeshore itself was booming: Duluth and its near neighbor, Superior, increased from a combined population of 45,000 in 1890 to 119,000 twenty years later.[4]

In his popular but impressive survey, *The Romance of Steel*, Herbert Casson summed up the situation. He seized on the possibilities of the Upper Great Lakes and explained both the reasons for their present lowly status in steel and the prospects of a way out in emotive language: "some day Duluth will awake and make her dream come true. . . . Now that less and less coal is needed to pro-duce a ton of iron, Duluth's opportunity to build profitable blast-furnaces is growing better year by year. Geographically, Duluth is located so that she can-not escape being an important iron and steel community. She stands at the western doorway to the Great Lakes—almost in the exact center of the conti-nent. One of the few possible sites for a grand city is hers, and she has her face towards the rising sun." So far, though, achievement of her destiny had been frustrated: "The Pittsburgh vikings sail up to the iron ranges and carry off the loot—millions of dollars' worth every summer week. And all the while, for some reason which no outsider can understand, the men of Duluth and Su-perior . . . have been satisfied to run errands and quarrel, like a couple of mes-senger boys."[5] All this apparently illogical, discriminatory flouting of commercial logic, which he presented so provocatively, was about to change; Minnesota realized that its ore deposits gave it the power to move even the greatest of steel companies. An important new steelworks project was not, like the frequent past failures, to be the outcome of hopeful paper calculations of relative production costs at various locations and still less of wild, romantic dreams of a great new steel center meeting the apparently insatiable needs of the rising West, but rather the price the United States Steel Corporation would eventually have to pay to maintain good relations with the state that controlled the price of its iron ore.

Throughout its earliest years, one of the impressive features of US Steel policy was the deliberate manner in which it set out to gain an ever more commanding control of mineral resources. In iron ore this meant further acquisitions in the Lake Superior region. The twin aim was to guarantee supplies to its constituent companies and to deny them to its rivals. Corporation furnaces required very large tonnages of ore; in 1902 its Lake Superior mining subsidiaries produced 16.1 million gross tons, over 45 percent of the national total. In this department, as in others, the huge size of its operations yielded operating economies and gave it bargaining power with ore railroads and independent lake carriers. It was not immediately obvious that they also might deliver the Corporation into the hands of those who controlled the states in which its mines were located.

The twin aims in controlling more ore, and the powers that its size gave to US Steel in these matters, came out well in the course of two spring 1902 meetings of the US Steel Executive Committee. Pres. Charles M. Schwab remarked: "I think we have been a little too conservative in taking mines; some that we have refused have doubled in value since. I think we ought to buy Mr. Hill's [James J. Hill of the Great Northern Railroad] ores." To this Gary replied: "We certainly have everything on the Vermilion; we bought everything on the Mesaba [*sic*] that is good, that is best, and that is first class, with perhaps one exception, which we could not get, and with the exception of beyond Hibbing; we cannot get there. Mr. Hill is in there now. The prospecting which has been bought is low-grade ore, and we have of that class of ore certainly 60 to 70 million not yet opened. We can afford to let our competitors get the low-grade ores if we have the better ones." Schwab visited the northwest before they again took up the matter two weeks later. Hill was found to be leasing properties to their competitors. Even so, Schwab believed they must come to an agreement with him and that they could do so without difficulty, for Hill had told him he was holding the ore for US Steel: "We have depended upon his assurances that he would let us have the properties, and we are now giving him a million tons of ore every year down his road. I think Mr. Hill ought to give us the leases of all his properties subject to a contract with him by which we would guarantee him a certain minimum of traffic over his road. All he wanted was the traffic. He does not want any royalty." Gary was understandably, and as it proved justifiably, doubtful about this.[6] In fact, things proved much less straightforward than Schwab anticipated. A US Steel committee negotiated with Hill for almost four years before they felt they were approaching an agreement, which was at last reached in October 1906, the lease being dated from 2 January 1907. Despite Schwab's earlier understanding, there was in fact not only a royalty but a high one at that. For 1907 it was eighty-five cents a ton as compared with royalties on some of their earlier leases as low as fifteen to thirty-five cents a ton. There seems to be reason to believe that the purpose of this lease was to deny these huge ore reserves to others.[7] From 1901 all US Steel supplies of ore had come from the Lake Superior region. Minnesota's share that year was 60.9 percent; by 1906 it was 76.8 percent. Future huge increases in tonnages required were signaled at the end of December 1905 in the decision to build the Gary works. The Hill lease was a further step in the same direction.

On the other side of the situation was the extraordinary significance of mining in the Minnesota economy. By 1909 the state's manufacturing output was worth $17.2 million; two years later its iron mines produced ore worth $48.4 million. The Minnesota Tax Commission estimated in 1908 that US Steel controlled 913 million of the state's 1.192-billion-ton ore reserves. Some 73 percent of all US Steel holdings of northern ore (from November 1907 it also possessed southern ores through its control of the Tennessee Coal, Iron, and Railroad Company) were there. In short, except for Alabama, the current operations, commercial success, and continuing prospects of this preeminent industrial combination depended on a continuing flow of ore from the iron ranges of one state. Not surprisingly, this state came to realize it could exact a high price for use of its natural resources. It wanted more income from its minerals and ultimately a US Steel integrated works.

From 1906 to 1908 state policy raised the taxable assessments of its mines to levels more nearly commensurate with their value as rated in the iron trade. When in May 1908 Theodore Roosevelt held a conference of state governors at the White House to discuss the conservation of natural resources, Gov. John A. Johnson of Minnesota revealed that a few months earlier Thomas Cole, superintendent of US Steel ore-mining operations, had told him that they had only just begun to scratch the surface of the Mesabi reserves. Addressing his fellow governors Johnson touched on the improvement of waterways, both in the Mississippi basin and on the Great Lakes, before disclosing that his main interest lay in another dimension, that of mining in relation to economic development: "It seems there are two problems here, not only the conservation of the resources, but the development of the industries. One is just as important as the other."[8] Procedures for attaining the latter were soon being perfected. During 1909 a bill was introduced to the state legislature proposing a tax of about five cents per ton on ore shipments. That was small in relation to that year's average commercial price of hematite iron ore throughout the United States of $2.17 per gross ton at the mine, but on shipments from Mesabi and Vermilion it would have cost US Steel over $900,000. One of the arguments used against the tax proposal in the legislature was that the Corporation had promised to build a plant in the state on the express understanding that no tonnage tax would be imposed. The bill was passed but thrown out by Governor Johnson. The fact that the threat had only been defeated by his veto was taken seriously, for as it was put shortly afterward, "as the governorship is not a life job it looked well to the Corporation to go ahead with the plant."[9] The point was well made, for in fact Johnson died in office in September 1909. Although, as suggested in the legislature, US Steel had by now committed itself to an integrated works at Duluth, it had done so without enthusiasm and carried the plan into effect with a leisureliness contrasting starkly with the urgency with which it pushed ahead at Gary.

In January 1907, five months after ground was broken for Gary, mounting Minnesota pressure caused the US Steel Finance Committee to resolve "that it is the sense of this committee that a substantial plant for the manufacture of iron and steel be constructed and operated in the vicinity of Duluth, Minnesota, *provided it is practicable and will be reasonably profitable.*" Gary, William Corey, H. C. Frick, and George Perkins were appointed as a special committee "to make an exhaustive study of the question."[10] Ten weeks later the finance committee recommended an integrated works consisting of a coal dock, one hundred byproduct ovens, one blast furnace (producing 150,000 tons of iron annually), a six-furnace open-hearth shop to make 200,000 tons of raw steel a year, and finishing mills capable of 225,000 tons annually of standard and light rails, shapes, and bars. The cost would be $5.5 million.[11] The press tried to build up the project. One claimed that the northwest was the likely region of most future extensions of rail mileage and would undergo a great growth in manufacturing. It further claimed that the future of the new works was considered by US Steel officers to be "very promising," cautiously adding that it would be extended "to care for all business . . . that can be better handled from the head of Lake Superior than from Gary or other works."[12] On the other hand, whether intended or not, a few months later the words used about the project in the US Steel Annual Report for 1907 managed to convey a mood of uncertainty and a lack of enthusiasm: "There has been purchased a site containing about 1,580 acres (of which 300 acres are now submerged) located in St. Louis County, Minnesota, ten miles from the center of Duluth, on which it is proposed to construct a moderate sized iron and steel plant. . . . The plans for the scope and construction of the steel plant have not yet been fully

developed."[13] For two years, apart from authorization of railroad connections
to the site and a small housing development, there was no further progress.
The sharp recession of 1908 was no doubt partly to blame, but Duluth interests began to suspect that reluctance to set things in motion had deeper
causes. In August 1909 Chester A. Congdon, a member of the Minnesota
House of Representatives who seems to have had interests in Mesabi ore, con-
tacted Frick about conditions there and the building of what was now referred
to as the Minnesota Steel Company. Corey was authorized to reply, expressing
surprise that "anyone should have questioned the intention of the Steel Corporation to erect a plant at Duluth, calling attention to the steps already taken
and stating that consideration of further plans would be taken up when a full
meeting of the Committee could be obtained."[14] Congdon's letter seems to
have stimulated action, however reluctant. At a meeting on 28 September, a
report from a special committee dominated by two representatives from Pittsburgh and two from Chicago agreed with the opinion of the original committee of two years before, "viz that 'strictly as a manufacturing and commercial
proposition, we cannot recommend the project,' but that on the understanding that the Corporation is committed to the building of a plant at Duluth,"
they proposed a rather larger project than their predecessors—130 ovens, two
blast furnaces, seven open-hearth furnaces, and various rolling mills. The
necessary capital expenditure was now put at $10 million.[15] Work started
that year. As the new mill began to take shape, company publicity and jour-
nalistic talents together contrived to build it into something bigger than had
been announced and certainly more than was intended. The site was reported
to be of about 2,000 acres (within US Steel it was still put at 1,500 acres) with
a two-mile frontage on the St. Louis River, and would include facilities for handling the larger lake carriers. As in Pittsburgh and Chicago, a belt line would
connect with every railroad coming into the steel center. It was expected that
the new works would daily make up to 1,000 tons of rails, shapes, and merchant products. (The company internally put its daily capacity at 1,000 tons
of *ingots.*) One account made what at first sight seemed a grand claim but on
further thought was modest enough, that as a result of experience gained at
Gary, the works might be "the most modern and economical in operation of
any steelworks yet built." In other reports the project took wings. One hinted
at long-term plans for a number of additional open-hearth shops that might
make the works second only in size to Gary; another, noticing the iron and
steel works would occupy only one-fifth of the site, looked ahead to the arrival
there, as at Gary, of other US Steel subsidiaries for which Minnesota Steel
might provide semifinished material.[16] Amid the euphoria, corporate officials
remained circumspect. In a summer 1911 speech to the Duluth Commercial

Club, US Steel's new president, James Farrell, who had come from the sales division, chose his words carefully: "It is the aim of the Corporation to build a steel plant here complete in every respect for the manufacture of every class of steel product for which a market can be found."[17]

The Duluth melting shop was at work by December 1915, and on 13 December its first billets were rolled. They were sent to Chicago for finishing until the merchant mill was at work in the middle of the next year. Duluth blooms, rounds, and billets helped supply the requirements of the war industries, but early operating experience was disappointing. There was no large inrush of other subsidiaries to colonize the remainder of the large site; even in a wartime boom steel-using industries did not find the area congenial. Whereas mushroom growth of metal-fabricating industries in Calumet brought about a temporary abandonment of "Pittsburgh Plus" pricing, as early as 1917 local interests were urging that Duluth should be made a basing point specifically in order to stimulate local consumption. This proposition was unattractive to the Corporation, for it would break up a system that bolstered the competitiveness of centers of production of far more importance to its overall success. Matters now began to come to a head. In summer 1918 residents of the area requested US Steel to commit itself to postwar extensions. The matter was referred to Gary and Frick. Corporate officers joined in a consultation with a committee representing a large number of Duluth citizens, resulting in "frank and somewhat extended conversations." Afterward, the US Steel Fi-

The Duluth works of the Minnesota Steel Company. *Courtesy of USX Corporation*

nance Committee proposed to deliver a letter "if it would be received as an adjustment of the questions involved." Taking up points made by the citizens, a new agreement was negotiated. Again it took the form of a bargain: some expansion by US Steel in return for a Minnesota willingness to back down on the basing point issue. Arguing that tight supplies of constructional steel and labor meant they could not then make extensions, US Steel intimated that when normal business conditions replaced wartime restrictions, it was the present intention of the Corporation, "on the assumption that it is the purpose of those who have been connected with the effort to establish a basing point at Duluth for the sale of iron and steel commodities to abandon the same," to build a wire mill, nail mill, tinplate works, and sheet mill "all up to the standard of our usual developments." They would also "make further additions and extensions at this point from time to time, whenever, in our opinion, the same are justified by the demands of the markets tributary to Duluth, together with the cost of production and distribution and other business conditions."[18]

Despite this commitment, after the Armistice was signed, steel making in Minnesota continued to lag. Sometime afterward, US Steel set up a substantial investigation into potential finished-product outlets in the twelve-state area that had been anticipated as Duluth's natural market. One major hope was completely defeated: its rail mill never turned out a single piece of track, for the boom in railroad construction was already over and the much bigger capacity at Gary could meet the needs of the time. The rail mill was put onto blooms and billets, but on these semifinished products margins over the raw steel price were lower than with fully finished products. The Western Association of Rolled Steel Consumers for the Abolition of Pittsburgh Plus took up the interests of Duluth. In February 1919 a local member asked a searching, potentially threatening question: "Furthermore, to us in Minnesota, a state which furnishes 80 percent of the ore that goes into the manufacture of steel, doesn't it really seem unreasonable for us to pay the freight on this ore clear to Pittsburgh and back to Minnesota, and be charged the Pittsburgh Plus Price?"[19]

Investigation now convinced US Steel that wire and small plates and shapes represented the best prospects for Duluth. In September 1919 it was decided to spend $9 million on new mills. By late 1920 their estimated cost had risen by over one-third. Yet the experience of the past year had been disillusioning. While there had been very high activity in the national economy, Duluth shipped only 108,000 tons of finished steel products. Less than 28,000 tons of this output went to the western north-central states; 10 percent of the total was exported. Even more upsetting, Illinois Steel had supplied 8,000 tons more steel to Minnesota than Minnesota Steel had shipped to all destinations.[20] By the end of 1922, finishing capacity was 120,000 tons, or about one-third the tonnage projected a decade earlier. From that summer on, the works had been equipped to produce both wire rod and wire but was still without either a sheet or a tinplate plant.[21]

Experience at Duluth highlighted another problem, the external costs involved in building a considerable steel plant in a remote location. In summer 1925 the vice president and general manager of Minnesota Steel outlined this situation in a rather rambling letter to the US Steel head office:

> In this particular institution we have to be very careful. Good men are scarce and stabilized men are scarce. . . . We have always been obliged to train men for every new operation we undertake. The principle was laid down in the early days of the property and we had to have men that could do more than one thing. In other words it has always been the practice here to take construction gangs on new departments and train them for operations. Naturally this policy was followed out in the erection of the wire mill. As a matter of fact during the boom times during the war we had 3,500 men in our employ and a careful analysis showed that only 366 had ever worked in any part of a steel works before or in any of its appurtenances. In other words in round numbers 3,200 were trained here. . . . The wire mill when running full has 572 men and I think we have only three wire drawers who did not learn the business here.[22]

In fact, by the early 1920s the prime inducement for US Steel expansion at Duluth had been overtaken by events. A second effort to impose a tonnage tax had been vetoed by Governor Burnquist (1915–1921), but then in 1921, against the opposition of legislators from the mining districts, a bill providing for a tax of 6 percent on the net value of the mined ore was passed. Giving evidence in the "Pittsburgh Plus" enquiry, Gary steered clear of the ore-tax issue, but he made it clear that Duluth was a high cost and generally undesirable location. "Duluth, it was stated, came along and objected to the Pittsburgh base and wanted a steel plant of its own. Judge Gary said the Steel Corporation established a plant there and that it had not been very profitable, but it has been possible to supply part of the demand. Others it was said couldn't afford to go into Duluth." Speaking to the Duluth Chamber of Commerce, Gary revealed that costs of finished products there were 13 percent higher than at Pittsburgh and 38 percent above those at the Gary works. From the other side, the Joint Committee of Civic Organizations of Duluth made clear the disappointment felt in the area: "it was fondly hoped that local industries would follow and that Duluth would be developed into a manufacturing center."[23]

The rod and wire mills at Duluth had been intended to supply not only the twelve-state area but also Canadian demand in the huge region from western Ontario to the Pacific Ocean. Competition from the steel plant at Sault Ste. Marie, Canada, and Dominion tariffs together largely excluded it from the Canadian market, and a slow down in the growth of the northern Great Plains states penalized it at home. Between 1900 and 1920 the five states of Minnesota, Wyoming, Montana, and North and South Dakota had added 1.6 million to their population; over the next twenty years they increased by only 450,000. Between 1900 and 1912 the rail networks of these states had been extended by 6,603 miles; over the following thirteen years no more than 1,616 miles were added.

Capacity changes within US Steel naturally reflected the unfavorable cost situation of this mill. After first stagnating, production began to contract. Between 1920 and 1930 rated ingot capacity at Gary increased by 45.7 percent, while that at Duluth remained unchanged. Through the 1930s the Great Plains were in deep economic distress. During the terrible trading conditions of its early years, Minnesota Steel was having to change rolls for as little as five-hundred-pound orders on its merchant and bar mills, "work that must have been done at considerable loss." Byproducts from its coke batteries could find no local outlet, and the benzol plant was closed.[24] In May 1932 the company suffered the indignity of losing its separate identity, being merged into American Steel and Wire (AS&W). Five years later, discussing the merger, the chairman of US Steel remarked, "We have not been entirely fair with the Wire Company because, when we wanted to dispose of some of our property in Duluth, we dumped it onto them."[25] In the mid-1930s, one of Duluth's blast furnaces and five open-hearth furnaces were abandoned; they were said to be "both obsolete and uneconomically located." When the consulting firm of Ford, Bacon, and Davis reported on AS&W in spring 1937, they did not chose Duluth as one of the five locations at which AS&W's multiplant operations ideally should be concentrated. They recognized that individual units of plant were better than the division's average but, because of the liberal layout of the works and its later failure to expand, they were inconveniently far apart. The report provided a cool yet damning summary of Duluth's beginnings and a minimalist justification for its future: "The original concept of the Duluth plant was greater than the development of any market that it could serve. . . . The investment in this plant practically dictates its continuation."[26] By 1938, capacity for iron had been halved and that for steel was 44 percent less than the 1930 level. To add a final touch of commercial reality to a project whose very rationale had been political, when US Steel abolished most of the price differentials at its basing points in 1938, it retained slightly higher levels for Worcester, Massachusetts, and for Duluth, explaining that production costs there were high partly due to frequent roll changes and in part because of high distribution costs, the cause for both being the scattered and varied demands they served.[27] After some expansion during World War II, the works survived another twenty-five years before most of it was abandoned. The life history of Duluth works had proved that allowing politics to play an important part in industrial location was unwise.

Minnesota Steel was built in an area in which, for various reasons, the scale and pace of further economic growth was overestimated. In the Southeast US Steel became involved in an area of arrested development whose horizons eventually expanded far beyond initial expectations. Duluth was an instance of plant construction and operation by a genuinely reluctant company. The apparent doubt that accompanied the acquisition of the Tennessee Coal, Iron, and Railroad Company (TCI) in 1907 reflected the fears of those in command that the federal government might outlaw what increasingly was seen as an attractive proposition.

"Manufacturers of steel in Alabama make very cheap steel on paper, but they have only made it there [on paper] yet." The skeptical tone of these words of Andrew Carnegie in fall 1898 combined the confidence of northern steel makers with an undertone of growing fear that before long southern steel might prove troublesome. Two years earlier Henry H. Campbell of Pennsylvania Steel had shown that, in theory at least, the cost of stock per ton of basic hearth steel in Birmingham, Alabama, should be no more than about two-thirds as high as in Pittsburgh.[28] By this time the metallurgical industry of Birmingham, and indeed of the whole of the South, was dominated by TCI.

Closely surrounded by large deposits of iron ore and coal in the southernmost section of the Appalachians, TCI made its first commercially produced basic iron in 1894. The next year it sold some of its product to Carnegie Steel, which with other northern sales helped open the eyes of TCI management to the prospect of adding steel to their iron products. Even so, by 1897–1898 they were initially unresponsive to a suggestion from the president of the Louisville and Nashville (L&N) Railroad that they should build a steelworks to make use of some of the pig iron they were dispatching to the North. When in 1898 construction of a ten-furnace open-hearth shop was begun at Ensley near Birmingham, the iron company, the L&N Railroad, and local citizens each contributed one-third of the capital.[29] TCI's first commercial shipment of steel was made to a Connecticut customer in January 1900. In December 1902 the plant began to roll steel rails, the first in the nation to be made of open-hearth steel. Yet, notwithstanding these signs of progress and success, by the first years of the new century TCI was in financial difficulties; generally, its plant was of inferior quality. In 1901 Don H. Bacon came to TCI from the presidency of the northern iron-ore company Minnesota Iron and was chairman for five years. He later recalled what he found. His new company was "an immense, loose-jointed concern . . . slung together in a very slip-shod way"; its equipment "a medley of make shifts." Despite its uniquely favorable minerals situation, pig iron costs were higher than average, the result of inefficiencies in the coal and ore mines, relatively poor coke ovens, and until 1904 total dependence on old blast furnaces. A good rail mill was built, but the open-hearth plant that supplied it, though recently constructed, was very poor. In the middle of the first decade of the century, steel consultant Charles P. Perin reckoned $20 million was needed to make TCI more or less equal in efficiency with northern plants; Bacon's own figure in 1906 was $25 million. Unfortunately, the company lacked adequate capital resources to upgrade its plant and could not finance it from profits, which were too small. In the five years to December 1905, $7.5 million was invested in improvements and enlargements; net earnings during the same period were $8.1 million.[30] Yet amidst all these problems, there were signs of innovation and still more of promise. In 1904 TCI began "duplexing" its steel: that is, processing the hot metal in an acid Bessemer converter before transferring it for finishing in basic open-hearth furnaces. Steel production that year was 155,000 tons; in 1906 it jumped to 402,000 tons. However, the large outlay on plant meant that as production rose, costs increased and profits fell.[31] During 1905 there was at last a glimmer of a way out. A syndicate lead by J. W. Gates, the northern wire and steel capitalist, disappointed in his hopes of becoming one of the leading figures in US Steel, gained control of TCI and consolidated it with the southern interests of the Republic Iron and Steel Company. In May 1906 Bacon, who had concentrated most on improving mining operations was replaced by John A. Topping, president of Republic, who, coming from Wheeling, had direct experience of northern steel making, and was better equipped to understand the various aspects of the business. This new controlling group also could call on larger capital resources. Prospects seemed to be opening up; as Topping expressed it to Herbert Casson in 1907, "The Tennessee Coal and Iron is now being run to make iron and steel, not for Wall Street purposes."[32] In spring that year, a completely new dimension of thinking and planning for the future was marked by the announcement that $25–30 million was to be invested in modernization and extension. The blast furnaces were rebuilt and the construction of more open hearths and an additional rail mill was begun with the intention of doubling capacity and "radically" cutting costs. By summer, over $6 million had already been spent. A breakthrough seemed likely, for as a trade journal put it, the "one great thing which the Southern iron business needs is the investment of millions of dollars in the building of the most modern furnace and steel plants to rival the great works in the East and the West."[33] The South could be expected to offer much keener competition.

There were occasional Wall Street rumors of US Steel interest in TCI. These were not wholly unfounded. Sometime in late spring or early summer 1907, the Corporation's first vice president in charge of raw materials and transportation, James Gayley, recommended that the acquisition would be a good investment from the standpoint of raw materials alone. Years later he recalled that his former Carnegie Steel colleague, Henry Clay Frick, who was much more influential than he himself was in US Steel's top decision making, "was absolutely opposed to it. He did not favor it at all," except at a very low price.[34] The two aspects of TCI of particular interest to US Steel were its iron-ore reserves and its place in steel-rail production.

The controllers of TCI seem to have been rather vague about ore reserves. Gates recalled they always reckoned they controlled 300 million tons, but that it might perhaps be as much as 500–700 million tons. However, the ore was poor, low in iron content (the ore used by TCI was of 38–39 percent Fe), and phosphoric. So far, the cheapness of its minerals had not been reflected in iron and steel costs. It was generally agreed that Birmingham should produce the cheapest pig iron in the United States, though inferior blast furnaces had prevented it from doing as well as it might. Bacon recalled their older furnaces made iron at ten or eleven dollars a ton, but that a new furnace built in 1904 had cut the figure below eight dollars.[35] Partly because little scrap was available in this relatively unindustrialized section of the country, its raw steel costs were higher than in Pittsburgh. There was dispute about the size of the differentials. A critical consideration in US Steel policy was whether, having tried to tie up Upper Great Lakes ore resources, the time had now come to control southern ores as well. The high royalties paid from the start of 1907 for the Hill leases on the Mesabi Range suggested to some outside observers that the strategic moment for such action had arrived. The rail market seemed to point in the same direction.

TCI pioneered large-scale use of open-hearth steel for rails. By 1906–1907 US Steel managers had recognized that the railroads were about to switch from Bessemer metal to more reliable material, rails made from which commanded a normal premium of $2 over the $28 a ton price for Bessemer rails. In 1906 it was decided that Edgar Thomson extensions should be in open-hearth steel; even more important, rail production at Gary works, under construction at the time, would be confined to open-hearth steel. Birmingham was reckoned to be at least as well located as Pittsburgh or Chicago to supply the rail requirements of about one-third of the national area. In spring 1907 the vaguely looming threat from the South seemed to be taking substance when the Harriman system ordered 157,000 tons of Ensley rails. Only later did it become known that their average cost of production was $29.48, or $0.72 a ton higher than the average price received. Some of the rails were returned as defective.[36] On grounds of competitive costs, US Steel had little to fear, but those who controlled it were unaware of this. In any case, it was necessary to think ahead, and when this was done TCI was seen as a possibly important challenger.

These issues were brought to a head in a depression that began in autumn 1907. Like the industry generally, US Steel suffered a sharp contraction of business. After peaking in October at $17.05 million, net earnings fell to $10.47 million in November and then to $5.03 million the next month. For TCI, much less firmly based and caught in the midst of a major investment program, the situation was more serious. The company had a floating debt of about $4 million, and the syndicate that had taken over two years before, and had since then spent about $6.25 million on its facilities, was unwilling to release all the money needed for both modernization and financial rehabilitation.[37] Gary later recalled that it was around 27 October when he was first consulted by J. P. Morgan on the danger of a collapse of the New York stockbrokering firm of Moore and Schley, which was heavily involved in TCI stock, at this time of difficulty not regarded as a sound investment. In fact, four days earlier a special morning meeting of the US Steel Finance Committee had resolved to exchange $1.2 million par value in gold bonds for $2 million in shares of TCI. But, as Moore and Schley's position worsened, it was made clear that attempts to undergird the southern iron and steel company were inadequate to meet the needs of the time and that only outright purchase could save them from failure, which in turn would have dangerous effects on the stock exchange. Gary pointed out to Morgan that TCI stock had been offered to them before and they had turned it down. Earlier in the year it had been Frick who was "absolutely opposed" to Gayley's idea that they should gain control of TCI; now, as the lawyer Lewis Ledyard later revealed to the Stanley Committee, "Judge Gary was the reluctant one."[38] Gary indicated they did not think TCI was worth more than $50 or so a share. But there were greater dangers than that of not getting value for money. As Gary and Frick told Morgan, "it would not help the situation if the President of the United States came out and started proceedings against the US Steel Corporation as attempting a monopoly by the purchase of the Tennessee Coal and Iron Company."[39] Yet the pressures for action increased, and it was indicative of the urgency of the situation that the finance committee met on Saturday morning, 2 November 1907. The minutes of that meeting recorded: "The subject matter of the purchase of the Tennessee Coal Iron and Railroad Company on terms suggested was fully discussed and thereupon the meeting adjourned without action." Late on Saturday evening and on three separate occasions during Sunday, the committee reconvened to consider the same subject. The Sunday meetings were followed by the dramatic episode of Gary and Frick traveling overnight to Washington, where at breakfast on the morning of Monday, 4 November, they pressed Theodore Roosevelt for his reaction to the suggestion they should absorb TCI. While they were at the White House, the president dictated a letter to his attorney general, which Gary was shown. It gave at least Roosevelt's moral support to the acquisition as a step toward checking further business decline. Some of the points that were made are worthy of note:

> Judge Gary and Mr. Frick informed me that as a mere business transaction they do not care to purchase the stock; that under ordinary circumstances they would not consider purchasing the stock because but little benefit will come to the Steel Corporation from the purchase; that they are aware that the purchase will be used as a handle for attack on them upon them on the ground that they are striving to secure a monopoly of the business and prevent competition—not that this would represent what could honestly be said but what might recklessly and untruthfully be said. . . . But they feel that it is immensely to their interest, as to the interest of every responsible business man, to try and prevent a panic and general industrial smash up at this time. . . . But they asserted that they did not wish to do this if I stated that it ought not to be done. I answered that while of course I could not advise them to take the action proposed I felt it no public duty of mine to interpose any objection.

More tersely, Frick later summarized the presidential attitude as one of "tacit acquiescence."[40] On Wednesday afternoon, 6 November, a brief meeting of the finance committee agreed that $30 million of US Steel bonds should be exchanged for TCI. In addition to the purchase price, the Corporation had to pay off some $3 million in TCI debts. Wall Street had been helped to avoid a more desperate financial panic, but "the general impression was that a bargain had been secured." Further progress on the takeover was held up for some time until opposition to the acquisition could be overcome.[41]

Notwithstanding the doubts of some directors, US Steel obtained good value for its money, above all in development potential. In 1911 John Wayne Gates claimed that the Corporation "got the best property in the country, and at a bargain price. I regard it as a sacrifice of stock worth a great deal more than the purchase price. . . . The iron ore and coal deposits of the Tennessee Company are worth many times more than the entire cost of the property to the Steel Corporation"[42] He had never been the most reliable source of information, and in this instance Gates was for a variety of reasons a suspect witness, but he had good first-hand knowledge of TCI. More significant was an assessment made by a man of unquestioned technical expertise and temperamentally quite different from the flamboyant Gates. Julian Kennedy had worked for Carnegie Steel before practicing as an independent steel engineer, in which profession he established a world reputation. His valuation of TCI ore and coal reserves and manufacturing plant as of 1907 was from $90–100 million.[43]

In order to improve efficiency at TCI, US Steel had once again to change the organization there and spend very liberally. John Topping was replaced as president by G. G. Crawford, a man whose extensive experience of blast furnace practice had been gained at Edgar Thomson and McKeesport. (His success in his new job was indicated by his retention of the post until 1930.) TCI operations were inspected by Corey after the takeover, and he reported back to the finance committee that investment of $5 million would be needed annually for a number of years in order to put them into shape, after which the southern operations might pay dividends. When the previous administration's modernization program had been completed in 1908, US Steel pushed through its own improvements. By 1913 Gary put the capital outlay there by the Corporation at $32 million. Some of this was for expansion rather than modernization, but so effective was the latter that Crawford claimed that within eighteen months of the acquisition the scrap yield in rail making had fallen from an extraordinary 35–40 percent of the tonnage of steel used to only 10 percent, and the mill cost of rails was cut from $29 to no more than $20 a ton. Radical changes in melting-shop practice cut the previous high rate of rejects from the railroads. The old melting shop was replaced in 1908, and four years later the new plant was duplicated. By 1912, annual steel capacity at TCI was 840,000 tons, 162 percent greater than Alabama output as recently as 1907. Iron quality had been improved by moving from mixtures of brown and red ore to complete dependence on the latter.[44]

After the upgrading of operations the extensions continued. In 1912 American Steel and Wire built rod and wire mills and a byproduct coke plant on a site southwest of Ensley, the new plant and associated town being called

Corey, though soon renamed Fairfield. The AS&W mills drew their steel from Ensley. By 1914 canalization of the Black Warrior River southward from what became Port Birmingham was underway with the expectation that it would improve facilities for Birmingham shipments. During World War I, plate and structural mills were built at Fairfield. By this time, in strong contrast to Duluth, southern prospects were promising. So impressive was the change that, though an apparently confirmed skeptic when negotiations for TCI began, before he died Frick was said to regard Birmingham as the nation's coming greatest center for steel production.[45]

Despite wide movements of raw materials, semifinished products, and finished products, steel production generally did not cross national boundaries in the first half of the twentieth century except where there seemed the assurance of special politically based oversight as in the British Empire. An impressive feature of US Steel in its first decade was the purposeful way in which it developed its foreign business, through a new organization that replaced the until then separate overseas sales departments of its constituent companies. In 1903 the US Steel Products Export Company was created under the control of James Farrell, who until then had directed the foreign business of American Steel and Wire. From 1902 to 1913 the overall value of products and services sold by the Corporation increased by 32.5 percent, and during the same period the annual value of its foreign business went up from $100 million to $305 million.[46] Notwithstanding this impressive growth in exports, in one or two instances serious attention was given to the idea of setting up operations overseas, in which US Steel would secure operations inside another national economy but inevitably also be subject to the vagaries of government policy over which it could hope to have little influence.

One idea that was considered quickly fizzled out. Even before the formation of US Steel, what were later to become its constituent companies had considerable success in selling steel to Great Britain. It was indeed a complaint of American shipbuilders that steel plate was being sold to Belfast yards at below the price domestic builders had to pay on the East Coast. Soon, semifinished steel from the United States was being delivered to rerolling mills in the Black Country of England at prices that excluded home producers of the same material. Under the leadership of the eminent politician Joseph Chamberlain, a Tariff Reform League was established to campaign for protective tariffs for leading British industries and for a new system of imperial preferences. It was at this time that US Steel gave some consideration to a proposal from Arthur Keen, who had close business associations with the Chamberlain family, to buy its way into a prominent position in the iron and steel industry of Britain.

Gary seems to have favored going further, but Schwab adopted a very practical attitude: "the only objection he [Schwab] sees is that in times of depression here when we desired to unload some of our product over there, it would then be against ourselves." Though the finance committee gave serious thought to the proposal, nothing came of it.[47] The situation regarding Canada was very different, and there US Steel made its most important moves toward foreign operations in the early decades of its existence.

In 1901 Canada was the third largest destination—after Britain and Germany—for United States merchandise exports, taking 7.1 percent of the total. Conditions for that trade were becoming more difficult, however. In 1897 Canada adopted a preferential tariff in favor of British goods, amounting to 25 percent less than that on goods from foreign countries. By 1902 the margin of preference had been widened to 33.33 percent. Soon, under the stimulus of bounties from the national government, Canada's own steel output increased rapidly, from only 27,000 metric tons in 1901 to 410,000 four years later. In February 1905 the US Steel Finance Committee decided to investigate the question of building or acquiring a Canadian plant. Two months later, having visited the Algoma works at the Soo and other sites, they decided to prepare plans for "a complete, modern, balanced and self-contained steel and iron plant." A one-thousand-acre site at Walkerville, in extreme southwestern Ontario close to the United States border along the Detroit River, was chosen. Next year a letter from Farrell prompted an investigation for a Canadian rail mill. In 1907 W. B. Dickson, second vice president of US Steel, called Gary's attention to the loss of business due to adverse tariff legislation. There was another suggestion that US Steel should buy the Algoma works before, in 1909, Farrell submitted a proposal for building a Canadian plant to make all lines of materials they were then exporting there.[48]

Eventually, bridge shops were built at Walkerville, but a 1912 plan for a fully integrated works with rail, structural, bar, and wire mills on a nearby 1,200-acre site with a 1.5-mile frontage on the Detroit River at Ojibway continued to languish. Not until World War I was there an apparent increase in the pace of development. Less than two weeks after the United States entered the war, the finance committee at last decided to spend $9 million at Ojibway to build plant foundations, docks, and make improvements to a townsite. In February 1918 an even bigger project seemed to be underway when the finance committee first deputed Gary, Farrell, Frick, and Roberts to consider installation of a plate mill of 150,000 tons annual capacity and quickly followed this with approval of the mill, open hearths, and one or more blast furnaces. This decision was made conditional on Canada allowing the import of materials, entirely free of duty, for building the works and making a commitment to buy 75,000 tons of plate annually for five years at a price $20 a ton higher than that prevailing in the United States. Work began on two blast furnaces. There was vague talk that Ojibway might eventually have a capacity comparable with that of Gary. However, changes in Canadian duties ensured that the blast furnaces were never completed.[49]

Much more modest use was found for part of the large Ojibway site. Late in 1928 it was agreed that the Canadian Steel Corporation should sell the land it had hitherto reserved as a townsite. Next year it requested almost $1 million to install tin and galvanizing plant, and this was approved on 15 October 1929 just as the great stock market crash began.[50] Meanwhile, Canadian steel production continued to increase, from 0.75 million metric tons in 1910, to 1.12 million tons in 1920, and to 1.40 million tons in 1929. Over the next few years, not only did the Canadian economy fall into deep depression, but between 1930 and the Ottawa Economic Conference in 1932 Canada also moved to a more pronounced policy of imperial preference. By November 1932 US Steel's general counsel was preparing a submission to the Canadian prime minister calling his attention to the "practically insurmountable difficulties" caused by the tariff and asking for "equitable revision of the rates." One way out of the situation seemed to be to procure supplies from within Canada, and early in 1933 Myron Taylor reported on a proposal to buy Algoma Steel of Sault Ste. Marie as a source of steel. No action was taken.[51] In 1935 a plan to sell Canadian Steel and the Canadian Land Company for $1.5 million was discussed, and two years later it was decided to dispose of the bridge and tower shops at Walkerville and the tinplate, galvanized sheet, and wire products operation at Ojibway. As Taylor explained to stockholders, "due to the prohibitive taxes imposed under the Canadian-Empire agreements, they were no longer outlets for semi-finished material." Canadian Steel, Canadian Land, and Canadian Bridge, along with the associated Essex Terminal Railroad Company, were sold to the Dominion Steel and Coal Company for $5.9 million.[52] The Canadian venture seems not to have cost US Steel a very large sum of money, but the experience had proved that foreign operations, even when so exceptionally well located in relation to the United States, placed even a most powerful corporation in too much dependence on the unpredictability of governments.

Entrepreneurial Failure?

Technological Backwardness, Constructional Steels, and the Universal Beam Mill

It might be expected that a prominent advantage of a giant corporation would be an ability to pioneer in the best technology, either by using the inventions of others or by applying the results of its own researchers. Carnegie Steel had led in the large-scale adoption of the basic open hearth and had always shown a readiness to scrap and rebuild in order to equip its mills with state-of-the-art plant and equipment. At its projected Conneaut works, it had planned to be technologically so far ahead of its rivals as to produce tubular products at far below the costs of the recently created agglomeration of units, National Tube. Benefits of this kind seemed logical further steps toward the possible processing cost economies that Charles M. Schwab had outlined on the evening of 12 December 1900. Ten years later, reviewing his own presidency, Schwab's successor, William Corey, claimed that US Steel still led the industry: "The history of the Corporation is marked, not only by . . . vast increase in capacity, but also by some radical changes in methods, such as the installation of gas engines, the displacement of steam engines by electric motors, and the building of by-product coke ovens." In fact, the first two advances implemented were only improvements in equipment and only the last could be claimed as a radical change in methods of production, even then in an activity that at that time had been peripheral to the making of steel and finished products. Corey continued: "As a general proposition, I believe that as a result of the expenditures and developments of the past ten years, our companies are at least five years in advance of their competitors in general efficiency and in the ability to serve the trade; and that we can produce on the average $5.00 per ton cheaper than the average competitor."[1] In those words he, like many others before and since, seemed to be underrating the significance of the word "average." In fact, though a giant from birth, US Steel in most respects failed to be a techno-logical pacemaker.

Did this sort of failure occur in spite of the size of the company involved, or was it rather a natural result of predominating size? Is difficulty in recog-nizing and implementing the most important new technological break-throughs an unavoidable cost of huge combinations that must be balanced against their inherent operating economies? It is obvious that such back-wardness cannot be due to their lack of technical expertise. Rather, it must be a consequence of other factors. A firm having a vast commitment in fixed cap-ital cannot justify its replacement as easily as an expanding firm is able, as it were, to "grow" into a new technology. In short, the timing of an innovation in relation to a company's life cycle is important. Additionally, a giant business unit, especially one with scattered operations and a division of power between a hierarchy of top decision makers and plant-level management not only pro-fessionally but also geographically, seems to be less responsive to changes in the market and signals from technological innovators than single or smaller operations with either one key leader or a close-knit group of founders who maintain close contact with both headquarters and mills. In a group as large as US Steel, any decision to strike out on a new line or process had to be taken after the desirability of change had been recognized at the plant level and when some one person or group had proved willing to fight the cause through the (unavoidable) bureaucracy of an extended chain of command.

These were inherent problems and accordingly give no grounds for ac-cusing US Steel of particular innovative failure. However, for many years its position was made worse by a desire to avoid upsetting the rest of the industry at a time when its own behavior was always under public scrutiny and for many years the subject of government antitrust enquiry. Possibly more im-portant still was the fact that for its first quarter-century, US Steel was led by an inherently cautious man trained in law and not in steel making. Although after his death there were important changes, the traditions and habits of mind that he had instilled could not readily be swept away—laggards cannot become leaders over night. In short, it is difficult to evaluate the various fac-tors that together ensured that US Steel did not pioneer the application of new technologies. What is unquestionable is that it paid a high price for most often being behind. This was first shown by its experience in the manufacture of heavy structural shapes.

One of the most dramatic increases in the use of steel in the years around the turn of the century was for construction. Wooden bridges were replaced by steel, there was a large expansion in the consumption of beams and angles in shipbuilding, and steel shapes were used in increased tonnages in rail cars. Most significant of all, heavy sections were now employed on an ever-extending scale in buildings, the multistory office or apartment buildings and, most spectacularly, the skyscraper being essentially a skeleton of steel beams on which the rest of the building materials were hung. Because of these new uses, demand for structural shapes increased rapidly. In 1879 these products, almost all of which were made of iron, composed only 3 percent of the total rolled iron and steel tonnage. Twenty years later their share was 8 percent of a vastly increased total finished tonnage, and 97 percent of the structurals were of steel. Ten years further on structurals amounted to 11 percent of all rolled products.

In the late nineteenth century, the rolling of structural shapes was dominated by a handful of firms. There were three small operators on the East Coast: Phoenix, Pencoyd, and Passaic. To the west, by 1890 the North Chicago works could roll up to 50,000 tons of structurals at the expense of rail production. At Steelton, Pennsylvania Steel had a small capacity, and from 1886 Jones and Laughlin produced Bessemer shapes. There were a few smaller operators. The new Erie shore works built by the Lackawanna Steel Company at the turn of the century was designed with some capacity for structural steel. But from the mid-1880s, the unquestioned leader in the trade was the Carnegie group, whose Homestead mill was as early as 1889 offering beams up to twenty-four inches in width. US Steel inherited the Carnegie dominance of the business, in 1901 turning out 62.2 percent of the nation's structural shapes. Yet within a few years, it had been surpassed in efficient technology and was beginning to experience erosion of its lead in output. This seems to have been a clear instance of entrepreneurial failure, a deficiency understandable in the decision making of such a huge and diverse agglomeration, but nonetheless painful in its consequences.

National production of structural shapes in 1901 was 19.1 percent greater than the previous record, achieved two years previous. Homestead was preeminent, and 60 percent of all production came from Allegheny County. Although the area's economy was then in its most dynamic phase, its annual consumption of structurals in buildings, bridgework, and other construction was estimated at no more than about 50,000 tons, less than one-twentieth of the national total. The main centers of demand were far away, above all on the eastern seaboard, where US Steel controlled one of the three local producers, Pencoyd. Each year, New York City used about 125,000 tons in buildings alone.[2] In this growth sector, the Corporation was inevitably interested in expansion, in spreading its capacity more widely, and if possible in new technology. Even so, in 1905, when offered the opportunity to purchase the Passaic

TABLE 6.1
Apportioning of sales in the structural steel pool, 1897 (percentage of all sales)

Carnegie Steel	49.3	Cambria Iron	5.0
Jones and Laughlin	12.7	Universal Construction Co.	4.5
A&P Roberts (Pencoyd)	11.5	Pottsville Iron and Steel	3.0
Passaic Rolling Mill	6.0	Cleveland Rolling Mill	3.0
Phoenix Iron	5.0		

Source: W. C. Temple to Stanley Committee, quoted in *Iron Age*, 17 August 1911, 367.

Note: Each firm made monthly returns of production and shipments. Excess shipments involved payment to the pool of 0.5 cents per pound; shortfalls received payments from the pool of 0.5 cents to the pound. Each month $5,000 was paid to the New Jersey Steel and Iron Company in Trenton for not producing beams and channels.

Rolling Mill Company, US Steel declined.[3] On the other hand, the fact that through American Bridge the Corporation was also very prominent as a fabricator was seen to have dangers. Less than two months after formation, Schwab told the executive committee that some important independent fabricators who had formerly bought large tonnages from Carnegie Steel now took the position that as American Bridge, also a US Steel subsidiary, was a competitor, they might instead obtain steel from other producers. Time confirmed his fears.[4] In June 1904 Henry Bope reported to fellow directors of Carnegie Steel Company: "A strong feeling against us is apparent in the attitude of the independent users of structural material on account of the Bridge company in taking about all of the work offering in their line. The McClintic-Marshall Construction Company are now buying steel at the eastern mills for shipment to Rankin, and they say frankly that this is done because to buy from us would practically mean to play into the hands of a competitor." Willis King of Jones and Laughlin had told him that some large fabricators had approached his firm in hopes of making arrangements to enable them to compete with American Bridge. A few years later W. B. Dickson suggested to Corey that to avoid the ill effects of this loss of customers, US Steel might need to expand American Bridge so as to take practically all of their structurals.[5]

The early pressures for increases in capacity came from the west. In 1902 a US Steel director, the Chicago dry-goods wholesale and retail dealer Marshall Field, urged the Corporation to build a structural mill there. After discussion the executive committee resolved to take no action on his suggestion.[6] However, within a few years major structural-steel capacity had been installed in South Chicago. Meanwhile, it was gradually recognized that even Homestead's large capacity was insufficient to meet demand and that to increase it would require a large investment from US Steel, at a time when other plants and districts were also pressing for funding. The gathering problems came out clearly in the directors' meeting of the Carnegie Steel Company in February 1905. Henry Bope reported that they were under general pressure to meet high levels of demand—in plate, except for just over five weeks of capacity, they were "sold up"; in bars they had already contracted for half their annual capacity—but the demand for structurals was their most serious problem. Already the division had 490,000 tons of orders on the books or under contract, and their monthly capacity was only 50,000 tons. Bope concluded: "We are sold for the year, as nearly as we can figure it today. And yet there has never been so much work on the architects' boards as there is at the present time." Alva Dinkey, the company president and until 1903 Homestead's general superintendent, suggested that a provision for a new and improved shape mill should be a first concern and made clear that technology as well as tonnage was on his mind. "We have done nothing on shapes since the big beam mill was put in at Homestead. Our 23" and 33" mills are obsolete, as they stand now pretty much as they were when built. I think Mr. Hunt [Azor R. Hunt] might as well proceed to have a study of a new mill worked up, to roll such sizes as we are now and have been congested on." To service such a mill and new sheared-plate capacity, they would need additional open-hearth capacity and a new blooming mill.[7] Soon the technological lag implied by Dinkey became an important influence on their inability to meet the demands of the time.

By July 1905 Bope had recognized that US Steel as a whole was not holding its own in structurals. More capacity was required, and the Corporation lacked an up-to-date mill except for the one then being built at Illinois Steel's South works, which would not be ready until the fall. W. Singer remarked that if they failed to increase output, "it is certain that other people will." Dinkey seemed now to be reconciled to upgrading rather than replacing existing mills: "We have made the structural business, and to see other people step in and take a good part of it from us would be too bad indeed. Now our 23" mill is practically as it was built years ago and Mr. Hunt could reset it and double the tonnage and decrease the cost in proportion. The 33"mill is about in the same shape and the 35" mill needs resetting." In one week that month, Bope sent two telegrams to Corey in New York pressing the case for additional structural capacity. What they had in mind was a conventional unit; they reckoned they could "get a mill together" in sixty days at an outlay of $200,000. It would produce 10,000 tons a month and, incredibly, pay for itself in only two months.[8] Within two years they were facing competition not only from newly built "conventional" plants but also from a radically new type of mill.

The costs of producing heavy structural shapes were relatively high, a fact reflected in prices: in October 1900, when standard rails were $26.00 a ton at Pennsylvania works, structural shapes were $33.60. These shapes were rolled on two-high mills in a laborious process. In turn the shapes were costly for the structural engineer to assemble. Since at least 1875 it had been realized that costs for both producers and fabricators could be cut and an improved product made if structural shapes could be rolled in a universal mill—that is, one with side as well as top and bottom rolls; the side rolls would shape the flanges while the width of the beam was determined by the horizontal rolls in the second stand of the mill. In such an operation it should be possible to produce wider and thinner flanges more efficiently than in conventional mills. Use of shapes of this sort would cut out much of the riveting, reducing the work and costs involved in constructing the frame of a major building.

Various attempts were made to design and operate a universal structural mill in the 1890s, most notably in what proved to be a commercially unsuitable location—Duluth. Early in 1902 the US Steel Executive Committee resolved to look into a process for rolling I-beams controlled by John Ennis Searles of New York through the American Universal Mill Company.[9] Nothing further seems to have come of this. Soon afterward, the true line of advance proved to have been marked out by a man who had worked on the ill-fated Duluth mill, Henry Grey. The Grey universal beam mill made it commercially possible for the first time to produce wide-flange beams. Grey's first mill was installed at Differdange, Luxemburg, in 1902. A few years later he claimed that savings in weight of material used and in labor meant beams could be produced by his method on the order of $9.75 a ton less than by conventional methods of rolling.[10] Such a saving, coupled with a product more convenient for the fabricator, pointed to large demand for the new product once mill-teething troubles and conservatism on the part of structural engineers had been overcome. The inventor began to look for entrepreneurs to take up his process in America. Given its dominance of the business and immense financial resources, he naturally turned first to US Steel.

Schwab, still president, had many years of practical experience in heavy structurals from his superintendency of Homestead and was impressed by the commercial possibilities outlined by Grey. He suggested that the Corporation purchase the rights, but this was turned down by the finance committee. The decision closed off a promising avenue for Grey. By November 1904 he was again in contact with Schwab, but the latter, having resigned his directorship in US Steel, was now actively engaged as owner and head of the newly reconstructed Bethlehem Steel Corporation. After he and his new colleagues had twice visited Differdange, in December 1905 Schwab contracted with Grey for a mill to be installed at Bethlehem. Some time later, on a point of honor, he is said to have decided that he must inform US Steel that he was adopting an invention first brought to his notice when he was at its head. When he did so, Judge Gary referred the matter to experts who recommended they should not adopt the new-style mill. This decision was communicated to Schwab.[11]

By spring 1906, well before the Bethlehem mill was built, competition in structurals was mounting. There were seven mills in the East, and Bope anticipated there would be nine within a year. Some of these were not of any great significance. Phoenix, Passic, and another operator, the Eastern Steel Company, were expected to fall behind in the race for business. Jones and Laughlin were reported as "very strong" and their new mill to be one of the best in the nation. A tidewater works was being built by Milliken Brothers. The other addition would be the Bethlehem installation. Bope described Bethlehem Steel as "putting up about the same class of mill" as Jones and Laughlin, but at this point in their discussions Dinkey interrupted to point out that the Bethlehem plant would be a universal mill.[12]

As a result of various problems, in part at least related to the acute commercial crisis, Bethlehem did not roll its first wide-flange beams until January 1908. The new product came onto the market at a difficult time. Trade was bad after the high levels of the previous year, and initial resistance from engineers had to be overcome. However, the universal mill soon made an impact in the temporarily shrinking market. In addition to its new, superior technology, Bethlehem had another major advantage, that of ready access to those major cities of the Mid-Atlantic Coast, which were the prime markets for heavy structurals. Already, transportation costs were rising, penalizing more distant suppliers. In 1898 the Pennsylvania Railroad rate on Carnegie structurals to New York City had been $1.54 a ton with a fifteen-cent-per-ton rebate to the steel company's Union Railroad; by 1908–1909, the rail rate Pittsburgh to New York was $3.58 a ton, a higher rate of increase than that of steel prices.[13] By late summer 1908 US Steel was beginning to feel the impact of the Bethlehem mill. In early autumn Corey wrote to Frick, "Gary seems considerably worried over something, don't know what." But, in fact, Corey thought he could identify the cause. "I believe it would be advisable to make an agreement with Schwab on a fair basis, say not over three years. This would give us time to build a mill, strengthen the structural situation, which is rather mixed up and Schwab will not make a great deal of money." Almost immediately afterward, Corey returned to the same theme but this time put it into a wider context both in relation to industrial and national circumstances:

"Gary may bring up the question of an arrangement with Schwab during my absence. Am very decided in my opinion that nothing should be definitely decided until after election for we shall want our hand free in case the unexpected happens [presumably the election of William Jennings Bryan]. At times I almost wish for a change for our present policy is building up competition all over the country, some of which is cutting our throat. We should make a good fair bargain with Schwab and am certain we can if we are only careful and patient."[14] In 1907 Allegheny County made 45.8 percent of the nation's structurals; the following year it managed only 42.8 percent.

During fall 1908 Henry Grey reported to Schwab that US Steel was planning a mill very much like their own. In response Schwab confirmed that Hugo Sack of Duisburg, who had designed another type of universal mill for heavy sections many years before, was looking for an American outlet for a mill that would obtain the same result as "our process."[15] By early 1910 Bethlehem competition had become more serious, though for a time those at Carnegie Steel attributed this to special circumstances. Bope admitted to fellow directors: "Bethlehem is cutting quite a swath in the east at present on their 'H' sections, and taking some business from us, but we have felt this condition is only temporary, and is due to the weather and the fact that they have not got as much business ahead as we have, and want to get enough to tide them over the spring." By mid-summer he had to admit that, whereas in the past, with a contribution from Pencoyd, they had about 51 percent of the structural business, US Steel was now getting no more than 25 percent. That December the new Gary structural mill, of a more conventional design, was almost ready for commissioning, with the prospect that Carnegie Steel would lose business farther west to Illinois Steel at the very time it was having a harder struggle elsewhere. They had just lost orders for 1,500 tons of structurals for five jobs in New England, in four of which Bethlehem sections were specified. Ruefully, Bope recognized they had been out-smarted. "These people have been very active in their work among architects towards having their sections specified, and we have been unable to counteract this work in many localities. . . . In this connection I might say that the structural situation throughout the country has been giving me a good deal of concern. There was a time when there was probably not a single user of structural material in the United States who was not on our books for more or less of his tonnage, but today we are not getting any of this business." He suggested, "we have got to keep our organization at the very highest pitch to take everything that we can." There was a short boom in early 1911, but in April a very sharp decline in US Steel's business in all lines, whereas Bethlehem was still running at about 90 percent capacity.[16] That fall the finance committee approved an expenditure of $250,000 for new rolls at Homestead, and by the following year Carnegie was marketing twenty-seven-inch I-beams in an effort to compete with the twenty-six-inch H-beams of Bethlehem.[17] As superintendent of the Homestead structural mills, they now had David Kennedy, son of the world-renowned steel engineer, but it remained an awkward fact that in Pittsburgh, Carnegie was now operating an improved plant with an old technology whereas Bethlehem, with a new process and new product, continued to make headway at their expense.[18] By 1913, a record year for most products, Allegheny County's share of structural-shape production was down to 38.1 percent.

After the special circumstances of wartime had passed, activity in construction and production of heavy structural shapes again advanced rapidly. As this happened, US Steel discomfiture in the field it had dominated for so long increased. Economies in fabrication possible with universal beams and impossible with standard beams meant the latter were losing more of their market. Competition between US Steel and Bethlehem came to a head. By 1922, with 650,000-ton capacity, the Bethlehem mill was considerably larger than Homestead or South Chicago, the two main US Steel structural mills. Because its product was superior, Bethlehem invaded the natural market area of the other mills, some 12 percent of its deliveries of structurals and plate now being made in areas west of Chicago and another 4–5 percent to Pacific Coast states.[19] The Corporation was driven to remedial action. In early fall 1923, its finance committee considered and referred to a subcommittee of key

TABLE 6.2
U.S. structural steel capacity by districts, 1922 and 1930 (thousand tons)

District	1922	1930
Pittsburgh/Johnstown	1,431	1,635
Philadelphia	1,177	1,538
Buffalo	140	520
Chicago	820	1,128
Valleys	n.a.	152
St Louis	n.a.	40
South	33	125
West	60	173
Total	3,661	5,311

Sources: Iron Age, 1 January 1925, 10–17; and 1 January 1931, 100.

directors capital appropriations totaling over $60 million for three major projects: two at Gary and the other tackling the Homestead structural-steel problem. At an estimated cost of $22.6 million, US Steel planned to build new blooming mills, a thirty-six-inch roughing mill, a fifty-two-inch universal mill, and ancillary plant to replace mills from twenty-three to as much as forty years old now and characterized as "obsolete and uneconomical to operate." The universal mill, able to roll the largest-sized wide-flange beams, would be able to make 744,000 tons, or, if steel was available, 50 percent more than that. On Monday, 24 December 1923, the finance committee resolved "to commence installation of the improvements at Homestead works under the general plan proposed."[20]

By the time the new Homestead mills began work, Bethlehem was improving its position for supplying Midwestern markets. Having acquired Lackawanna works four years earlier, in spring 1926 Bethlehem Steel began to build two new Grey mills there. Their first shipments were made in April 1927. Some 38,000 tons of material for the new Mercantile Mart, Chicago, were rolled there. During 1928 Lackawanna cut further into Midwestern markets with the introduction of a weekly lake boat to Chicago and Milwaukee. That year Homestead shipped 81,000 tons of wide-flange beams into Illinois Steel territory. In order to try to recapture more of the local market, US Steel invested $15 million in a new Chicago beam mill.[21]

Understandably, in Atlantic Coast outlets US Steel structural mills were having an even harder time. From the beginning of the decade, only small deliveries of Carnegie sections were made to the prime New York City market. After that, they were further reduced by rail-freight rate increases and the abandonment of "Pittsburgh Plus" pricing. In 1920 rail charges were increased by 40 percent in the Eastern Territory, a bigger rise than in the other three rate classification districts. New basing points set up in 1927 enabled Bethlehem to increase its command of East Coast outlets. By that time the rail charge per hundred pounds of steel from the Bethlehem mill basing point to New York was 14.5 cents as compared with 34 cents from Pittsburgh; to Philadelphia, the respective rates were 13 and 32 cents.[22] A 1924 report from the Carnegie Steel sales department summarized their increasing difficulties: "The tonnage lost on account of competition with Bethlehem cannot readily be estimated, but it is an ever growing tonnage and if not lost outright we are obliged to sell our standard sections at unusually low prices in order to compete." In spring 1927, as Lackawanna rolled its first sections, J. A. Coakley of US Steel pointed out that in the last seven years, during which national output of structural shapes increased 50 percent, Carnegie Steel deliveries of heavy structural shapes, plates, and bars to eastern points had fallen by almost

160,000 tons. Of 969,000 tons of heavy structurals sold in the East in 1926, his company supplied 74,000.[23] It was under these crisis conditions that US Steel set out to boost sales and cut costs by producing the same sort of sections with a very similar technology used by Bethlehem. Unfortunately, they were attempting this in secrecy. The project was uncovered in embarrassing circumstances.

During March 1926 Schwab visited Homestead to mark the twenty-fifth anniversary of his departure from Carnegie Steel to head the newly forming US Steel. In the course of his celebratory visit, he toured the Homestead mills of which for so many years he had been a distinguished superintendent. To his amazement he found that seven of the old structural mills were being demolished and as much as $31 million was being invested in a new, fully electrified mill, which later that year would begin turning out what was effectively the patented Bethlehem beam. Challenged about this, Gary apparently at first and unconvincingly maintained that the mill was an experiment and not a Grey beam mill. A later meeting ended in an angry outburst from Schwab; Eugene Grace, president of Bethlehem Steel, also became involved. At the end of November, the question of obtaining a license from Bethlehem was brought to the finance committee and it agreed to pursue this—"if practicable"—with the advice of the legal department. Farrell and others negotiated with Grace and his lawyers "in an effort to come to an agreement."[24] The first structurals were rolled on the new fifty-two-inch mill five days before Christmas 1926, and from the next month Homestead was selling what it called "Carnegie beams." Early this year, as the eighty-year-old Gary became ill, he made his peace with Schwab, and US Steel prepared to settle the possible infringement of patents. Discussions were still underway in the middle of the year, but after Gary's death that August, US Steel seems to have gone back on the agreement. Eighteen months later Bethlehem Steel filed a suit alleging infringement of patent rights. A year later the suit was withdrawn, the Corporation agreeing to take out a license to use the Bethlehem process.[25]

In addition to its lead in production, at this time Bethlehem increased its share of national capacity for fabricating structural shapes, gaining bigger outlets for its own steel. In the construction of the Empire State Building, US Steel's American Bridge Company had split the contract with its keenest competitor, the independent McClintic-Marshall, which controlled about one-seventh of national fabricating capacity. Forty percent of McClintic capacity was in the Pittsburgh area, a major part of it at Rankin across the river from Homestead, and the 58,000 tons of steel used in the Empire State Building were rolled by US Steel. Then, early in 1931, Bethlehem took over McClintic-Marshall.[26] By that time decisive, even if heavy-handed, action by US Steel had to some extent checked the slide in production. By the mid-1930s, US Steel had a heavy structural-shapes capacity 55 percent greater than Bethlehem. Homestead was again the nation's leading structurals producer, though, because its mill had been squeezed into an old site, it had staggered roughing and intermediate stands and lacked space after the finishing stands.[27]

Early delay and later controversy over the universal-beam mill showed US Steel in a bad light over much of the quarter-century of Gary's control. First, it had failed to recognize a decisive innovation in what was a major section of its business. Second, when this innovation had been proved a conspicuous technical and commercial success by greater initiative on the part of a competitor, its own top decision makers spent another fifteen years before they took purposeful steps to close the gap. Third, having decided on action, they muddled the situation by acting unlawfully. All in all, it was a serious blot on the Gary age. In the latter years of his chairmanship, another even more momentous rolling-mill revolution was taking shape. As with structurals, US Steel had the chance to pioneer with continuous rolling of strip and sheet but instead lagged behind firms having much smaller financial and engineering resources but more enterprise. Fortunately in this instance, though belated, a new regime took quicker and more effective steps to make up the lost time.

The Interlude of World War I

On 28 June 1914, while a US Steel Finance Committee meeting was underway in New York City, the Archduke Franz Ferdinand of the Austro-Hungarian Empire was assassinated in Sarajevo. The aftermath of this far-away event marked a major stage in the fortunes of US Steel. World War I more or less coincided with a watershed in the American industrial economy—in Walt W. Rostow's terminology a time of transition from the drive to economic maturity to the age of mass consumption, a change over from an economy of coal, steel, and railroads to one dominated by petroleum, steel, and the internal-combustion engine. Unfortunately, in relation to these long-term trends, the urgency of wartime requirements for heavy steels provided the wrong signals to those looking for the best direction for advance and thereby delayed the inevitable changes called for by the shifts in the structure of the national economy. Older steel firms were ill placed to deal with this challenge. Many of their leaders had been reared and trained in the traditional product ranges—rails, plates, structurals, wire, and bars. Some of these lines had already peaked, while some reached new high levels during the war. By the time of the post-war boom, it was already clear that other sectors of the industry were growing more rapidly. This was especially marked with sheet and blackplate for tinning and with tubular products.

During World War I steel mills throughout the United States reached record levels of production; the highest annual war output was exceeded in only eight other years to and including 1940. For US Steel the situation was even more extreme. Its 23.4 million net tons of steel in 1916 was its highest

TABLE 7.1
Production of hot rolled steel in the United States, 1905, 1910, 1917, 1920
(thousand gross tons)

	1905	1910	1917	1920	1920 as % 1910
Heavy steel products					
Rails	3,376	3,636	2,944	2,604	71.6
Plates	2,041	2,807	4,158	4,755	169.4
Structurals	1,660	1,912	3,110	3,307	173.0
Long-term "growth" products					
Wire rods	1,809	2,242	3,137	3,137	139.9
Tubular materials	1,436	1,828	2,674	3,220	176.1
Sheets	1,491	2,147	4,110	4,582	213.4
All finished products	16,840	21,621	33,068	32,348	149.6

Sources: AISI annual statistics.

to date, and except for 1929 production would not be topped until 1941. Such exceptional activity followed a serious slump. The record outputs of 1913 had given way the next year—in the middle months of which the fighting began in Europe—to the most depressed period since 1908. Addressing the Iron and Steel Institute on 22 May, Elbert H. Gary pleaded for a continuation of cooperation and restraint from fellow leaders: "I ask you to consider not only the propriety, but the necessity and more than that the pleasure of being fair, reasonable, and generous in our treatment of our employees and in our treatment of one another. It is not necessary at this time to speak of being generous to our customers, for they are taking care of themselves."[1] Throughout 1914, not a single new blast furnace was blown in across the nation, and at the year's end 287 of the existing 451 furnaces were idle. In the boom of 1913, US Steel had expanded less rapidly than its rivals, having thirty-one open-hearth furnaces either built or still under construction, whereas other companies had sixty-six furnaces underway.[2] Now, its output underwent a decline almost identical to that of the industry generally. Shipments of products were 74.2 percent the 1913 level and, as one of its own documents records, by the end of 1914 "the Corporation's operations had reached their lowest point since the organization of 1901."[3] Rapidly rising demands from home and export markets the following year quickly swept national production back to high levels and then on to new highs. In January 1915 pig iron production was running at an annual rate of 19 million tons; before the end of that year it had doubled. By mid-1915 all mills in the Pittsburgh area were operating day and night to meet wartime demands. Extensions to plant were soon underway. Across the industry 17 new open-hearth furnaces were completed in 1914 and 29 in 1915. The next two years recorded additions respectively of 103 furnaces (with 4 million tons capacity) and 101 furnaces (5.4 million tons capacity). During this same time national ingot capacity increased by 25 percent. Expansion of iron plant was smaller but still impressive: in 1917, 14 new blast furnaces, amounting to 2.2 million tons annual capacity, were completed. In some fields the growth was even more dramatic: electric furnace capacity, built or under construction, doubled in 1915. Tonnage increases in 1916 more or less matched those of the previous year. Directly or indirectly, the fighting in Europe was the prime motivating force for this unprecedented industry expansion. By the second half of 1917, some 75 percent of iron and steel output was supplying war demands.[4]

As industry leader, US Steel played a major part in the high activity and expansion. Demand for steel exports in 1915 was the highest in the history of the Corporation. Production throughout that year was about 85 percent the capacity of the mills and reached 100 percent during the last quarter.[5] Yet in some important respects, the Corporation's experience contrasted with that of steel makers generally. In both raw steel and steel shipments, its activity peaked in 1916, whereas for the industry as a whole 1917 was the year of highest output. In 1916 output of raw steel at US Steel was 25.5 percent and steel shipments 27.8 percent higher than in 1913; for the other companies 1917 outputs of raw steel and steel products were respectively 69.2 percent and 39.9 percent above 1913. Even so, it is a striking indication of its stature that from 1916 onward, US Steel alone delivered more steel each year than all the plants of the German and Austro-Hungarian Empires.

TABLE 7.2

US Steel's unfilled orders in hand at the end of each quarter,
1912–1919 (thousand tons)

Year	31 March	30 June	30 September	31 December
1912	5,305	5,807	6,551	7,932
1913	7,469	5,807	5,004	4,282
1914	4,654	4,033	3,788	3,837
1915	4,256	4,678	5,318	7,806
1916	9,331	9,640	9,523	11,547
1917	11,712	11,383	9,833	9,382
1918	9,056	8,919	8,298	7,379
1919	5,431	4,893	6,285	8,265

Source: US Steel Board of Directors minutes, USX.

In expansion, the Corporation lagged badly. Its 1914 raw steel capacity, according to its own records, was slightly less than half that of the national industry; by 1920 other companies aggregated 40.4 percent more capacity than US Steel. Statistical evidence does not suggest that in 1914 the other firms had more unutilized capacity than the Corporation, indeed it was quite the reverse situation. Why then was growth at US Steel less? Was it a case of an ingrained cautiousness, of a situation in which others were always more responsive to opportunities? Certainly it was not because of a lack of income. In the iron and steel industry, wartime experience had quickly dispelled the effects of the 1913 depression. During the years 1916–1918, US Steel paid high dividends on common stock; in the peak year for distribution, 1917, these amounted to 7.12 percent of the value of products and services sold as compared with only 4.53 percent in the excellent peacetime year of 1913. Should more have gone into extension of plant, or, as at Bethlehem Steel, into major acquisitions? Apart from the fact that the Corporation already had a plethora of units, a decisive factor against additional absorption of companies was the fact that the government's dissolution suit was still pending. Not until March 1920 was a final decision in favor of US Steel delivered by the Supreme Court.

Though it expanded less than the rest of the industry, US Steel did make some major extensions during World War I. There was a substantial increase in its capacity to make electric steel at the South Chicago works with one new furnace built in 1915 and two bigger ones during 1917. Above all it made important investments in order to contribute to the apparently insatiable demand for shipbuilding material. In 1916 Congress authorized a three-year building program for sixty-three warships. After the United States entered the war, emphasis switched to merchant-ship building. Under very effective leadership, above all from Charles M. Schwab, the Emergency Fleet Corporation achieved exceptional results. In April 1917 there were 61 shipyards in the nation; by the end of 1917 there were 132 yards and when the fighting ended 223.[6] Bethlehem Steel was already a major factor in shipbuilding; now for a time, though on a much smaller scale and against its own inclinations, US Steel went into this alien trade. Responding to a government request, it organized the Federal Shipbuilding Company. In summer 1917, corporate officials decided to build two yards, one on the Hackensack River, New Jersey, and the other at Chickasaw, Alabama, near Mobile.

The most important production changes in the war were in coke making and heavy steels. The brief period 1913–1919 saw the long-delayed triumph of the byproduct coke oven over the older and extremely wasteful beehive method of production. In 1913, 27.5 percent of American production of coke and 33.6 percent of that within US Steel came from byproduct ovens; six years later the proportions were 56.9 percent and 61.6 percent. When the war in

TABLE 7.3

**US Steel earnings per share of common stock and
common stock dividends, 1913–1918**

Year	Earnings per share	Distribution in dividend (millions)
1913	$11.02	$25.4
1914	below $0[a]	$15.2
1915	$9.96	$6.4
1916	$48.46	$44.5
1917	$39.15	$91.5
1918	$22.09	$71.2

Sources: Earnings data from F. L. Allen, *The Lords of Creation* (London: Hamish Hamilton,
        1935), 207; distribution figures from D. A. Fisher, *Steel Serves the Nation: The Fifty
        Year Story of United States Steel* (New York: US Steel, 1951), 225.
[a] Earnings failed to completely cover the preferred stock.

TABLE 7.4

**Raw steel capacity of US Steel and of the rest of the U.S. steel industry,
1914–1920 (thousand net tons)**

Year	US Steel	All other companies
1914	21,262	21,416
1915	21,528	22,926
1916	23,280	25,986
1917	24,721	29,193
1918	24,868	32,215
1919	25,018	34,156
1920	25,047	35,173
1920 as % of 1914	117.80	162.24

Sources: US Steel information calculated from company production and operating rate
        data; national totals come from U.S. Department of Commerce, *Historical Statistics of
        the United States,* series P234 (Washington, D.C.: GPO, annually), 418.

Europe began, US Steel had 1,452 byproduct ovens; by early April 1917 it had
added 1,118 more. Now for the first time provision was made for the Pittsburgh-area plants to be supplied from a local coke works rather than draw
supplies from Connellsville. The first phase of a new installation on the banks
of the Monongahela River at Clairton included 640 ovens, which also made
available large supplies of coke-oven gas for corporate steel operations and for

TABLE 7.5
Shipbuilding in the United States, 1910–1920 (thousand gross tons)

Year	Tonnage	Year	Tonnage
1910	342	1915	225
1911	291	1916	325
1912	233	1917	664
1913	346	1918	1,301
1914	316	1919	3,327
		1920	3,881

Source: Encyclopedia Britannica, 12th ed., s.v. "Shipping (United States)."

power generation. Further building during the next few years made Clairton far and away the largest coke plant in the nation.

US Steel played an important part in providing new ordnance capacity to the armed forces. It put up 155-mm gun and 240-mm howitzer plants at Gary. In spring 1918 a government contract was received for a 12-inch and 18-inch gun and projectile plant. The site chosen was on Neville Island, in the Ohio River just below Pittsburgh, but the planned works was abandoned with the Armistice in November 1918. The new shipyards and remarkable increases in tonnages launched boosted demand for plate and angles. For the four years to 1912, production of plate averaged 2.6 million gross tons. Early in 1915 plate was selling for about one dollar per ton less than most forms of rolled steel; by the later months of the year its price was as much as fourteen dollars per ton higher. Prewar prices for ship plate were under 2 cents a pound, but by 1917 they had climbed to 12 cents or more. (Even so, when in late 1917 its "normal" selling price was 4.25 cents a pound, costs of manufacturing ship plate were no more than 2.5 cents.)[7] Plate output reached 4.2 million tons in 1917 and 5.1 million tons the next year. To serve its own Mobile yard and other new southern shipbuilding operations, late in 1917 US Steel began installation of structural and plate mills at its recently built Fairfield works, hitherto largely confined to wire products. There were important extensions of steel-making plant at Ensley to supply these new mills. New sheared-plate mills were also built at Gary and South Chicago, and the most dramatic of all Corporation responses to the demand for ship plate was the building in only six months of a new 110-inch plate mill, the "Liberty mill," at West Homestead. This boom in pricing and production reinforced both the heavy-steel bias and its concentration in what before the war was already being seen as a

less-suitable location. Yet notwithstanding such major extensions, in the period straddling the war years, there was a dramatic shift of both the national
industry and of US Steel from the Pittsburgh area. In 1912 Allegheny County
produced 24.4 percent of United States rolled products but by 1920 only 20.6
percent. Alabama's proportion crept up from 2.16 to 2.77 percent during that
time, though there were many more years of southern economic backwardness and slow growth ahead. Interestingly, in this period the Valleys made impressive headway, with US Steel breaking ground for the McDonald rolling
mills outside Youngstown in 1916. Even more striking was the 45 percent increase in output in Illinois and Indiana; in 1912 they produced 16.7 percent
of the nation's steel and in 1920 18.5 percent. These extensions were accompanied and followed by other important changes. Under the pressures of the
times, the War Industries Board made Chicago a basing point for plates, structural shapes, and bars from September 1917 to June 1918.

At the same time, an especially large expansion of steel processing took
place in the Midwest and above all in the Chicago area. In Calumet alone it was
reckoned the number of industrial plants increased by more than half from
1917 to 1919.[8] In January 1919 the Western Association of Rolled Steel Consumers for the Abolition of Pittsburgh Plus was formed. Its first complaint to
the Federal Trade Commission (FTC) was designed to secure the restoration of
Chicago as a basing point. Within a few months the FTC had decided that it

TABLE 7.6
**Production of all finished rolled iron and steel products
by areas, 1912–1920 (thousand gross tons)**

Area	1912	1915	1920
USA	24,656	24,392	32,347
Allegheny Co.	6,015	5,733	6,654
West/Cent. Pa.[a]	2,574	2,604	2,876
E. Pennsylvania	2,489	2,419	2,856
Valleys	2,937	3,310	4,142
Ohio[b]	2,564	2,654	3,640
Illinois	2,253	1,889	2,487
Indiana	1,873	2,104	3,499
Alabama	532	556	896
All other areas	3,419	3,123	5,297

Sources: AISI annual statistics.

[a] Excludes the Shenango Valley, which is included in Valleys district.

[b] Excludes the Mahoning Valley, which is included in Valleys district.

could not assume jurisdiction, but in September 1920 rail-freight charges were advanced by as much as 40 percent and the Western Association took up the matter again. The writing was now on the wall for a pricing system that had helped bolster the primacy of the Pittsburgh district. Brought to a head by wartime expansion, the effects of the changes now set in motion were to work themselves out through the interwar years and beyond.

Some of the developments required by war soon proved an embarrassment in the very different conditions of peacetime operations. The Chickasaw yard delivered the last of its fourteen ships in September 1921 and was then closed, executives deciding a few months before that "the investment has proved to be of no economic value or utility; and . . . there is no prospect of it ever being of much, if any value to the Corporation."[9] But the yard had already stimulated expansion at Fairfield, and this not only survived but was carried further, work beginning there in 1918 on installation of its first open-hearth and blast furnaces. Construction of the new Homestead plate mill had increased that works' capacity for sheared and universal plate to 1.23 million tons, or almost 17 percent of the national total by 1922. This new plant was grossly underutilized. Even in the boom year 1920 the whole of Carnegie Steel deliveries of this product within the United States—including small contributions from other Pittsburgh mills, which cannot be disaggregated—represented a 67 percent utilization of Homestead capacity. The following year, a time of depression, the rate was no more than 30 percent.[10] Another instance of the inappropriateness of wartime outlay was provided by the gun-forgings plant built by American Bridge at Gary. In June 1922 the US Steel chairman pointed out to the finance committee that "since the conclusion of the war the American Bridge Company had endeavored to find some practicable use for the plant in the manufacture of commercial forgings, but the plant having been constructed to manufacture a specialized product, after practical experience it had been found impossible to economically operate it for general forging work."[11] The challenges and opportunities of the 1920s would be in very different directions.

Labor Conditions and Relations
during the Gary Years

From the start, the United States Steel Corporation employed a huge work-force, and its numbers for many years grew roughly in proportion to the increase in output. Average employment in 1902 was 168,127 men. Highs were reached at times of high activity—210,180 employees in 1907, 228,906 in the boom of 1913, and a World War I annual peak of over 268,000. In times of difficulty, men were thrown out of work, though the declines were much smaller than the shrinkages of output. In 1903–1904 the utilization rate for raw steel capacity averaged 77.3 percent as compared with 97.2 percent in 1902. Employment over those two years averaged 93.7 percent of the 1902 figure. In the severe depression of 1908, employee numbers were 45,000 men, or 21 percent down from the high of the previous year, and the 1914 figure was almost 50,000 below that of 1913. In 1921 76,000 fewer workers were employed than in 1920. Gradually, a policy was worked out for necessary layoffs. As a US Steel manager explained to Ida Tarbell in 1914, single men or those last taken on—or the least experienced or efficient—were laid off first; older or more experienced workers were retained as long as possible. Some plants and mines were then working half time, jobs available being divided up between more employees. Ten years later the president of Carnegie Steel confirmed this had remained their practice.[1]

Given the presuppositions of the society of the time, it was inevitable that there were immense inequalities in the material rewards from the activities of the Corporation, for employees, managers, directors, and stockholders. Skilled workers were well remunerated, but they were a small minority of the

whole, and year-by-year some previously skilled positions were downgraded
—deskilled—by mechanization. Overall, the massed ranks of the workers—
"common labor"—received a comparatively small share of the returns from
production. In 1910, for instance, payments to the average of 218,000 employed throughout that year amounted to $175 million; $50.7 million was
distributed to holders of preferred and common stock. In the early years, US
Steel directors received $20 for attending the once-a-month board meeting in
their Broadway offices; they were paid ten cents compensation for every mile
traveled to that meeting place. The average hourly wage for their workers was
twenty cents in 1902 and did not rise above twenty-six cents until 1916.
Sometimes there was a very obvious paternalistic element in the way the Corporation dealt with wage matters—for instance, in 1912 it chose 24 December to announce increases for all its employees.[2] Normal hours of work were
long and showed no clear tendency to fall. Indeed, the average working week
was 67.5 hours in 1902–1904 and 68.5 hours ten years later.

Even more dramatic and depressing than the distribution of the fruits of
production was the way in which labor was used, its uncertain employment,
everyday working conditions, and the state of the still rapidly swelling mill
town communities in which the workers and their families lived. Between
1900 and 1910, the number of people in Allegheny County increased 31.4
percent and in the city of Pittsburgh (including Allegheny City) by 18.2 percent, but in Braddock, Homestead, Duquesne, and McKeesport, the rise was
38.9 percent—an extra 27,000 people squeezed into their often already
crowded housing areas. The previously insignificant towns of Munhall and
West Homestead together contained another 8,000 residents by 1910 and
two new mill towns, Donora and Clairton, whose sites had only minute settle-
ments in 1900, were home for 11,500 inhabitants ten years later. As the
renowned Pittsburgh Survey of this first decade of the century made clear, for
most of those within the mill towns, life was poor, drab, and a continuing
struggle. The Homestead findings of one of these pioneering social scientists,
Margaret Byington, were later summarized: "The men toiled long hours,
nearly all working a 12-hour shift, with a 24-hour stretch every two weeks
when they exchanged day and night shifts. There was no leisure, little family
life, and little civic spirit; there were only hard work, poor food, and wearied
sleep." A year or so earlier, the Monongahela Valley was visited by an Englishman, Arthur Shadwell, who was familiar with the heavy industrial areas of
Germany as well as those of his own country. Of Homestead, he wrote with a
despair matching that of Byington and even more emotively: "Never was
place more egregiously named. Here is nothing but unrelieved gloom and
grind; on one side the fuming, groaning works where men sweat at the fur-

naces and rolling mills twelve hours a day for seven days a week; on the other, rows of wretched hovels where they eat and sleep, having neither time nor energy for anything else." In fact, at this time working the Sunday shift was increasing, National Tube at McKeesport starting it after a strike in 1901. Long afterward, William Hogan wrote that after the 1909 strike, there was "contentment" and "cooperation" between the workers and US Steel, but the evidence suggests this was a gross distortion of the reality.[3] Much nearer the time and commenting on the first two decades, Charles A. Gulick of the University of California came to a very different conclusion: "For the first 17 or 18 years of its existence the Corporation did not pay to the average common laborer in its employ sufficient wages to enable him to support a normal-size family in health and decency. . . . Such a conclusion receives considerable support from the fact that cigar factories and similar 'complementary' industries in which women and child labor can be utilized flourish in steel districts, and from the further fact that a large proportion of families in which the man is a common laborer take lodgers."[4]

Even this mean existence was precarious. Sometimes the very basis of employment and the life of a community could be swept away by a single decision from upper management. W. B. Dickson, who as vice president worked in the 71 Broadway offices, recalled the effect of Judge Elbert Gary's 1907 decision to close eight of US Steel's plants rather than cut prices in the hope of increasing their share of the reduced business available. The eight works were idle for an average of almost thirteen months, "most of them were the mainstay of the communities which had been built up around them. . . . The trade of these towns was paralyzed, some of the merchants ruined, and the workmen, many of whom owned their own homes, had to seek work elsewhere. . . . [N]ot one of these persons so vitally interested were even aware of the contemplated action, nor were they consulted in any way. They had lost the old American status of family economic independence, the sine qua non of our boasted American democracy. By the arbitrary act of a man in a New York office, the devotee of an unworkable trade theory, their means of support were ruthlessly destroyed; and they had no redress as the strictly legal right to close these mills could not be questioned."[5]

As well as the exhausted state of the workingmen and their prevailing material poverty, there was a pervasive oppression, an overriding company preeminence not only in the mills it owned but even in the everyday affairs of many of the mill towns. Looking back from 1920 over the almost thirty years since the Homestead battle, for two-thirds of which the town had been controlled by the Corporation, John Fitch reckoned that along the Monongahela Valley "working men have lived in an atmosphere of espionage and repression. The deadening influence of an overwhelming power, capable of crushing whatever does not bend to its will, has in these towns stifled individual freedom, initiative, and robbed citizenship of its virility."[6]

Despite these appalling conditions, studied, publicized, and denounced by a new generation of social scientists, US Steel showed little inclination to improve things. Its predecessor companies had been harsh with their labor. This is well known to have been the case with Carnegie Steel, whatever the visions and protestations of its chief stockholder and ever-watchful overseer. It applied to the other constituent companies as well. One example was provided at a US Steel meeting as late as 1945, when a seventy-seven-year-old employee recalled starting work in the late 1880s at the Pennsylvania Iron Works of Percival Roberts. At that time newcomers had to sign their acceptance of a book of rules. "This book stated that the applicant did not belong to a labor union, and that he would not organize a labor union and that he would have nothing to do with labor unions." The weight of this oppression was not lessened by the fact that in this case the speaker added, "which exactly fitted my character."[7] Roberts transferred his prejudices from his family firm to his role as a board member of US Steel. Asked during the Stanley Hearings about excessive hours of work, he replied with the traditional formula of classical economics: "Who shall say [what] is the proper limit? There is no doubt that the minimum number is the pleasantest; but in the economies of this world, how shall we determine what the limit may be?" Thus, Percival Roberts found his own answer in "the laws of nature."[8] Under such a regime there was little hope for an unassociated workingman.

Most important of all, as far as Corporation policy was concerned, Elbert II. Gary seems to have been emotionally distant from "ordinary" workers and therefore woefully deficient in his capacity to understand their aspirations. Not only was he implacably prejudiced but also insufferably paternalistic. He resisted the abolition of Sunday work although one of his vice presidents, William Dickson, argued for years that nothing would so commend them to the workers and the public than action on these lines, especially since US Steel's dominant position in the industry would compel its competitors to follow suit.[9] In a May 1910 address to the newly formed American Iron and Steel Institute, Dickson suggested that failure to act on abolition of the seven-day week and twelve-hour day might result in a radical response from state or federal governments. Closing these same Institute proceedings, Gary slammed the door on the prospect of change, observing, in an apparently not unusual convoluted way, "that for himself he did not favor taking up sociological questions, such as the one they had considered, because of any feeling that a public sentiment compelled them to do so. Fidelity to the interests of those whom they represented was to him an important consideration. The highest type of an honest man is not the one who leans backward in his effort to be fair, but the one who is not afraid to decide in favor of his own friend if that decision would be right. The American Iron and Steel Institute should never be in the position of being pushed to take a certain stand because of a public sentiment that might overwhelm it, without regard to whether that sentiment was right."[10]

Gary's approach to labor also came out well in a statement that he issued in January 1912 in response to an accusation from the liberal progressive jurist Louis D. Brandeis that US Steel maltreated its workers.

I believe, taking everything into account, that the treatment accorded by our Corporation to its employees compares with that of any line of industry in this or any other country at the present time or any period in the history of the world. We are paying 25 percent higher wages than we were when the Corporation was organized [in fact later figures from US Steel indicated that the average hourly earnings were $0.201 in 1902 and in 1911–1912 averaged $0.236[11]], and we have spent and are spending millions to prevent accidents in the works, to improve the sanitary conditions, to furnish voluntary relief in case of accident, regardless of legal liability, and for pensions for superannuates. We have standing committees constantly engaged in welfare work. We have largely abandoned seven-day work and have, to a large extent, eliminated 12-hours-a-day work. It is true there is considerable of the latter still in force, but this is largely because the employees prefer twelve-hours a day in order to receive a larger compensation. This question, therefore, largely comes back to the amount paid for labor. We could, of course, reduce the hours of work by reducing the wages. Whether or not we are paying as much as we ought to pay is a question always up for consideration. It is well known we have in the past declined to reduce wages when many of the others were making reductions. Above all we are trying to satisfy our employees that they are receiving fair and liberal treatment, and I think they appreciate our efforts.[12]

In spring 1914 he claimed that the elimination of a seven-day week meant the Corporation had lost many of its best men, over four thousand during a short period of 1913.[13]

Even friends recognized that Gary had a touch like that of "the velvet glove over the iron hands of right and justice." Much more important than the velvet was one's conception of the values that lay beneath. He was clearly biased against the new immigration that had provided so much of US Steel's workforce. Indeed, in his expressed opinion that immigration should act as a pool of labor for American business, he played into the hands of those Marxist propagandists he so much disliked by essentially regarding the newcomers as a reserve army for the capitalist industrial machine. As he stated during a press conference in 1923, "Measures for limiting the number of immigrants to those who are healthy, morally, politically, and physically, ought to be clear, strict and enforceable, but the number allowed to come here should be equal to the necessities of our industries."[14]

It is important to recognize that opposition to concessions for workers extended far wider than the top leadership of US Steel. In 1911 stockholders resolved that an investigation should be made into conditions of work. Notwithstanding Gary's response to Brandeis, the resulting report showed that 50–60 percent of the Corporation's employees were working twelve-hour days. The committee of enquiry recommended a reduction in hours, but this was rejected by the finance committee. Then in March 1912 a stockholder, Charles M. Cabot—"a cultured, high-minded gentleman" as Dickson recalled him—commissioned a study by John Fitch of the Pittsburgh Survey, "Hours of Labor in the Steel Industry." His report was sent to 15,000 stockholders. Cabot, who was in favor of abolishing the twelve-hour day, asked fellow stockholders to write Gary. Their answers provided a fascinating insight into the mind of the investor, though it must be assumed that the responses that were printed by US Steel above all emphasized what to its officers seemed positive rather than negative points of view. (Naturally, the present selection is even more partial.) One stockholder wanted to eliminate the seven-day week but retain the twelve-hour day, for "I believe any other arrangement would be detrimental to the best financial interests of the corporation." A letter from "TC" of Philadelphia was brief and uncompromising; one must assume he had not visited Homestead or walked the newly laid out gridiron streets of Gary: "In my experience of forty years I have never known any one to suffer from over work or long hours of labor." "MJA" of New York had equally definite opinions: "As a business man, I would not care to have an outsider interfere with my business. I have invested in the securities of the United States Steel Corporation because there is competent management, and I will leave the matters of details to their best judgment. There is too much meddling with business. It would be wise if everyone of us would quiet down for a period of five to ten years, and attend strictly to our own private matters." Gary himself described John Fitch as "an advocate rather than a judge."[15] In late May 1912, the finance committee agreed to eliminate the seven-day week, recording that it was "feasible and practicable" to eliminate the "long turn," but on the twelve-hour day its members were much less definite, resolving that Gary, James Farrell, and Percival Roberts be appointed a committee "to consider what, if any, arrangement with a view to reducing the 12-hour day, in sofar as it now exists among employees of the subsidiary companies, is reasonable, just and practicable."[16] Ten years later the twelve-hour day still existed. In spring 1923 Gary spoke optimistically to the annual meeting about the Corporation's good treatment of its men, though he gave minimal explanation for this approach. "He said that the Steel Corporation treated its men better than any large corporation in the world. This was not due to a spirit of benevolence, but was due to two reasons. First that it was because they ought to be treated so and second because it paid."[17]

In terms of management-labor relations, the Gary years saw little advance toward harmony. The formation of US Steel offered what seemed to the Amalgamated Association of Iron, Steel, and Tin Workers its best opportunity to reestablish itself. To gain public confidence, US Steel might be expected to wish avoiding early bitter battles with labor; on the other hand, when firmly established it might be too powerful for this already diminished union or others.[18] In fact, from its earliest days the Corporation set its face against unionization and did its best to stamp out any effort to organize. On 17 June 1901 the executive committee resolved "That we are unalterably opposed to any extension of union labor, and advise subsidiary companies to take firm position when these questions come up and say they are not going to recognize it, that is, any extension of unions in mills where they do not now exist; that great care should be used to prevent trouble, and that they promptly report and confer with this Corporation."[19] Early the following month, the Amalgamated Association entered into a dispute with three US Steel subsidiaries, American Sheet Steel, American Tin Plate, and American Steel Hoop. Its aim was not to secure shorter hours or higher wages but to establish the right to unionize the mills; in short to establish a principle. At two meetings on 1 and 2 July, the members of the US Steel Executive Committee revealed a diversity of attitudes. Both meetings were attended by Gary as chairman of the committee, Charles M. Schwab as president, Edmund Converse of National Tube, William Edenborn of American Steel and Wire, and Percival Roberts of American Bridge. Charles Steele of J. P. Morgan was at the first meeting and Daniel Reid of American Sheet Steel, American Tin Plate and American Steel Hoop at the second. The Amalgamated had tried to ensure that all sheet, tinplate, and hoop mills, whether unionized or not, signed the wage scales. The American Tin Plate mills accepted, but the others did not. As a result, on the morning of 1 July much of those divisions' capacity was idle. Schwab, Edenborn, Converse, and

Steele thought no action should be taken because the men had a weak case. Gary was less sanguine. Roberts felt they could not sign the scales for half of the mills (those already unionized) and not for the rest: "he states that if the union is a bad thing, wipe it out and that it is a sign of weakness to do otherwise." At the next day's meeting, Gary and Reid were conciliatory to union demands, and Converse reckoned that in this case to give way was "a good way out of a bad situation." Edenborn had now decided that he would not concede an inch. Robert's position seems to have become particularly confused: "Mr. Roberts stated that he would either agree with Mr. Edenborn or would be willing to concede any non-union mills in companies that now have any union mills; or he would vote in favor of wiping out the whole thing in all of the mills, but he believes under the conditions to-day, it would not be advisable to do so." Edenborn contributed the additional reflection that "a concern operating with unions is pretty badly handicapped."[20] In fact, the Corporation, through a powerful committee made up of Morgan, Gary, and Schwab, offered to compromise. The union, convinced time was on the company's side and believing it could count on support from the whole labor movement, refused an accommodation. Morgan and others made their antiunion attitudes clear. In August the dispute widened when a general strike was called in US Steel mills. Another 16,000 employees joined the 46,000 men already out. The workers in the mills of Illinois Steel could not join because they had a no-strike clause in their contracts.[21]

US Steel now resorted to the threat of closure and relocation of mills to put pressure on the workers. At Canal Dover in central Ohio, the managers of the tinplate and sheet works told the strikers that they had been given orders that unless the mills were at work again in a reasonable time, they were to begin dismantling them. In such sections of the trade, whose plant was relatively cheap, relocation was always a possibility, leaving small communities desolated. In two cases the threat was carried out. Alleging that the townspeople were largely in sympathy with lawlessness and the mayor refused to use the police to protect industrial property, American Sheet Steel decided to move its large Dewes Wood sheet mill from McKeesport to the Kiskiminetas Valley twenty-five miles to the northeast. Local striking workers having rejected a call to return to work, American Sheet Steel's Chartiers plant, a feature of the industrial scene in the Pittsburgh suburb of Carnegie since 1884, was also moved to the same area.[22] By September the dispute ended with the workers defeated. US Steel claimed it had suffered no great loss. (There seems reason to doubt this claim; at the time, one economist reckoned the cost to the Corporation was $20 million. On the other hand, US Steel could claim that it had made some headway, for it had managed to convert fourteen previously unionized mills into nonunion operations.) The press generally chose to present the company case in a favorable light; as the Philadelphia correspondent of a leading British trade journal noticed at the end of August: "The newspapers give all points of encouragement favorable to the steel combination, and cut out all favorable reports for the workers' side."[23]

Eight years later, and in the same branches of the Corporation's holdings, the humiliation of the Amalgamated Association was taken further. On 1 June 1909 American Sheet and Tin Plate announced wage cuts of 2–8 percent and that from 20 June all plants would be "open," that is, employ nonunion as well as union men. Company spokesmen busily advanced the simplistic argument that any requirement of union membership was at odds with its employees' rights of personal liberty. Despite support from the American Federation of Labor, the Amalgamated had been weakened by its own diminished membership and failure to collect dues. The dispute dragged on for fourteen months before it was called off. Effectively, this marked the end of craft unionism in steel.

In his retirement speech at the 1911 President's Annual Dinner, William Corey made clear that he fully shared Judge Gary's opposition to organized labor. Even though, ironically, his speech was written by William Dickson, whose own views were much more liberal—and some of the fruits of whose wide reading came out in its phrasing—the sentiments conveyed had to be those of the man who delivered the address.

> If there has been any one subject in which I have been intensely interested, it is that of what I am pleased to call "free labor" as against so-called "union labor." The Company in which I passed the early years of my business life had to face this question many times, and decided once and for all in 1892 that however beautiful in theory, as a matter of practical operation the intervention of any third party between a company and its employees could not be tolerated. No sane man will question the abstract right of the workmen to organize. It is a "condition, however, and not a theory which confronts us." Until organized labor has demonstrated its ability to deal with economic problems in an enlightened and progressive spirit, and abandons its reactionary attitude, as indicated by its pernicious practices of restriction of output, dead level of wages regardless of efficiency, and the closed shop, we must deal with it as a hindrance to progress and steadfastly refuse to be hampered by its unreasonable demands.[24]

After the defeat of Amalgamated, various smaller strikes had mixed outcomes. In June 1912 National Tube employees in Pittsburgh struck for higher pay and "greater conveniences" in their working places. Most of their demands were met. Five months later there was a serious strike at Homestead and Braddock. A wage increase was granted, but those who had led the men were discharged. At the end of that month an attempt to restart the mills failed, and at the same time the Industrial Workers of the World tried to call a general strike.[25] On 1 May 1916 martial law was declared in Braddock after riots at Edgar Thomson. When the United States entered the First World War, the unions dissuaded working men from involvement in labor disputes.

Over the years, increases in welfare provisions—US Steel was said to have spent $65 million on this between 1912 and 1919—and the idea of forming company unions seemed to pose threats to further movements for industrywide unionization. On the other hand, World War I gave some hope of better things, in 1917 the War Labor Conference Board going so far as to assert the rights of employees to organize unions without interference from employers.[26] In many respects the conditions of work remained deplorable. The average workweek for the 268,000 employees of US Steel in 1918 was of 66.1 hours. Now, as the end of the war approached, there occurred a very purposeful push for unionization. After the June 1918 American Federation of Labor convention, the twenty-four unions that then claimed jurisdiction over the steel industry were urged to cooperate in a movement for improved rights. A central organization was formed, the National Committee for Organizing Iron and Steel Workers. After a start in Chicago, attempts to organize workers centered on Pittsburgh. Twelve demands were submitted, including an eight-hour day and increased wages, but once again the real issue was the recognition of the right to unionize the labor in the works. The steel company response was to fire some of the union members. As the industry's leader, Elbert H. Gary, recognizing that to act differently would be to tacitly acknowledge the right of unionization, ignored a request for a meeting from the veteran union leader Samuel Gompers. In other words, he let slip an opportunity for a grand conciliatory gesture. In contrast to this snub, Gary did extend the common courtesy of a written refusal to meet the men's organizing committee, even though his letter to them ended with his typed name rather than a signature. Altogether, the judge showed himself woefully out of sympathy with current labor realities when he declared during this dispute, "We stand firmly on the proposition that industry must be allowed to proceed untrammeled by the dictates of labor unions or any one else except the employer, the

employees, and the government."[27] It was indicative of the anachronistic nature of Gary's views that Henry Clay Frick, now within two months of his
death, was among those who expressed approval of his actions . On 2 October
Frick congratulated his chairman on his refusal to negotiate with Gompers: "I
was glad to see . . . that there would be no compromise. . . ; that's the kind of
stuff."

When on 22 September the call to strike went out, it was necessary to pub-
lish it in seven languages. By this time the industry was busily taking action to
protect its material interests. Pittsburgh mills were once again fortified and
provisioned, and the sheriff of Allegheny County deputized large numbers of
men who had been hired and armed by the local steel companies. These
"deputies" were backed by state troopers, local police, and vigilante groups.
On the eve of the strike, rioting broke out and arrests began. At Clairton, a
large open-air meeting, which included women and children, was attacked
by mounted police. The next day, a majority of the nation's steel workers
came out on strike. (Figures of those involved vary widely: 365,000 people,
284,000, and 250,000 being variously cited. It seems likely that the number
at the beginning was at the lower levels with a maximum of perhaps 350,000
reached by the end of September.) The workers' dispute was coordinated from
Pittsburgh. In the newspapers of that area, between 27 September and 3 October alone, the industry ran thirty full-page advertisements advancing its
case, and as a research worker later recognized, "A perusal of the newspapers
during the strike could leave the reader with no other impression than that
held by the Corporation officials."[28] Men drifted back to their jobs, and by mid-
October a Senate committee found that of a normal workforce of 28,200 at
Homestead, Duquesne, McKeesport, and Clairton, as many as 23,514 men
were at work. In contrast, Chicago, Cleveland, Youngstown, and Wheeling
were almost fully out and the works in those cities closed. At Gary, the company town that, from his distant perspective, Farrell was to characterize as "a
beautiful city," the walkout on the day the strike began was nearly complete.
On 7 October the town was occupied by federal troops under the command of
Gen. Leonard Wood.

As the weeks went by, a variety of circumstances undermined the men's
position. The organization of the strike was in the very capable hands of
thirty-eight-year-old William Z. Foster, but his radical past as a member of the
Socialist Party meant that it was easy to represent a dispute conducted under
his direction as threatening civil order at a time when the background of labor
relations was particularly complicated and embittered by nationwide fears of
Bolshevik activity. In fact, Foster's key role was used by US Steel to help persuade public opinion that the strike was the cutting edge of communism.

There were even suggestions that a tract by Foster on Syndicalism, written five or six years earlier, was now circulated by the steel companies in order to turn public opinion more decisively against him and the strikers. In western Pennsylvania, gatherings of more than three or four strikers were outlawed. The start of a strike of 435,000 coal miners on 31 October helped reinforce the general impression that the nation's whole economic and social fabric was facing imminent collapse. Between 1 and 3 January 1920 in raids in thirty-three cities, federal agents arrested 2,000 "Reds," 233 of them in Pittsburgh.[29] The companies turned to scab labor, bringing in over 30,000 black workers that fall; some men were also brought in from Mexico. Meantime, the strike dragged on into midwinter, always a hard time for idle men and their families. On 7 January, after occupying the works for three months, federal troops were at last withdrawn from Gary. A week later another factor was added that undermined the strikers' position when it became known that Henry Ford had traveled to Pittsburgh to place orders worth $15 million with US Steel subsidiaries. Men were already returning to work by then, recognizing that the cause was lost; the National Committee authorized the 100,000 still out to return on 20 January. One estimate was that the workers lost between $87 and $112 million in wages; some twenty strikers had been killed. Those employees who were taken back were required to surrender their union membership cards. As Gary summed up the industry's achievement in this crisis, it had been saved from "the closed shop, Soviets, and the forcible distribution of property."[30]

Yet, although their experiences during fall 1919 seemed to represent another crushing defeat for the men, in some ways at least their prospects were beginning to improve. The problems of the industry had been exposed to wider scrutiny. The public impact of press reports was heightened by the insights provided by a commission of the Federal Council of Churches, which recorded the full range of serious defects in the industry: "the average week of 68.7 hours . . . and the underpayment of unskilled labor, are all inhumane. The 'boss system' is bad, the plant organization is military, and the control autocratic."[31] Now at last there came progress toward a more reasonable working day. In the middle of the strike, a Senate committee of investigation had concluded, "laborers in the steel mills had a just complaint relative to the long hours of service on the part of some of them." Although a historian of US Steel has argued that from 1911 Gary was an advocate of the abolition of the twelve-hour day, though his taking this step was delayed partly because of the increase in production costs that would result and partly by the urgency to maximize output during the war, the evidence indicates that the chairman temporized. The president of Carnegie Steel estimated that in 1919 about 60 percent of their 55,000 employees were still working a twelve-hour day.[32] In March 1921, at a time of depression when the loss of production would not be so important, US Steel announced the end of the seven-day working week and the "long turn" at the changes of shifts. The men now looked ahead to a time when the twelve-hour day might go as well. However, two years later as president of the Iron and Steel Institute, Gary read out the report of an investigating committee, which recommended that this last reform should be delayed on the grounds that it was likely to increase costs by 15 percent, making it necessary to employ an additional 60,000 men throughout the industry. The committee went so far as to express doubts that the twelve-hour day was injurious to the workers. But the state of public opinion meant that this sort of line could be defended no longer. On 18 June 1923, President Harding wrote to Gary about the

"US Steel and the Twelve-hour Day," a newspaper cartoon of 28 May 1923. *Archives of USX Corporation.*

abolition of the twelve-hour day. A special meeting of the finance committee eight days later agreed to recommend acceptance of Harding's opinion. By August the industry had at last conceded the twelve-hour day.

There is little evidence that in the remaining four years of his term as US Steel chairman Judge Gary softened in his attitude toward organized labor. Two years after Gary's death, James Farrell wrote a sympathetic account of his longtime senior colleague. He almost contrived to present Gary as a liberal in his labor attitudes: "He advocated and established many pioneer measures for the welfare of the employees of industrial corporations, including stock ownership by them and participation in profits, high wages and safe, sanitary, and pleasing [*sic*] surroundings. He was always a strong advocate and a firm upholder of the 'open shop.' During his chairmanship the seven-day week and the 12-hour day for labor in the steel mills was abolished."[33] Farrell failed to explain to his readers that this last reform essentially had been forced on Gary and the rest of the industry. By the time Farrell wrote these words, circumstances for steel makers and their workforces were approaching a dramatic change. Fortunately for US Steel, another corporate leader was emerging who would prove capable of a more flexible and imaginative response to both pressures and opportunities.

Part Two

The 1920s, Depression, and Reconstruction

The Changing Shape of Competition and the End of the Gary Years, 1919–1927

When in the course of 1920 the postwar boom began to turn toward depression, Elbert H. Gary was already in his seventy-fifth year. Under his guidance US Steel met these new challenges in time-honored style. Whereas other top steel producers, led on this occasion by the Midvale Steel and Ordnance Company—headed by his former Carnegie Steel colleagues William Corey, Alva Dinkey, and William Dickson—decided to cut prices, the Corporation was reluctant to do so. Whereas shipments and employment fell sharply, it firmly maintained dividend payments. During this first postwar depression, it even managed to increase its share of United States output.

Gary continued to dominate US Steel, and in public perception at least the

TABLE 9.1
Depression and the performance of US Steel, 1920–1922

Year	Average U.S. prices (per gross ton)			United States Steel Corporation		
	Pig iron[a]	Rails	Prods and services sold (millions)	Prods and services sold (per t.shipped)	Employees	Dividends paid (millions)
1920	$42.05	$53.83	$1,291.00	$83.08	268,000	50.6
1921	$21.87	$45.65	$726.00	$82.89	192,000	50.6
1922	$23.89	$40.69	$809.00	$61.63	215,000	50.6

Sources: US Steel annual reports for 1920–1922; U.S. Department of Commerce, *Historical Statistics of the United States* (Washington, D.C.: GPO, 1957).
[a] Price at Valleys furnaces only.

whole steel industry, through most of the generally good years of the 1920s. Although his speeches revealed an awareness of disturbing economic forces below the surface of prosperity, he continued to address his fellow steel executives and the public at large as if both the industry he represented and the company he led were still unquestioned leaders in the national economy; neither was the case. Speaking in Cleveland in the spring of 1927 as Judge Gary's career closely approached its end, Eugene Gifford Grace, now effective head of the Corporation's nearest rival, Bethlehem Steel, neatly if rather unkindly demonstrated how much things had already changed when he pointed out

TABLE 9.2
Assets of largest steel companies, 1917 and 1930
(millions of dollars)

Company	1917	1930
US Steel	2,449	2,394
Bethlehem	381	
Midvale	270	
Lackawanna	117	720[a]
Republic	123	308
Youngstown	97	
Brier Hill	46	258[b]
Jones & Laughlin	160	219
Armco	30	148
Inland	57	103

[a] By 1930, Bethlehem, Midvale, and Lackawanna were considered as one
   company.
[b] By 1930, Youngstown and Brier Hill were considered as one company.

TABLE 9.3
Net profits of leading steel companies, 1926 and 1927
(annual average in thousands)

US Steel	$102,282	Inland	$6,977
Bethlehem	$18,036	Wheeling	$4,517
J&L	$13,194	Republic[a]	$4,041
Youngstown S&T	$11,086		

Source: Iron Age, 22 March 1928, 845.
[a] Republic Iron and Steel, which in 1930 after having been extended, became Republic Steel.

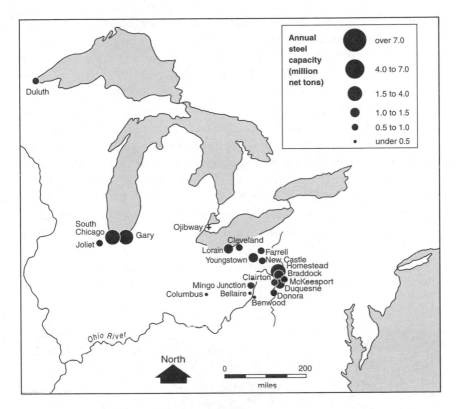

US Steel Integrated Works in the Northeast, 1922

that during the previous year, the net return per $100 of capitalization had been $6.54 at US Steel but $29.20 at General Motors.[1]

Throughout the decade US Steel leadership of the industry continued as measured by most criteria, but there was unmistakable evidence that its relative standing was falling. This was clearest of all from its divisions' shares of national output; there was no catastrophic collapse, merely a steady, apparently unstoppable erosion. In utilization of capacity, it did better than the industry as a whole. Whereas capacity was growing generally, at US Steel the increase was a good deal smaller than at its main competitors or for the industry as a whole, in which a series of important mergers increased the sizes of many companies. There were still major differences in efficiency between the various US Steel plants. This was shown in Federal Trade Commission data of mill costs at a large number of the Corporation's operations in August 1918. It must be stressed that the figures excluded transportation charges, depreciation, and interest on invested capital as well as administrative and sales expenses—they were in fact merely processing costs. Given such diversity, it is understandable that US Steel was still pruning the overextended structure it had already been struggling with for over twenty years. For instance, in the middle months of 1925, the finance committee approved the dismantling of the Emma Furnace at the Newburgh works, Cleveland, and agreed that the Carnegie Steel Company should abandon and dismantle five furnace plants, one furnace at another works, and five rolling-mill operations.[2] In 1922 US Steel had twenty-one fully integrated works with a combined raw steel capacity of 25.4 million net tons, or a 1.2-million-ton average per plant. In fact, only nine of these mills held capacity of over 1 million net tons; four were smaller than 500,000 tons. Each of the latter—in ascending size, Columbus, Benwood, the American Steel and Wire Pittsburgh works, and Bellaire—were closed during the 1920s. The only new integrated works built during the decade was Fairfield, in such close proximity to the existing Ensley works as to be what a later generation would call a "brownfield" rather than a "green-

The US Steel Finance Committee in 1925. The members are, from left to right: James A. Farrell, Richard V. Lindabury, George F. Baker, Elbert H. Gary, J. P. Morgan, Percival Roberts, W. J. Filbert. *Courtesy of USX Corporation*

TABLE 9.4

The range of mill costs at various operations of US Steel, August 1918

Product	Number of plants	Highest cost (per ton)	Lowest cost
Open-hearth rails	7	$48.27	$32.46
Sheet bars	11	$54.98	$31.30
Wire rods	23	$78.10	$36.41
Black plate for tinning	14	$99.17	$89.12

Source: Federal Trade Commission study quoted in N. R. Lamoreaux, *The Great Merger Movement in American Business, 1895–1904* (Cambridge: Cambridge University Press, 1985), 143

field" venture. Closures, additions, and the one new plant resulted in a consistent theme—the Corporation was growing less rapidly than its rivals. During the six years to 1920, it had added 3.8 million tons of raw steel capacity, while other companies created an additional 13.7 million tons; through the next five years the disparity was even greater—0.86 million at US Steel and more than 4.8 million tons for the others. Addressing the American Iron and Steel Institute in spring 1929, James Farrell's figures were that, over the last five years, whereas US Steel capacity had increased 1.7 million tons, the independents had added 8.25 million tons.[3] The gap was almost as great in production. Between 1919 and 1927, national steel output increased 29.6 percent; for US Steel the growth was 7.5 percent.

By this time US Steel was facing competition from both rapidly expanding older rivals and new pacemakers such as Inland, Youngstown Sheet and Tube, Armco, and Weirton. Even as compared with Bethlehem, the Corporation seemed to be dragging its feet. Between 1923, by which time it had completed its major acquisitions, and 1927, Bethlehem's net income increased 9.7 percent, while that of US Steel fell by more than 19 percent. Much more serious in the long run than a declining share of capacity was the fact that US Steel was not keeping pace with the rest of the industry in product development or diversification. Incompetence had been and was still being shown in its trade in universal beams; even more, it now fell behind in sheet and wide-strip technology. As before, the cause for many of these problems can be attributed to its huge size and unmanageable structure. Apart from the abolition of single-basing-point pricing, the decreases cannot be blamed on government intervention or fear of such action, for the 1920s was a decade whose three successive administrations were unusually favorably inclined toward big business. Additionally, it is difficult to avoid the conclusion that in an age of growth and change, the United States Steel Corporation was also hindered by the capacity and inclinations—in short, the quality—of its top executives. Wall Street rather than steel making was still preeminent in the makeup of its board. Too many of its directors were now veterans. Most important of all, at the top, Gary remained both the public face and the unquestioned prime decision maker. Taking a longer view, it is seen that in almost all respects a steady decline in relative standing characterized his entire tenure. Whether or not it might have been practicable to avoid this decline is a matter for debate. When US Steel was formed, Gary had been a lawyer and business organizer in a world of aspiring steel men; by the 1920s he was also a survivor from another generation, compared with the leaders of most of the other companies, men such as Eugene Grace (Bethlehem), Tom Girdler (Jones and Laughlin), George Verity and Charles Hook (American Rolling Mill Company, or Armco), James Campbell and Frank Purnell (Youngstown Sheet and Tube), and Ernest Weir (Weirton). At the beginning of the decade, Gary was seventy-four; the average age of the other seven executives was forty-seven. In 1925 Gary's biographer, Ida Tarbell, eulogized him by writing, "He has made a lasting contribution to our difficult and often baffling problem of substituting in American business balance for instability—mutual interest for militarism—cooperation for defiance—frankness for secrecy—good will for distrust. No man in contemporary affairs has more honestly earned the high title of Industrial Statesman."[4] These were indeed qualities that, from the collective point of view, deserved praise, but as far as the Corporation's interests were concerned, Gary was now being outmaneuvered by nimbler operators who took advantage of the opportunity to shelter behind the industry leader. Another, later, and more critical writer of a short biographical note identified a fatal deficiency in Gary's range of abilities: he was "restricted in imagination." Such a characteristic is a liability at any time, but amidst the changes and opportunities of the 1920s it was a dangerous quality. Over the years Gary's dominance at US Steel had caused either a flight to other companies or a withering of top managerial talent. As two historians of a slightly earlier period of his control summed up, his "pursuit of 'stability' had drifted into an obsession with maintaining the status quo—an ominous development for any firm, no matter how powerful."[5]

As he had done in-and-out every day, month, and year for more than a quarter-century, on Tuesday, 28 June 1927, Judge Gary came to the Broadway head offices of US Steel and chaired that day's regular meeting of the Board of Directors. It was his last visit. On 26 July his illness made it necessary for J. P. Morgan to take his place in the chair. On 15 August, the eighty-two-year-old chairman of US Steel died. Warm appreciations were made by his colleagues. President Farrell referred to his "lovable, patient, and helpful character. . . . [H]e builded [*sic*] well—his constructive ideas and principles of

TABLE 9.5

US Steel share of U.S. iron and steel production and annual profit per employee
through the Gary era, 1901, 1913, 1920, and 1927

Year	Pig iron	Raw steel	Finished rolled products	Profit per employee
1901	43.2%	65.7%	50.1%	$537.09 (1902)
1913	45.5%	53.2%	47.8%	$354.73
1920	39.4%	45.8%	41.6%	$409.32
1927	37.7%	41.1%	37.7%	$379.62

Sources: H. R. Seager and C. A. Gulick, *Trust and Corporation Problems* (New York: Harper and Brothers, 1929), 258; US Steel, annual reports for 1901, 1913, 1920, and 1927.

Note: All the years except for 1927 were "good" years for national production.

business conduct will maintain to his credit and honor." Among the many tributes from colleagues within the wider industry, one from a relative unknown, Severn P. Ker of the Sharon Steel Hoop Company (later Sharon Steel), indicated the never-to-be-matched stature Gary had achieved and the outward appearance of harmony that resulted: "[N]ot only has the great steel industry in the United States, of which he was the acknowledged leader, suffered a great loss, but all industry in America has lost a leader whose clear thinking and fair attitude towards all interests have done more than any other to establish the present high standard of corporate management. The relations existing between competitors, between capital and labor and between all these interests and the general public, are undoubtedly largely due to Judge Gary's influence."[6] When the US Steel directors next met on 30 August, the fulsome tribute they ordered to be recorded in their minutes included the sentence, "He will be sadly missed, but he builded [*sic*] so well, and established in the organization methods and policies so firmly founded, that we look into the future with confidence and courage."[7] Already, there was speculation as to his successor. Given his inheritance, the next man's task would be a formidable one.

In the late 1920s, US Steel outwardly seemed an epitome of the unshakeable power of big business. It now straddled the map of industrial America to an even greater extent than in its earliest days. This was partly because in autumn 1929, it gained a strategic position in the American West through the acquisition of the Columbia Steel Company. Though the number was smaller than in 1901, by the end of the decade US Steel controlled 132 manufacturing works containing 101 blast furnaces, 335 open-hearth furnaces, 29 Bessemer converters, 552 rolling mills, 72 wire mills, 46 pipe and tube works, 13 bridge and structural plants, and other plant. It operated iron-ore mines in

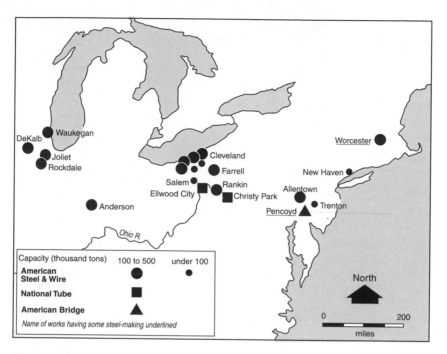

US Steel Nonintegrated Operations in the Northeast, 1922 (excluding American Sheet and Tin Plate)

the Great Lakes region and Alabama; had coal mines, beehive coke ovens, gas, and oil properties in Pennsylvania, West Virginia, Kentucky, and Alabama; and through Columbia now controlled smaller mineral operations in the West. The Corporation also held manganese properties in Brazil, owned or leased 1,117 miles of railroad in addition to the rail tracks inside its plants, and possessed over six hundred lake and river vessels and thirty-one ocean-going steamers.[8]

The head of this great agglomeration of plant and productive power had died, but the membership of the board that he left behind pointed to a strong thread of continuity in its affairs. For some, it may have given a reassuring sense of stability. J. P. Morgan Jr. was chairman and James Farrell had almost completed twenty years as president and chief executive. Gary's friend, banker George F. Baker, a director since the foundation of the Corporation, had now been joined by his son, G. F. Baker Jr. Eugene J. Buffington, a top executive from the old American Steel and Wire and Illinois Steel days, and Thomas Morrison from Carnegie Steel remained directors. Thomas Lamont, who replaced George Perkins at J. P. Morgan's in 1910, had joined US Steel at that time as

Morgan's representative; twenty years later he was still on the board. William J. Filbert, controller, had been auditor for Federal Steel before the merger. In summer 1927 the average age of these eight senior colleagues was 63.75 years. The need for new blood and a thorough shakeup in corporate affairs was not yet generally apparent to the public or stockholders since year after year US Steel had continued to pay solid dividends—in 1926–1927 an annual 7 percent, which in 1929 was increased to 8 percent. Even so, the foundations of the immense superstructure were less sound than they appeared. The commercial conditions that soon followed would threaten its collapse.

ERW Pipe and the
Wide Continuous Strip Mill
Instances of Delayed Innovation

In its extensive 1936 coverage of US Steel, *Fortune* magazine recalled that the Corporation's policy had once been summarized as "No inventions: no innovations." If true, then the interwar years provided a most inappropriate setting for the practice of such a business article of faith, because change, efficiency, and competitiveness were now more important than inherited size and previous growth. As Charles M. Schwab confided to a press correspondent sometime during 1935 or 1936, the chairman of US Steel had recently admitted to him that the Corporation, in fact, had missed every "new thing" in steel.[1]

At the time when US Steel was building its own universal-beam mill at Homestead in an attempt to make good an earlier lapse in technology, it was also being out maneuvered in two other important fields of manufacture, tubular products and thin flat-rolled products. During the 1920s national output of pipes and tubes increased slightly more rapidly than that of finished rolled products generally, tonnage rising from 10.2 percent of the total in 1920 to 11.3 percent by 1929. Although National Tube remained by far the biggest producer, its share of total production fell more in this period than did that of US Steel in all products. The Corporation's loss of standing was related to changes in markets and technology. The 1920s saw a huge expansion of oil production, especially in the West and Southwest. Line pipe for the oilfields and then for longer distance pipelines grew in significance. With this was associated a shift to the manufacture of seamless pipe, the production of butt- and lap-weld pipe falling away. In 1928 US Steel approved National Tube expenditures of $30 million to replace lap-weld by seamless mills at the

National, Lorain, and Gary works. This was fortunate for seamless products were to dominate the 1930s business. In 1936 these categories made up 57.5 percent of sales but contributed 94.3 percent of National Tube net profits.

In 1927 there occurred another important change in pipe and tube technology. The A. O. Smith Corporation of Milwaukee, which dated only from 1909 but had in the 1920s developed a highly efficient and rapidly increasing manufacture of electric-welded automobile frames, introduced the electrical welding of large-diameter pipes, above all for oilfield use. The new process used steel of higher tensile strength and consequently produced a lighter pipe than with older welding processes. For its new line, A. O. Smith bought plate from Chicago steelworks, tonnages used amounting to about 70,000 tons a month by mid-1929, by which time it was making up to fifteen miles of ERW (electrical resistance weld) pipe daily. The need to supply A. O. Smith was an important consideration in the installation at this time of a new continuous-plate mill at the US Steel South Chicago works. So efficient was the Milwaukee firm that National Tube was unable to compete in large welded pipe even though the latter's operations were linked with steel making and those of the former were not.[2]

Meanwhile, US Steel, like other major producers such as Youngstown Sheet and Tube, continued to make large-diameter seamless pipe, the former building a new plant for twenty-four-inch seamless at Gary, though its product was expected to cost more than the ERW pipe. By the early 1930s the older producers were reacting to the challenge, Youngstown Sheet and Tube and Republic Steel both building ERW mills. In 1930 US Steel installed the ERW process at the Christy Park works near McKeesport. At the same time, it tried to improve its outlets for pipe by acquiring the Oil Well Supply Company. Unfortunately, the following year the pipeline business collapsed, and Christy Park, after making only a small tonnage of ERW pipe, turned out none at all

TABLE 10.1

National Tube Company sources of profit (loss) by process of manufacture, 1930 and 1936 (percentage of net profit)

Process	1930	1936
Butt-weld	40.1	9.9
Lap-weld	7.5	(3.7)
Seamless	35.7	94.3
All other processes plus exports	16.7	(0.5)

Source: Ford, Bacon, and Davis, "Reports," vol. 159, "National Tube," 18 November 1937.

for several years.[3] In 1932 R. L. Smith proposed that US Steel should buy his company, but by this time both the industry and the Corporation were in the depths of depression, and after consideration by a special committee, the finance committee decided "that the proposals submitted for purchase of plants and engineering research achievements and service of the A. O. Smith Corporation do not, under the terms and conditions attached to such proposals, interest the Corporation."[4] A. O. Smith remained an important competitor in the oil-pipe field. Most thought-provoking of all, there were suggestions that, years before, the ERW technology had been offered to US Steel, which had shown a lack of interest.[5] If true, this was very much like its reaction to the Grey-beam mill. More practically relevant, by the time the ERW process was taken up by A. O. Smith, a far more important instance of technological lag was taking shape.

The development of continuous rolling in the field of thin flat-rolled products and specifically the successful introduction of the wide continuous strip mill were without doubt the most important advances in rolling-mill technology in the first half of the twentieth century. As far as the impact of change went, whereas the production of heavy structurals in old-style mills had been concentrated in a few large operations, some of which successfully made the transition to the new Grey-mill process, the new strip and sheet processes were deadly in their effect on that sector's previously small, usually unintegrated, and widely scattered mill operations. When these failed or were replaced by strip mills closely tied to large-scale iron and steel plants, serious problems of localized distress were inevitable. These problems coincided with the radical thinking about worker and community issues associated with the New Deal, and consequently the new process also brought the social implications of the actions of private enterprises into the foreground of public attention.

US Steel lagged seriously in recognizing and acting to correct its backwardness in sheet, tinplate, and wide-strip technology even though, from the first, it had been the leading producer of thin flat-rolled steel products. Even more telling, the Corporation had been involved in some of the first experiments designed to improve the sheet and black-plate trades. After this phase US Steel was left behind by technological pioneering of much smaller rivals, whose experiments, backed by persistence and large capital outlay, eventually led on to commercial success. Whereas others had pressed on, the Corporation had withdrawn. From the late 1920s, it was trying to make up for its slowness by licensing others' technology.

A continuous mill is one in which the piece of steel is rolled in at least two successive stands of a train of rolls at the same time. There are two main technical problems: the coordination of the contours of the rolls in order to ensure an even reduction in the dimensions of the original mass of steel in the direction that is ultimately desired, and the coordination of roll speeds. If the latter problem is not tackled successfully, the piece will be subjected to either compression or stretching between successive stands of rolls; the thinner and more fragile the product, the greater these dangers and therefore the more difficult to develop a technically successful continuous-rolling process. Quite apart from these practical problems, a commercial prerequisite for a continuous mill is a large order book for standardized products. Given favorable circumstances in all these respects, the tremendous advantage of continuous rolling for the operator is that a largely increased output can be secured with a great economy in handling during manufacture and consequent major reductions in the workforce. Such savings enable the company operating such an installation to more than cover the initial high capital costs and to reduce unit costs of production, which in turn, through lower prices, may help expand the market.

Continuous-rolling operations were introduced as early as 1862 in the Bedson rod mill. Wire-rod manufacture was a branch of the steel trade in which two of the essentials for successful continuous rolling were present: large tonnages of standard-gauge wire were consumed and the wire rod itself was fairly robust so that it could withstand the pressures or tensions of a less than perfectly coordinated succession of rolls. Building on the Bedson mill, in the 1870s the Morgan continuous-rod mill managed to sidestep problems involved in achieving exactly coordinated gearing by running the down-mill stands slightly faster than what was thought to be the ideal speed, thereby putting a slight tension on the rod. By the 1890s the Morgan Construction Company was also building continuous billet and bar mills, including plant for rolling sheet bars or the slightly smaller tin bars, which were then sent on or sold to other mills that rolled them down into sheet or a variant known as "black plate" for the tinning dipperies. In what at that time was a completely separate line of business, by the early twentieth century steel was rolled in the form of strip in continuous mills, but this was only of narrow widths, usually from six to eight inches.

Soon after 1900, continuous-mill operations were still very limited in their range of application. At that time their commercial situation was summarized by two experts: "The output of such a mill is, therefore, very high, and the labor cost very low, but the first cost [capital cost] of so many stands of rolls is a serious matter. The cost of the numerous rolls which are needed is very large and much time is wasted in changing them, so that it is only practicable to use a mill of this kind where a continuous demand exists for one spe-

cial section."[6] For the British industry, with which consulting engineer John
W. Hall was especially concerned, operating conditions were generally unfavorable for mills of this type. By this time, however, American companies commanded a much larger overall demand, and as this was overwhelmingly
centered in the home market, it was also more standardized. A new field of favorable circumstances for continuous rolling seemed to be emerging in thin
flat-rolled products, but for many years serious technical problems dogged all
attempts to produce a practicable continuous sheet or wide strip mill.

Into the twentieth century, tinplate- and sheet-mill operations were labor-
intensive, low-productivity sections of the rolling-mill industry. Yet demand
for the products of both was growing rapidly. Output of sheets of thirteen
gauge or thinner increased by over 400,000 tons, or 133 percent, in the ten
years to 1900, during which period the rise for all-rolled iron and steel products was only 57 percent. By the early twentieth century, sheet production
was about 6–7 percent of total rolled output. Tinplate tonnages were 361,000
in 1899 and 577,000 six years later. Much of the increase in both sections of
the thin flat-rolled product industry was for standardized specifications. Given
these conditions, most producers continued with the well-proved methods,
merely building additional mills when favorable opportunities seemed to justify the outlay. The widely spreading amalgamations of capacity, which produced the American Sheet and the American Tin Plate Companies of US
Steel—in December 1903 merged as American Sheet and Tin Plate—meant
they incorporated a good deal of mill capacity that from the start was recognized as surplus to requirements. This situation as well as the importance of
this division of the Corporation provided opportunities not only for the usual
sale or demolition of redundant plant but also for some of it to be used for experiments designed to improve efficiency in the continuing operations.

The first attempt to roll sheet steel in a continuous mill was made in the
Bohemian industrial area of the Austro-Hungarian Empire in the 1890s. It
was an unpromising setting for such an innovation, and the experiments
never led on to anything important. Soon after its formation, US Steel began
to investigate the possibilities. Its explorations extended over fifteen years. In
1902 Charles W. Bray began to try to improve operations at the Monongahela
works in Pittsburgh by breaking down tin bars into tinplate stock mechanically. To this end he arranged for four continuous bar-heating furnaces to feed
a roll train of eight two-high black-sheet mills arranged in tandem. After passing through these mills, the "pack" was mechanically doubled and went on to
the regular tin mills to be finished in ordinary two-high mills. Economy in
labor costs as compared with normal practice was put at about $2.00 a ton. In
May 1901 the average price of black sheets was $71.68 and of tinplates

$93.86 per gross ton so that, comparatively speaking, the saving was small, but, given the large tonnages of these products, significant in total. But gains in labor costs were cancelled out by losses due to an increased number of "seconds" and extra scrap. As a result it was decided to give up the experiments and to try instead with the rather heavier-gauge material used in standard sheet mills. This attempt was carried on from early 1904 to July 1905. Then in November 1905 another experimental mill was put up at Mercer works, South Sharon in the Shenango Valley of northwestern Pennsylvania, using nine rather than eight stands of rolls, but this too ended with mechanical doubling and finishing on an ordinary two-high mill. The Mercer plant operated intermittently from November 1905 to October 1907, was idle throughout 1908—presumably because of that year's depression—and partially at work again in 1909 and into 1910. A tendency for both bars and sheet "breakdowns" to stick in the mills caused problems. It is said that savings as compared with old hand-mill operations were this time about $1.80 a ton and that on the other side of the balance sheet there were a number of extra costs: $2.00 for maintenance, $1.00 for rolls, $1.00 for a greater yield of scrap, and $0.80 for other costs. The innovations therefore increased costs by about $3.00 a ton over conventional practice. In May 1910 the Mercer mills were shut down. A final American Sheet and Tin Plate experiment was made in 1917 at its Vandergrift works, an unusual establishment in two respects: its large capacity and the fact that its rolling mills were backed up by an open-hearth plant. Such conditions might have been expected to lead on to success. The new procedure was given the name of the "Tipperary" system, presumably reflecting the wartime song or perhaps also the long production sequence. This time three two-high mills were linked. Operations yielded useful experience, but the output was inferior to that in some old-type mills so that the trials went on for only a short time.[7] By now the Corporation was a smaller factor in this field than it was in the steel industry generally. Even so, sheet and tinplate tonnages amounted to 11.3 percent of its 1920 production; next year, one of extreme depression, they were 13 percent. The decision to abandon experiments in this field eventually proved to have been unfortunately timed. The initiative in the development of improved flat-rolled product technology passed to others.

The Monongahela and Mercer—and presumably also the Vandergrift—experiments and their outcomes were widely known so that later workers built on the experience gained. Now the pace of advance quickened. The first major breakthrough was the continuous two-high sheet mill installed in 1923 at the newly acquired Ashland, Kentucky, works of the American Rolling Mill Company (Armco). More important in the long run was the grad-

TABLE 10.2
**US Steel's share of production of all U.S. rolled products
and sheet and tinplate**

Year	Of all rolled products	Of sheet and tinplate
1902	64.8%	60.6%
1905	47.3%	62.6%
1909	48.9%	53.0%
1920	42.9%[a]	31.8%

Sources: US Steel annual reports; *Iron Age,* 19 February 1925.
[a] US Steel products shipped.

ual widening of the dimensions of strip steel that could be rolled continuously. With roots in experiments made by two separate companies at Massillon and at Elyria in Ohio and drawing on important new insights into roll design and the coordination of mill speeds powered by electric motors, this approach to the problem of continuous rolling of thin flat-rolled steel came to fruition after the 1926 acquisition by the Elyria company of the steel works at Butler, Pennsylvania. Butler was equipped with a wide continuous strip mill able to roll up to 400,000 tons of strip and sheet a year. This works proved to be the pioneer of a veritable revolution in the sheet and tinplate trades. There was now rapid progress in the application of the new technology. In 1926 only one in every 35 tons of sheet, strip, and black plate was produced in a continuous mill; within three years the proportion was one in every 6 tons.[8] The Ashland sheet mill and the Butler strip mill, by now both owned by Armco, were the only continuous operations at the end of 1926. In 1927 two more wide continuous strip mills were installed, another in 1928, and two more in 1929. Until 1929, expanding demand meant that both new and old-style mills could find employment, but with the onset of the Depression it became obvious that, because of higher standards of quality and productivity, for ordinary grades of sheet, it was only a matter of time before the old rolling-mill technologies would become redundant.

In spring 1929 the US Steel Finance Committee was informed that their patent counsel had examined the Armco patents and had concluded they were probably sound; the Corporation would have to pay others to use the new processes.[9] In fact, only one of the first thirteen continuous strip mills was owned by US Steel. This was a 42" mill with a capacity of 360,000 tons that American Sheet and Tin Plate started at Gary in 1929 to supply existing tinning operations onsite; it was expected to increase tinplate capacity by at least

50 percent.[10] Throughout the Depression, an industry-wide switch from heavy steels to lighter flat-rolled products continued. Nationally, 1936 was a somewhat better year than 1927; the overall slight increase during these nine years concealed a dramatic change in the structure of finished-steel production.

With increased tonnages of thin flat-rolled steel, there went not only a major improvement in quality but also an accompanying reduction in prices—generally, sheet-steel prices fell during 1926–1938 by 30 percent; in 1937 automobile fender sheets cost 46 percent less than in 1923 and 29 percent below the 1929 level.[11] Both higher quality and reduced price were only made possible by the extended use of the strip mill. From 1930, US Steel was forced to respond to challenge of the new processes of manufacture not only to retain its market in this sector but also to compensate for the sharp reduction in demand for a number of other, mainly heavier, products, which resulted in a decline of crude-steel requirements and calls on its existing blast-furnace and minerals capacity. To do so, the Corporation had to abandon a good deal of old-style plant and build strip mills by licensing Armco technology. In April 1929 the finance committee approved an agreement under which US Steel paid Armco $50,000 a year for the patents already at use at Gary and a royalty of thirty cents per ton of product rolled by this equipment at any other of its mills.[12]

TABLE 10.3

Light flat-rolled steels and other products in the United States, 1922–1926, 1927, and 1936 (percentage of total finished steel production)

	1922–1926 avr.	1927	1936
Light flat-rolled products			
Sheet	12.8	13.1	21.7
Black plate for tinning	5.0	5.1	6.8
Strip	3.1	4.1	9.5
Subtotal	20.9	22.3	38.0
Heavy rolled products			
Rails	8.9	8.7	3.6
Plate	12.2	11.5	7.5
Structural shapes	11.1	11.6	8.6
Subtotal	32.2	31.8	19.7
All other finished products	46.9	45.9	42.3

Source: C. M. White, "Technological Advances in Steel Products," in AISI, *Yearbook* (1937), 115.

Over ten years to 1938, US Steel closed twelve of its twenty-six sheet and black-plate mills, installations which in 1922 had represented 540,000 tons of its total of 2,111,000 tons of capacity for those products. Starting with Gary in 1927, it installed eight hot-strip mills having a combined annual capacity of 4.79 million tons. The distribution of these new plants was interesting for inevitably it reflected the changing balance of locational advantages and above all the importance of access to the prime sheet and strip markets of southern Michigan. Three of the first four plants were at Gary, which by 1936 had 930,000 of US Steel's 1,230,000 tons of strip-mill capacity; previously, the Chicago area contained no more than about 25 percent of the Corporation's capacity for these products. One strip mill was built in Youngstown, but none in the neighborhood of Wheeling, the district that suffered the greatest loss of hand-mill capacity. Fairfield was provided with a strip mill in 1936. In Pittsburgh, previously the leading flat-rolled product district, almost 39 percent of the capacity in old-style mills was abandoned. US Steel built two strip mills there, but only in the latter years of the program. One of these installations—at Homestead—was designed for light-plate rather than normal-strip mill gauges. In 1922 the Pittsburgh district contained 29.3 percent of US Steel's sheet and black-plate mill capacity; by 1939 the area contained 30.8 percent of its strip mill capacity. But the apparent balanced replacement of the old by the new plant was deceptive in the sense that—as shown above—not all the new capacity was for strip and sheet, the new plant was built later than other US Steel strip capacity, it represented a smaller net increase on what had gone before in terms of old-type mill tonnage, and its installation was only achieved with considerable difficulty.

After spending some years considering possible alternative locations, US Steel announced at the beginning of 1937 that it was to spend $60 million on new facilities or modernization of old ones near Clairton and at the Edgar Thomson works. Ground was broken for a new rolling mill located at Dravosburg, some two miles west of Duquesne, in May. The 600,000-ton-capacity 80" hot mill, two cold reduction mills for sheet, and another to produce black plate for tinplate covered fifty-two acres. Construction involved removal of 4.3 million cubic yards of material to produce a man-made plateau—the deepest ravines that had previously crossed the site being 250 feet from the top of the highest cut to the bottom of the deepest fill. Slabs were produced in a new mill at Edgar Thomson, railed across the Monongahela River, and carried by US Steel's Union Railroad some six miles to the new hot mill built 227 feet above river level. The 3,750 employees of the new operation were recruited from the cream of the workforces of the old-style sheet and tinplate mills that had been closed as a result of the new technology. The new strip mill was opened in De-

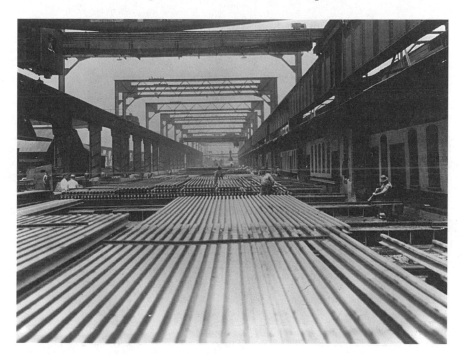

The shipping yard for rails at Edgar Thomson, Pennsylvania, between the world wars. *Courtesy of USX Corporation*

TABLE 10.4

US Steel sheet & tinplate mills, 1922; closures, 1928–1938; and strip mills to 1939 (capacities in thousand tons)

District	Sheet and Tinplate mills 1922		Sheet & Tinplate mills closed 1928–1938		Strip mills built to 1939	
	No.	capacity	No.	capacity	No.	capacity
Pittsburgh	10	619	6	239	2	1,478
Valleys	4	384	n.a.	n.a.	1	470
Wheeling	9	483	6	302	n.a.	n.a.
Cleveland	1	27	n.a.	n.a.	n.a.	n.a.
Chicago	2	598	n.a.	n.a.	4	2,340
Birmingham	n.a.	n.a.	n.a.	n.a.	1	504
	26	2,111	12	541	8	4,792

Sources: *Iron Age,* 1 January 1925; US Steel Annual Meeting, 4 April 1938, Hoboken, New Jersey, USX; Association of Iron and Steel Engineers, *The Modern Strip Mill* (1941).

cember 1938.[13] In engineering terms, the Irvin works, named after the US Steel president who had spent most of his working life in their sheet and tin-plate operations, was a triumph. Less positively, its construction indicated the importance of locational inertia, involved much increased capital outlay—the whole project cost about twice as much as Jones and Laughlin's recently completed 96" mill—and pointed to continuing increased operating costs as a consequence of the local physical difficulties.

TABLE 10.5

Productivity at the Shenango blackplate and tinning operations and at Irvin works, March 1942 (man-hours per ton)

Operation	Shenango	Irvin
Rolling	24.910	5.225
Tinning	9.332	6.672
Packaging and shipping	0.535	0.502
Total per ton	34.777	12.399

Source: US Steel, paper prepared for the U.S. Senate Judiciary Committee's Subcommittee on Antitrust and Monopoly, August 1957.

Before World War II began, US Steel had made up for a good deal of its slow start in the newer thin flat-rolled products field. Its own experience with both the newer and the older methods of production showed conclusively how wide was the advantage of the strip mill. Even so, twenty years later the Corporation still lagged far behind some of its leading rivals in the emphasis it gave to this sector of the business. By then there was also more evidence that US Steel had not fully learned the dangers involved when an industry's leader allows itself to fall behind in technology.

Crisis and Response

The Achievements of Myron Taylor, 1927–1938

In the late 1930s, students of business realized not only that the national economy was at last emerging from the greatest of depressions but also that the ways of doing business were and would always be different from what had gone before. The director of the Pittsburgh Bureau of Business Research recognized something of what this implied for the leading local industry: "The past ten years or so may be looked on as years of transition for the American iron and steel industry—transition from youth with its vigorous growth and rapidly expanding markets to the stage of economic maturity, attended by slow growth in productive facilities and intense rivalry for slowly expanding or stationary markets. Profits no longer accrue more or less automatically, but must be garnered as the fruits of competitive efficiency." He might have added that this major divide crossed in the nation's economic structure meant a permanent change in the types of steel most in demand. In fact, as hindsight was to recognize, although economic maturity seemed for a time to be associated with crisis and slow-growing markets, the nation had been moving into the age of high mass consumption. The United States was leading the way into a consumer-oriented society.[1] In such circumstances it was particularly important that US Steel, which for so long had showed little flare or sparkle, should at last bestir itself.

The Great Depression tested the Corporation's massive structure as it had never been challenged before. At its outset US Steel seemed to be holding on to its role as an industrial stabilizing force, but later it did less well. As Wall Street faltered in the latter part of October 1929, the Corporation was involved in a symbolic act designed to rally confidence in the soundness of industrial Amer-

TABLE 11.1
US Steel and other leading steel companies, ratio of net income
to invested capital, 1927–1936

US Steel	2.42%	National (1929–36)	6.28%
Inland	7.53%	Armco	3.90%
Republic	2.62%	Wheeling	3.60%
J&L	2.62%	Crucible	2.48%
Youngstown S&T	3.29%	Bethlehem	2.69%
		Avr. of the 9 independents	3.39%

Source: Ford, Bacon, and Davis, "Reports," 200:5.

ica. Midway through Thursday, 24 October, the vice president of the New York Stock Exchange put in a bid of $205 per share, the same price as in the most recent sale, for 10,000 shares of US Steel. After that he did the same with some fifteen or twenty other stocks, but it still seemed that US Steel, his starting point, was the hub of the business universe, at least to those most concerned to prevent its collapse.[2] This gesture of confidence in the essential soundness of the economy did not work. Bit-by-bit, the situation settled into the terrible conditions of the early 1930s. US Steel decided to pay only 1.75 percent on its common stock at each dividends distribution in June, September, and December 1930. Shipments were only a little over three-quarters the level of the previous year. Its steel-operating rate was 90.4 percent of capacity in 1929 but only 67.2 percent in 1930; profits fell from $197.5 million to $104.4 million. This was bad enough; fortunately, it could not then be foreseen that over the next decade profits would not again reach the 1930 level.

The problems were general to the industry, but US Steel felt them more severely than many companies. Key indicators emphasize some remarkable features of these years. First, the Corporation did not reduce its capacity to match the long-term decline in production. Second, in the early part of this cycle of bad years, it was unexpectedly slow to prune the workforce. Third, even when output of crude steel and shipments of products crept back to high levels at the end of the 1930s, by various criteria productivity showed no strong improvement on ten years previous. But, despite such apparently negative features, US Steel was very substantially changed and was much better fitted to face the challenges of a new age. The size of the task was increased by its inheritance from the Elbert Gary years. The Corporation's ability to tackle the decade's problems owed a great deal to the qualities of the man who succeeded him.

The obituary notice that appeared in *The New York Times* on the day after

TABLE 11.2
US Steel crude-steel capacity and production, shipments (million tons),
employment, and income (loss), 1927–1940

Year	Crude steel		Steel products shipped	Employment	Income (loss) (millions)
	Capacity	Production			
1927	25.9	20.7	14.3	231,500	$87.9
1929	27.1	24.5	16.8	254,500	$197.5
1930	27.9	18.8	12.8	252,900	$104.4
1931	30.1	11.3	8.4	215,700	$13.0
1932	31.2	5.5	4.3	164,300	($71.2)
1933	30.6	9.0	6.3	172,600	($36.5)
1934	30.6	9.7	6.5	189,900	($21.7)
1935	30.6	12.5	8.1	194,800	$1.1
1936	29.9	18.9	11.9	222,400	$50.5
1937	28.9	20.7	14.1	261,300	$94.9
1938	28.9	10.5	7.3	202,100	($7.7)
1939	28.9	17.6	11.7	223,800	$41.1
1940	27.8	22.9	15.0	254,400	$102.2

Source: D. A. Fisher, *Steel Serves the Nation: The Fifty Year Story of United States Steel* (New York: US Steel, 1951), 224–25.

Note: Steel capacity figures are calculated from production and operating rates.

TABLE 11.3
US Steel indices, 1927 and 1937

	1927	1937
Steel products shipped per man-year (tons)	61.80	53.95
(A) Avr. value of products/services sold per ton shipped	$67.12	$72.95
(B) Employment costs per ton of product shipped	$28.84	$31.71
Margin above employment costs (A - B)	$38.28	$41.24
Employment costs per $100 of products/services sold	$42.97	$43.47

Source: US Steel production and financial data from D. A. Fisher, *Steel Serves the Nation: The Fifty Year Story of United States Steel* (New York: US Steel, 1951).

Gary's death contained the subheading "Myron C. Taylor looms."[3] More than four months later a new chairman was at last announced: sixty-year-old John P. Morgan Jr. Long before he in turn handed over leadership of US Steel four years later, it was clear that the corporate reins were held by the man the *Times* had identified. Born in upstate New York in 1874, by the turn of the century Taylor had graduated in law from Cornell. Gary's practice of law led him during his middle years into creating and then running large corporate combinations; at a much younger age, Taylor began work on the wholesale rationalization of overextended long-established firms. Gary was concerned to maintain the stability of an existing corporate order, Taylor to upset it in pursuit of economic efficiency. While Gary was consolidating his position at the head of the steel industry, Taylor was involved in cotton textiles, initially in Lowell and Newburyport, Massachusetts. He was much exercised about output levels in relation to capacity; "he saw the problem of fixed overhead costs clearly," as one later appreciation summed up the central thread of Taylor's work. Modernization in textiles was tackled along four lines—a strategy similar to that he would later apply in steel. His four priorities were simplification of financing, installation of modern equipment, progressive management, and improved conditions for workers (including three eight-hour instead of two twelve-hour mill shifts). In 1912 the reorganizations he helped carry through culminated in a major consolidation of northern and southern mills as the International Cotton Mills. Taylor moved on, before World War I disposing of his interests in ordinary textiles, to turn to other lines of business including tire fabrics. Goodyear Tire and Rubber Company was in difficulties by

Myron C. Taylor, chairman of US Steel, 1932–1938. *Courtesy of USX Corporation*

1920–1921, before he took over its reorganization. Taylor was also involved in financial reconstruction at AT&T, among railroads, and later that decade in bank reconstruction. He acquired a reputation for combining ample talents in high finance with intellectual honesty.

In September 1925 Taylor was elected a director of US Steel and a member of its finance committee. He was spending much of his time traveling at first, but eventually two senior directors, G. F. Baker and J. P. Morgan, induced him to take a more active part in corporate affairs. By December 1927 he was chairman of the finance committee. He, and possibly those who had pushed him forward, already saw his role as that of a reformer of the hallowed institution. In retrospect, Taylor acknowledged that he entered active service "to undertake a readjustment of [US Steel's] basic organization, and of its plants, its facilities and its personnel, so that it might continue in the future to justify the splendid history of its past."[4]

Taylor set out to equip himself more fully for a leading part in the corporate hierarchy, spending much of the late 1920s learning first hand about conditions throughout the various divisions. He began to reveal his working philosophy. In October 1928, speaking for steel to the Conference of Major Industries held in New York City, he focused on an unresolved conundrum of a modern competitive industry. The total investment represented by the industry was reckoned to be $4.75 billion; in terms of replacement costs, it was worth at least $6 billion. There was need for vast outlays on modernization, requiring profit margins to encourage and reward this investment, yet keen competition threatened the price levels on which profits depended. He summed up: "Modernization of plant and facilities to lower cost of production is a prevailing practice throughout American industry. If this practice is not followed, survival cannot be assured. Hundreds and hundreds of millions of dollars of capital outlay have been made for this purpose, and in the older and more established industries these outlays have not produced to the owners a sufficient profit on the new investment."[5] In March 1932 Taylor was elected chairman of the US Steel board and chief executive officer, posts he retained until spring 1938. He now had the position and power to put his ideas into practice. In fact, his whole period of service, at the finance committee and as board chairman, may be considered as a ten-year campaign to overhaul US Steel. Given the times and his inheritance in the corridors and meeting rooms of 71 Broadway and in mines, plants, and other facilities across the nation, he did remarkably well.

It was natural for a man of Taylor's background that his first major contribution involved finance. Fortunately, he tackled this issue before the Depression occurred. During 1929 he carried through a reorganization that lowered the Corporation's bonded debt. As a result, whereas average annual payments of interest and other costs on debt averaged $27 million in the seven years to 1929, for the next seven they were only $5.2 million. This made a vital contribution to corporate survival in a period when profits first shrank and then, for a few years, turned into major losses; interest payments on debt would have had to be maintained, and holders of common stock could have gone unrewarded for much longer.[6] Above all, Taylor inspired and helped push through a major rationalization of organization, production, and labor relations. After his first two years as head of the finance committee, this work was done against a backdrop of almost continuous business uncertainty and for much of the time acute depression—at the depths of their fortunes, four months after Taylor became chairman, production of raw steel was as little as

11.8 percent of installed capacity. As a successor from more placid days recognized, Taylor "guided" US Steel through "that great and universal malady" with "consummate skill and vision."[7]

He pushed through important changes in corporate structure. In 1935 the formerly separate operations of Carnegie Steel and Illinois Steel, three-quarters of all the Corporation's steel capacity, were merged into the new Carnegie-Illinois Steel, removing some duplication of production and control and bringing important operating economies. Two years before, it had been decided to separate control of the making and selling of steel from the financial administration of the group. To this the end, United States Steel of Delaware was formed. The powers of the finance committee were reduced, financial matters were left at 71 Broadway, and control of manufacture and sales was moved to Pittsburgh, though the latter was not completed until early 1938.

Another line of restructuring was in terms of personnel; in turn this was divided into concern with the introduction of better top executives and improvement in relations with the work force. A notable feature of the first was the introduction of new blood in the form of younger men whose quality had been proved in other industries or, for the first time, at other leading steel companies. In 1934 the thirty-four-year-old Edward Stettinius was brought in from General Motors as vice chairman of the finance committee; next year he became its chairman. The accountant Enders M. Voorhees, who had already made important contributions to the reconstruction of a number of important firms, was made vice chair of the finance committee in 1937; he was forty-five. There were at least equally important changes on the production side. William Irvin had been associated with the sheet-and-tinplate industry since the 1890s, and after 1901 he rose gradually through management ranks until in the 1920s he was vice president of American Sheet and Tin Plate. In 1931 he was made vice president of US Steel and the following year succeeded James Farrell as president, a post Irvin retained until 1938. His competence went beyond an intimate knowledge of what had now become both the industry's and the Corporation's key growth sector. He helped draft the NRA code for the industry and was widely respected by labor for a constructive wages policy.[8] Provision for the next stage of the succession on the production side began in September 1935, when a forty-five-year-old executive vice president of Republic Steel, who had done well with sales even during the Depression, Benjamin F. Fairless, was brought in as president of the new Carnegie-Illinois Steel Corporation. This appointment caused some tension in the formerly separate companies. At Illinois Steel, Eugene Buffington, president since 1899, had been asked to retire in 1932, at age sixty-nine, to make way for younger men. When Fairless came in, Buffington's successor, George Thorp, and I. Lamont Hughes, president of Carnegie Steel, were given the option of resigning or retiring to make way for him. Three years later, having proved himself, Fairless took over from Irvin as president of US Steel. In short, to at least as great an extent as Gary had done in the course of the sometimes-painful maneuvers of the first years of the Corporation's history, Taylor now managed to secure the men he wanted for the key posts in finance, general organization, and production. Stettinius remained with US Steel only until 1940, when he resigned to join the Roosevelt government. Fairless occupied one of the two leading positions until 1955, and Voorhees chaired the finance committee until 1956.

As a result of over a decade of analyses by their own engineers, painstaking surveys by outside consultants, the commercial pressures of the times, and Taylor's own sense of direction for advance, the chairman was able to force through a program of plant and process rationalization far greater than any in the previous history of US Steel. It was on a scale that was not matched again for almost another half-century.

The process of weeding out superfluous, smaller, less efficient, or badly located operations and greater concentration on better endowed units had been going on since the formation of US Steel. Plant closure and abandonment had continued into Gary's last years, and after his death there were signs of even more comprehensive reviews. A leading instance was the March 1929 report by Farrell to the finance committee—of which Taylor was now chairman—of a "Complete physical analysis of the plants and properties of the Corporation." The details are unfortunately unknown, but the minutes indicate something of the scope. The report contained

> charts illustrating such properties, recommendations of the executives of the several companies covering improvements, replacement and extensions which in their opinion are either (a) necessary, or (b) desirable to be made to equip the various plants and properties with the most efficient machinery and appliances, and to round out production along lines necessary to meet market requirements as same now appear; also the estimated cost for same, together with a program for the accomplishment thereof, spread over periods of one, two, and three years, part of such program having already been authorized and undertaken. There were also presented summaries and charts indicating the effect of these expenditures upon productive capacities, both in semi-finished and finished products. . . . [T]hese reports and recommendations were received and ordered filed for

further consideration and reference. . . . The President reported and presented to the Committee, [a] statement prepared by the executive staff, listing the properties of the Corporation no longer needed for corporate purposes and recommended for sale together with the estimated value of such properties. A similar statement, prepared by the Comptroller from the corporate records, was also presented to the Committee.[9]

Apart from the involved and painfully obscure form of the statement, it is important to note the limitations of the analysis it described. First, it was an internal study made by constituent companies; that is, it came from below rather than representing an overall evaluation from above or outside the huge range of well-entrenched self-interests within US Steel. Second, its program was short term. Third, the beginnings of the economic collapse that autumn soon made the whole basis of the study redundant. In April 1930 the finance committee approved an outlay of up to $200 million for a three-year program of "expansion and modernization," but by August 1932 the severity of the Depression caused the cancellation of the program after little more than one-seventh of this budget had been spent.[10]

Along with his personal program of better understanding the Corporation's structure, in 1928 Taylor initiated a seven-year investigation by US Steel engineers of what was needed to overhaul the entire straggling assembly of mineral resources, transport facilities, and plants. These and other studies continued throughout the Depression and were periodically mentioned in committees. For instance, early in December 1933 with the finance committee, "Chairman Taylor reviewed the various phases of the reorganization of the Corporation's financial, plant, and personnel structure which had been in progress during the last five years." Six months later, he "called attention to a proposed construction program developing from a study of plant locations and costs of production which had been prepared with a view to replacing obsolete equipment and installing modern continuous methods. After discussion and on the affirmative vote of all present, a special committee consisting of the chairman [William J. Filbert], president [Irvin], Taylor, Roberts, and Junius S. Morgan was appointed to consider all the facts and report to the finance committee at the earliest practical date." In September 1935, Filbert reported that studies of four plants had revealed savings that could be made amounting to $1,211,878 annually. Similar analyses were then underway in other plants. Executives decided at this time to call on independent expertise, and within a few weeks Taylor reported that consultants were engaged to take part in what he described as a cooperative study of "the facilities, operations, business, management, and policy control of the Corporation and its subsidiary companies." The Corporation would be represented by a "control committee" of Stettinius, Watson, Vogt, and the presidents of each subsidiary company whose affairs were involved.[11] Looking back on this enquiry, he would explain to stockholders that well before their own studies reached full term, the board realized that the implications of their findings "were so far-reaching and involved expenditures of such magnitude that we decided that it would be wise to gain an outside opinion." The Corporation did so "to reassure itself with the results of these investigations and plans."[12] in short, at a cost of some $3 million, US Steel engaged the industrial engineering firm of Ford, Bacon, and Davis, which hired the expertise of other specialist organizations, to survey, report, and make recommendations for all corporate operations.

The Ford, Bacon, and Davis study was initiated in a meeting held at US Steel offices on Wednesday, 23 October 1935, attended by all principal executives and the heads of most divisions. Despite the size and importance of the venture being set in motion, the tone seems to have been relatively relaxed and generally different from the self-conscious dignity of the Gary years. Even so, in his remarks Taylor allowed himself one rather whimsical comment on their exceptional status: "[S]ome say that we have been very old-fashioned in many ways. I think that you can take that as an unpleasant criticism in some ways, but I think that it is rather a pleasing reflection nowadays to be called a little old-fashioned. Certainly some of the modern ways are not ways that this group would care to follow." Irvin made a brief statement. Then the chairman of the consultants, G. W. Bacon, spoke. He ended with a simple but thought-provoking sentence: "Don't get the idea that we are going to tell you how to make steel. What we expect to tell you is, [by] making steel, how to make money." For a company that over the previous three financial years had lost almost $130 million, it was no soft-option target.[13]

Over the next two years, the reports containing the findings of the painstaking studies made by the consultants were submitted to US Steel as each was completed. Eventually, there were two hundred printed volumes of widely varying length incorporating innumerable tabulations, diagrams, and maps. Most dealt with individual mineral and company divisions, while some looked at broader issues such as Report 100 on "Organization." The whole survey was summarized in Report 200, produced in April 1938. In addition to their extraordinary detail, the reports contained striking insights and made important recommendations. A necessarily cursory summary of their findings may be grouped under two heads: the analyses and recommendations for the various divisions, and the more general comments on the problems associated with the giant size of US Steel.

An important criticism concerned the Corporation's inadequate response to changes in demand and technical innovation. The indictment was blunt: "A general review of the three and a half decades constituting the Corporation's record fails to confirm that it has met or anticipated as it might have, in our judgement, the reasonable demands of certain of its most important customer requirements. . . . We refer particularly to (1) Its failure to introduce years before it did the continuous mill, although the mill was presented first for its consideration." Having once supplied 67.3 percent of the sheet market, by 1933 US Steel's share was 14.4 percent. The second failure listed was in ERW and seamless pipe, and the third in wide-flange beams. "In steel alloys, particularly in stainless steel, it followed where it should have led." In 1929 US Steel accounted for only 0.6 percent of national output of stainless steel, though by 1936 its share had increased dramatically to over 15 percent. The Corporation also had lagged in introducing the electric furnace. South Works in particular had important capacity, but as late as 1930 electrically produced steel had been of such slight significance that it was not listed separately in the annual report's summary of production.[14]

The consultants provided detailed analyses of the present situation of each division including the national setting for its product or group of products, consideration of the ideal pattern of production, and examination of the ways in which the nearest practical approach could be found to that ideal. It is impossible to deal systematically or anywhere near fully with the immense complexity of these reports, but the situation of American Steel and Wire (AS&W), covered in Report 32 of 20 April 1937, gives a good idea of the consultant's approach. AS&W had not had an illustrious record: "For years it bulked too large in the industry and was voluntarily allowed to decline in market participation. . . . [I]t suffered heavy losses during the depression along with other subsidiaries; it has not recovered as rapidly as most of the others. Its plants by their types of construction and scattered locations did not and still do not lend themselves to modernization without major expenditures and removals and major expenditures accordingly have not been made on them; as a result, they now show distinct obsolescence and are badly handicapped in comparison with competitors' plants."[15] In 1901 US Steel produced 77.6 percent of the wire-rod output of the United States. Much of this came from AS&W, but some was from former steel and semifinished firms such as Carnegie and Federal.[16] By 1921 AS&W accounted for only 38.8 percent of the national output, and in 1935 this was down to 34.1 percent. From 1929 to 1932, its net income before federal taxes fell from a $14.3 million profit to a loss of $7.2 million. The consultants attributed almost one-seventh of this fall in gross mill profits to loss of market share. In turn, some of this was due to shifts in the geography of the national economy. AS&W, which had begun life with a ramshackle assembly of plants and clearly had too many of them, had not been well placed to react positively. In the early 1930s, A. W. Rutt-kamp from that division had recognized this: "A period of thirty years necessarily brings about plant obsolescence. Plants become poorly located due to population changes and shifting trade centers. The construction of new facilities overlaps the old and to some extent brings about at least a temporary duplication of capacity until the old facilities are scrapped. The wire industry is going through exactly that sort of evolution."[17] In fact, response had been so slow that far too much of the old survived. Closures had been underway for many years—Craig listed sixteen plants or major units of AS&W plants abandoned between 1901 and 1930—but their place was not taken by new works in strategic locations. The slimming process continued. In 1931 the H. P. works in Cleveland and the Braddock works in Pittsburgh closed, the first a wire-nail maker, the other a wire works. In 1935 and 1936 AS&W abandoned its works at Farrell, two blast furnaces at the Central works, and the obsolete Newburgh steelworks, the last two in Cleveland. However, with the closing of Newburgh, AS&W lost its only local source of ingots for further Cleveland operations. The plan was to replace these ingots with billets from Lorain, but growth there in orders for tubular products made Lorain an unreliable source of supply; the Cleveland works were thereby "compelled to resort to excessive cross shipments of billets." By the time Ford, Bacon, and Davis reported in spring 1937, AS&W still had twenty-two active plants, but they were a very mixed bunch; "a series of acquired works, very few of which are satisfactory from the standpoint of proper economical flow of material." In fact, as their tabulation of company's mills showed, the division remained after over thirty years a collection of very differently endowed operating units, most lacking modern equipment, and nearly all having poor layouts. They were widely scattered, "involving excessive freight charges for the assembling of raw materials, excessive overheads due to maintaining too many organizations, and other unfavorable factors incident to this condition." In some cases, as before 1901, AS&W still bought semifinished steel from separately controlled integrated works—in 1935, for instance, Illinois Steel had 130,000 tons of wire-rod capacity. Only two AS&W works were integrated, and they were ill located. An "ideal" AS&W, "if there were no practical considerations involved," would have had five of what the consultants called "integrated" plants of a composition and in locations very different from the existing ones. There would be two fully integrated works with blast furnaces, steelworks, and mills, one in the Cleveland the other in the Chicago area; a scrap-based steel, wire-rod, and wire-products plant on the New England coast near New Haven, Connecticut;

a plant in the South or Southwest; and a works on the Pacific Coast making wire from brought-in rod. The consultants recognized that this ideal scheme made no provision for a Pittsburgh-area plant even though at that time the integrated Donora works had, by a considerable margin, the company's largest rod and wire capacity. Nor could US Steel easily withdraw its commitment through AS&W from either Worcester, Massachusetts, or Duluth, Minnesota. They therefore came up with a compromise, a reconciliation of the ideal and the actual, suggesting seven future foci—Worcester, Pittsburgh, Cleveland, Chicago, Duluth, Birmingham, and somewhere on the Pacific Coast.[18]

One of the most interesting aspects of the analyses was the revelation of the extent to which generalizations about the various districts had to be qualified by reference to the very different state of efficiency of individual plants. The studies of the consultants confirmed what had long been obvious: the Chicago area possessed major advantages over the Corporation's old heartland in the Pittsburgh-Youngstown-Wheeling region. Even so, as compared with the excellences of Gary, the layout and flow of materials at South works was "very poor as units have been added without any well defined plan for expansion." Pittsburgh works, generally burdened by the effects of their age and the local topography, varied widely in technical efficiency. Homestead was unavoidably limited by site conditions, the impact of which had been accentuated by piecemeal development: "additions have been made wherever it was possible to fit them in rather than as part of a preconceived general plan." Their comments on aspects of its operations would have appalled Andrew Carnegie or Charles Schwab: "The equipment, with the exception of the structural mills and the new 100 inch plate mill, is all old and requires replacement by modern capacity. . . . [C]ontrol of quality is good, but the control of cost is rudimentary." Of its Number 2 open-hearth plant, the report ran, "Housekeeping in this department is exceedingly poor and reflects very lax supervision." Comments on Clairton were crushing. Open-hearth furnaces and mills were "old and not in good condition. . . . The plant did not operate for several years and general plans were made to abandon it. The policy seems to have been to defer all maintenance and the equipment and buildings reflect that determination." Duquesne was no better: "The facilities of the entire works, with but few exceptions, are old and in poor condition. . . . With the plant in the condition it is, a study should be made of the feasibility of placing this capacity in some other locality . . . with equipment capable of meeting market conditions profitably." Though established earlier than any of them, Edgar Thomson, which contained much old and some poor plant, in other respects contrasted sharply. Of its blast furnaces, the survey commented, "the entire department is exceedingly well maintained and the housekeeping is excel-

TABLE 11.4

American Steel and Wire Company works, spring 1937 (annual capicity, thousand tons)

	Pig iron	Steel ingots	Wire rods	Plain wire
East Coast				
Worcester, North works	–	–	–	68
South works	–	190	134	60
New Haven	–	–	–	27
Trenton	–	–	–	27.5
Pennsylvania				
Allentown	–	–	105	115
Donora	380	627	320	235
Rankin	–	–	–	80
Cleveland				
Cuyahoga	–	–	150	105
Consolidated	–	–	75	90
Central furnaces	473	–	–	–
Newburgh	–	–	–	95
American	–	–	40	75
Chicago area				
Scott Street	–	–	–	90
Joliet	–	–	220	–
Anderson	–	–	105	75
Waukegan	–	–	219	195
Rockdale	–	–	–	70
DeKalb	–	–	–	75
Duluth	164 (idle)	300	100	97.6
Totals	853[a]	1117	1468	1580

Source: Ford, Bacon, and Davis, "Reports," vol. 32, "AS&W," 20 April 1937, exhibit 53.

[a] "Dormant" blast-furnace capacity at Duluth is not included in the total.

lent." Throughout the rest of the plant, "the work has been intelligently supervised in every respect, and . . . the maintenance in particular has been carried out in such a manner that it might well serve as a model for the district." Summarizing their analysis of Carnegie-Illinois, the consultants emphasized a lack of uniformity throughout the company: "The facilities range from the very old to the most modern and the use and care of the facilities, as indicated by operating conditions and maintenance, range from very poor to very good."[19]

As well as examining and recommending changes to the physical plant of US Steel, Ford, Bacon, and Davis also tackled organizational structure, which they defined as "the medium through which the three major elements of industry—capital, labor, and management—function in the public interest."[20] Outside critics had been sure that the heart of the painfully obvious difficulties of US Steel was the fact that it was unmanageably large. In summer 1932, when the Depression was about at its worst, *Fortune* pointed out that, despite all the plant closures, US Steel still consisted of twenty manufacturing subsidiaries and 143 works. Four years later, in an important series on the Corporation, the same journal described it as engaged in too many fields of operation and having too much capital tied up in plant. It concluded: "The trouble with the United States Steel Corporation can be briefly stated: it has been too big for too long."[21] The consultants recognized there might be a problem in this respect and faced it directly. For them the real question was not "Is the Corporation too big," but rather "a question of balance between the integrated parts." They filled out their meaning: "The problems of mere bigness which beset the Corporation, however, do not hold for us the terrors which seem to concern many. The greater the number of men involved in an organization, the less the effect upon it of failure or even the loss of the individual. Bigness makes possible far-reaching policies, plans for the selection, the training, and the succession in the ranks of human material, denied lesser operations. Bigness implies a commanding position which should, properly handled by those concerned, become a synonym for service to the public of the highest order."[22] Their answer to the "problem" of bigness may have been correct in principle, but it was not of much practical help. First, was "public service of the highest order" really the priority or should it be maximum income? Second, and more practically, how could one ensure a corporate structure which made these sizable assets realizable and would, above all, guarantee direction of the highest quality throughout the whole group? Any solution would need to coordinate the whole vast structure and take the initiatives necessary for continuing commerical success. The consultant's theory was not exceptional; the practice would need to be just that. Meantime, having received the two hundred reports, the Taylor regime did the best it could to implement some of their recommendations as the industry began to emerge from years of the greatest commercial difficulty.

There was one final proposal from Ford, Bacon, and Davis that US Steel seems to have allowed to founder without follow-up, probably much to its long-term cost. They recommended a "planning board . . . to carry on a continuous program of long-range planning." The need for such an "agency" had long been recognized, but they now reckoned it was essential, and would be relatively easy to finance: "In an organization the size of the Corporation and its subsidiaries the problem of coordination of effort, and of placing before those in final authority the necessity for changes or improvements becomes a

LOOKING SOUTH SHOWING GENERAL
EXCAVATION FOR HOT STRIP MILL
IRVIN WORKS
OCTOBER 20, 1937
NEGATIVE No. 6-B-73

Excavation for the hot-strip mill at Irvin works, Pennsylvania, 20 October 1937. *Courtesy of USX Corporation*

complicated one and delay in action generally results. The Corporation, due to its size, is not only in greater need of a long-range planning agency than a smaller corporation, but is better able to carry the burden of its expense. Such an agency would probably have led to avoiding some of the mistakes of the past. . . . The functions of the Planning Board would be to plan the way ahead for the Corporation and its subsidiaries to the end that their products shall keep abreast of competition and be manufactured and distributed by the most economical methods." Such past failures as those with H-section beams and continuous rolling might still have occurred, but "the probability is that with the Planning Board properly set up, the advances in the art in connection with these products would have been so forcefully brought to the attention of the management that earlier action would have been taken."[23]

In the early 1930s, steel consumption had fallen so much as to make reorganization difficult but, particularly in the long run, its nature was changing. Such conditions presented the whole industry with problems, but for no other company than US Steel were they more serious. As Taylor acknowledged, the Corporation was ill equipped to meet the needs of a nation moving rapidly from a capital-goods to a consumer economy: "The trend of consumption was away from the heavy products where the Corporation was strongest to the lighter flat-rolled products where the Corporation did not possess the most up-to-date facilities." In 1934 about 20 percent of the business of the industry was in the automobile field, but for US Steel it was only 9 percent.[24] It was a sign of the needs of the times when in June 1936 the American Sheet and Tin Plate Company, concerned in large part with the automotive sector, was merged into Carnegie-Illinois. It then had a capacity for 2,318,000 tons of finished products but for only 340,000 tons of steel. Most of its requirements had been supplied by Carnegie-Illinois, whose rolling-mill capacity was 14,666,000 tons of semi- and fully finished products but which had a total steel capacity of 20,218,000 tons.[25] To make things more difficult, changes in demand had called forth the highly productive but costly new technology of the wide continuous strip mill. Under the new regime, US Steel sheet and tin-plate operations would no longer draw semifinished steel from integrated works, but would be conducted increasingly as part of these works. Speaking to shareholders, who were mainly laymen like himself, Taylor simplified the issue so as to drive home the lesson. "Into a settled industry, which, on the whole was losing money, there came an almost instant demand for the laying out of great sums to meet the revolution in the science of steel making which transformed it almost overnight from a rough industry into a precision industry." In fact, for a large producer that was a laggard in this field and yet had a major commitment to it in older technologies, such innovations were particularly disruptive. A program of large-scale closures had to be carried through, in many cases the plant and equipment being of little more than scrap value: "In certain of our plants the obsolescence, taking that word in its broader significance, was such that their abandonment was scarcely a matter for discussion." Over the years Taylor was in charge, US Steel abandoned almost 5 million tons of finished rolled-steel capacity—approaching one-third of their total when he came into the Corporation—and built 6 million tons in new capacity, of which more than two-thirds was in hot-strip mills.[26] This revolution cost about $600 million—over the same period, net income amounted to little more than two-thirds as much. Two other considerations in the rationalization program were concentration of capacity and substantial change in its geographical distribution.

In the 1930s there appeared a series of analyses that purported to show the relative costs of manufacture for pig iron in various producing districts.[27] However accurate—and they varied a good deal—these failed to recognize that, in practice, the critical factor was not the expense of iron but the overall cost of making steel. The predominance of open-hearth furnaces meant the price of scrap became a vital consideration in the comparative-cost analysis.

TABLE 11.5
*** Shares of steel rails and flat-rolled products in U.S. steel production,
1901, 1928–1929, 1936***

Year	Steel rails	Sheet and strip	Tinplate
1901	23.0%	6.5%	3.3%
1928–1929	8.3%	19.0%	5.0%
1936	3.5%	31.0%	8.0%

Source: M. Taylor, "Ten Years of Steel," statement at the annual meeting of US Steel stockholders, 4 April 1938,
Hoboken, New Jersey.

Carroll R. Daugherty showed that in 1934, the costs of assembling the materials for a ton of iron were currently over ten cents per ton more in Chicago than in Pittsburgh. But in the same year the average price of "No. I" heavy melting scrap was $12.36 per gross ton in Pittsburgh and $10.16 a ton in Chicago.[28] When better site and internal-movement conditions with superior market access are also taken into account, finely tuned calculations of assembly costs for ore, coke, and limestone were shown to be of limited relevance. Even so, for various reasons US Steel's reaction to the reality of production costs was not completely logical.

In 1920 US Steel operated twenty-one integrated works whose average ingot capacity was 1.15 million tons. (Sharon last operated in 1920.) During the next ten years, it created one more such operation by backward integration at Fairfield. By 1938 the Corporation had closed or disposed of seven integrated works, and the average capacity of the fifteen retained had increased to 1.83 million tons. At the various dates when closed, the seven abandoned operations averaged only 515,000 tons. Apart from size, another factor involved in winnowing old plant was technology. Five of the seven works had been wholly dependent on Bessemer converters. The revolution in thin-flat-rolled-products technology also encouraged concentration. In the early 1920s rated capacity at American Sheet and Tin Plate was 1.21 million tons of sheet and light plate and 1.15 million tons of black plate for tinning. Twenty-six works were involved, of which Gary alone produced both products. Only 587,000 tons of the sheet and light-plate capacity was in two works backed by steel manufacture, Gary and Vandergrift. The remaining 623,000 tons was divided between eleven rerolling operations. Gary accounted for 208,000 tons of the black-plate capacity; thirteen works without steel making for the remaining 946,000 tons. Between 1928 and 1937, seven of eleven unintegrated sheet works and five of the thirteen uninte-

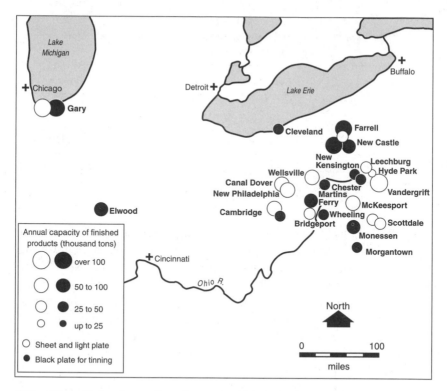

US Steel Sheet, Light-plate, and Tinplate Works in the Northeast, 1922

grated black-plate works were closed. Their production was taken over by a handful of strip mills located at or dependent for slab supplies on integrated works in the immediate vicinity of the rolling mills. Though very much a minor theme as compared with closure and reconstruction, there were a few acquisitions. Columbia Steel Company was absorbed in 1929. US Steel also acted to control slightly more of its outlets, acquiring the Oil Well Supply Company in 1930 and the Virginia Bridge Company six years later.

In April 1938 Taylor was replaced as chairman by Stettinius. Voorhees took over the finance-committee chairmanship. That year, like the industry at large, US Steel was severely buffeted by a sharp return of the Depression. All parts of the Corporation were severely affected. Early in 1937 a visitor had found that the Ohio Steel works, Youngstown, was averaging 13.5 of its 15 open-hearth furnaces active each week. When the same man was there in early spring 1938, only 3 or 4 furnaces were in use. For over two months in 1937, Gary averaged 85,000 tons of steel a week; in late March 1938, one week's output was only 27,000 tons.[29] Much had been done to rehabilitate the Corporation, but by some criteria it was still performing less well than leading competitors. Taylor and his colleagues had not reversed its relative decline. In the year of Gary's death, US Steel made 46.1 percent of the nation's raw steel; by 1937 its share was 41 percent, and the harsh conditions of 1938 brought it down further to 37.1 percent. In other respects too the overhaul of both structure and operations was incomplete. There was no uniform, comprehensive system of accounting covering all subsidiaries and as a result costs or profits for particular products could not be compared. Fairless later admitted the seriousness of these defects: "[C]osts were not available to the management in the way they should have been and the system of cost and production control was antiquated and far below those considered everyday practice in other industries."[30]

As detrimentally important as the survival of out-dated business practices was that of old attitudes and assumptions. Delusions of grandeur remained within and created problems for the Corporation. At the end of 1937, in a meeting of the control committee for the Ford, Bacon, and Davis studies, Taylor even alluded to the Great Pyramid in his remarks on the importance of the company they served. "This Corporation is so big and so wonderful it takes on, I think, the kind of atmosphere that makes you forget some times that it is a business." A few months later, in a valedictory address to shareholders, he went further, and in doing so exposed a continuing dilemma: "The Corporation is much more than a commercial enterprise. It is a national institution and its pulse throbs with that of the nation. And so it cannot be successfully managed solely and restrictedly as a commercial enterprise. But, at the same time, it is a private institution in that it must stand on its own feet." Taylor put limits to their ambitions: "The objective has been efficiency. We desire only to keep pace with the national market."[31] In other respects, as he had recognized, they seemed to have responsibilities beyond those of a limited public company, however great. A revealing instance of this concerned the South. In their study of Tennessee Coal, Iron, and Railroad (TCI), Ford, Bacon, and Davis had been upbeat. They had noted that the universal plate mill installed at Fairfield in 1932 had had little use, but generally TCI's recovery from the Depression had "more than kept pace" with that of US Steel as a whole. The consultants believed Birmingham might be further developed as a low-cost source of tinplate for both the Atlantic and Pacific Coasts.[32] Yet eighteen months after Taylor's resignation of the chairmanship, when giving testimony to the Temporary National Economic Committee, Ben Fairless revealed how strong the biases were in US Steel's developmental thinking against treating the South on equal terms with the North. In a different field, when the committee was considering the elimination of price differentials at some bas-

TABLE 11.6

**Steel capacity and production, operating rates, and profits of US Steel
and other leading companies, 1936–1938 (thousand gross tons)**

Company	Year	Ingot Capacity	Ingot Production	Operating rate (%)	Net profit ($000s)	Earnings per ton of ingot capacity ($)
USS	1936	25,772	16,908	65.6	50,583	1.96
	1937	25,790	18,532	72.1	94,944	3.68
	1938	25,790	9,397	36.2	(7,717)	(2.99)
Bethlehem	1936	9,360	5,994	64.0	13,901	1.49
	1937	10,042	7,270	77.1	31,820	3.40
	1938	10,042	4,348	43.3	5,250	0.52
Republic	1936	6,053	4,292	70.9	9,587	1.58
	1937	6,453	4,014	62.2	9,044	1.40
	1938	6,500	2,465	37.9	(7,998)	(1.23)
J&L	1936	3,671	2,375	64.7	4,130	1.12
	1937	3,671	2,472	67.3	4,789	1.30
	1938	3,671	1,359	37.0	(5,880)	(1.60)
Inland	1936	2,340	2,164	92.5	12,800	5.47
	1937	2,340	1,750	74.8	12,665	5.41
	1938	2,760	1,490	54.0	4,916	1.78

Sources: Iron Age, various issues, 1936–1938.

ing points that had been carried through in June 1938, Fairless stated: "If a railroad had to pay a dollar or two dollars more per ton for car material in Chicago than in Pittsburgh, the result of that would be it wouldn't build its railroad cars in Chicago; it would build them on the other end of the line, *and of course we couldn't permit that.* We were interested—as we are—in the development, industrial-wise, of every district in which we operate our properties, and whether or not we operate properties we are interested in the general overall industrial development of America."[33] These were heavy burdens to carry for a company whose prime objective might have been assumed as the maximization of profits from the manufacture and marketing of iron and steel in the interests of its multitude of stockholders.

Ford, Bacon, and Davis not only analyzed US Steel, but they also tried to rouse those in command with a rallying cry. They pointed out that in its early days, the Corporation had the best plants in the industry, but by the end of the 1920s and in the early 1930s, competitors such as National and Inland often operated better mills. As they put it: "The Corporation must never again permit itself to be led into the valley of declining participation in declining markets." Defense was not an effective antidote to shrinking market share, "The remedy for this is to be found in one word—aggression." Competition must be met on quality and price, but it would be a hard struggle: "[I]t must not expect to reap results overnight. Literally years are involved in breaking down the opposition which has grown steadily and which has entrenched itself largely because of past complacent policies of the Corporation."[34] Much was done in the late 1930s to improve the situation along lines that the consultants had advised. The outbreak of World War II, first in Europe and a little over two years later engulfing the United States as well, swept the steel industry and US Steel to new heights. In doing so, as later became painfully obvious, it helped halt the only partially completed improvement of plant, organization, and commercial attitudes.

The newly completed Irvin works, Pennsylvania, 1938–1939. The Monongahela River is down to the left. *Courtesy of USX Corporation*

12

Labor Relations under Myron Taylor and Philip Murray

In the 1930s there were major developments in the relations between US Steel and labor. During the worst part of the Depression, the Corporation introduced the policy known as "share the work." This was sensible and humane, though take-home pay was inevitably reduced severely. Steel shipments in 1932 were only a little over one-quarter those of 1929, but the average monthly number of employees was 64.6 percent as high and average weekly earnings about half as much as three years before. Employment costs in 1932 totaled a little over one-third the 1929 level. By the boom of 1937, output, employee numbers, and earnings were roughly back to 1929 levels. A positive change of the middle of the decade was the major step of accepting unionization. However, for reasons beyond the control or responsibility of US Steel, or of any single company, there came a new urgency in technologically induced unemployment, a problem that was especially serious in some of the more isolated steel communities in which the Corporation operated.

Many in the union movement resented the well-intentioned work-share scheme that Taylor had introduced in 1931. When he first came into the chairmanship, Taylor as had his predecessors resisted unionization. National circumstances caused a gradual shift in these positions. New Deal legislation began a general swing toward a more general acceptance of unions so that preparations by some industries to resist the trend began to look increasingly out of tune with the spirit of the age. The National Industrial Recovery Act of 1933 provided guidelines for industry—the NRA Codes—and also required advances in collective bargaining. US Steel made positive moves but was reluctant to go as far as its workers wanted. Nonetheless, Taylor recognized the

TABLE 12.1
**US Steel shipments, productivity, employees, hours of work,
and weekly earnings, 1929, 1932, and 1937**

Year	Shipments (th. tons)	Shipments per man-year	Employees, avr. monthly no.	Avr. hours weekly	Avr. weekly earnings
1929	16,813	66.06	254,495	46.2	$31.67
1932	4,324	26.31	164,348	25.4	$15.58
1937	14,098	53.95	261,293	37.6	$32.51

Source: US Steel annual statistics from D. A. Fisher, *Steel Serves the Nation: The Fifty Year Story of United States Steel* (New York: US Steel, 1951).

need for change. Having introduced a retirement age of seventy for management, in 1934 he brought in a fifty-three-year-old labor expert, Arthur H. Young, as the Corporation's first vice president for industrial relations. Young was charged with the responsibility of introducing a company union plan to conform to the requirements of the code. For US Steel, employee representation was expected to involve gathering together representatives from the various plants—a way of satisfying demands from workers and the administration for a machinery of collective bargaining. Despite his wide experience and rather more liberal view of workers than had been traditional in US Steel counsels, Young was not a man for imaginative thinking or groundbreaking practices. For instance, in the year of his appointment, he spoke of employee representation plans (company unions) as a means of developing "sound and harmonious relationships between men and management [comparable to the] sound and harmonious relationships between a man and his wife." Next year he made it known that he would "rather go to jail or be convicted as a felon" than accept "any formula of the conduct of human relationships in industry imposed on us by demagogues."[1] That year US Steel duly opposed the new National Labor Relations Act (the Wagner Act) and its creation of a National Labor Board to supervise the right of workers to organize and bargain, but the pressures on the Corporation to go further were now increasing. There were mixed signs as to how it would react. On the one hand, as Frances Perkins, secretary of labor discovered when she brought labor and steel industry leaders together in the same room, most of the latter shrank from direct contact with the former, but William Irvin, US Steel president, was a happy exception.[2] On the other hand, in spring 1936 it became public knowledge that some US Steel subsidiaries had been buying guns, grenades, gas masks and other riot-control gear. The long retired but ever-watchful William

Dickson wrote in horrified tones to Young. He recalled how forty-four years before he had sat on a bluff overlooking the Monongahela River as the Homestead steel workers made their bloody assaults on the hired Pinkerton detectives, concluding: "In the name of God, have the men in control of the Corporation no memories? 'Whom the gods would destroy they first make mad.'"[3] Soon afterward, the initiative was taken out of the hands of the companies.

In summer 1936 the Congress of Industrial Organizations (CIO) set out to unionize the industry through the Steel Workers' Organizing Committee (SWOC) headed by Philip Murray, vice president of the United Mine Workers. The first meeting of a recruiting drive was held on a plot of open ground by the banks of the Monongahela in McKeesport on Sunday, 21 June. Early the following month, the American Iron and Steel Institute published a spread advertisement in 375 of the nation's newspapers in which it committed itself to fight to the finish what it characterized as John L. Lewis's one big union. Speaking in Worcester, Massachusetts, Young pointed out that he did not think much of the CIO drive and said he thought the workers felt likewise.[4] Even so, by the beginning of 1937, SWOC claimed 125,000 members in 154 local unions. The first of April seemed likely to mark the start of a damaging strike. At this point, Myron Taylor took the lead in finding a way out for US Steel.

Gradually, he had recognized the futility of a continuing, implacable resistance to unionization. During summer 1936, while staying in his Mediterranean vacation home in Florence, he worked out the main principles for a recognition of the rights of the men to bargain through their own freely chosen representatives. Bruce Seely has suggested that Taylor not only realized that workers were now at last well organized, but also that under the influence of New Deal thinking, sentiment at federal, state, and local levels would now be in favor of unionization. By early 1937, he was also eager to avoid the disturbance of a possible extended strike for workers' rights at a time when, at long last, order books were rapidly filling. Above all Taylor was wise enough to recognize the drift of the times and conclude that unionization would soon be irresistible. By chance or contrivance, on Saturday, 9 January, he met and spoke briefly to John L. Lewis in the Mayflower Hotel, Washington, D.C. The next day there was a fuller, less public conversation and then a number of further meetings. Amazingly, it seems that it was only on 27 February that Benjamin Fairless, as Carnegie-Illinois president, first learned of these talks. Two days later he was in negotiations with Philip Murray, a situation which it is said—whether true or merely as an apocryphal but appropriate story—led to a decision to remove a portrait of Henry Clay Frick from the room in which the two men were meeting because they "did not think he could stand it," "it" being a signed agreement between management and labor.[5] The preliminary arrangement between Carnegie-Illinois and SWOC was signed on 2 March by Fairless, Murray, and three other organizing-committee representatives. It provided for full union representation with SWOC as the bargaining agent, an established hourly wage of five dollars, an eight-hour day, and a forty-hour week.[6] A few months later, after Irvin had "delegated" him to review recommendations of Ford, Bacon, and Davis's Report 37 on industrial relations with their various subsidiaries, and his own comparatively conservative ideas for developing labor relations having been superseded, Arthur Young left US Steel. Meanwhile, as a result of coming to this accommodation with its men, US Steel escaped the strikes—the so-called Little Steel Strike—that began in late May 1937 at the mills of the Bethlehem, Republic, Inland and Youngstown Companies. These were often violent confrontations in which fifteen to twenty strikers lost their lives. From 35.4 percent in 1936, the US Steel share of national steel output increased to 36.6 percent in 1937; the respective figures for its shipments of steel products in relation to national *production* of rolled iron and steel were 31.4 percent and 34.2 percent.

After the 1937 strikes, SWOC continued its recruiting drives, and by the time it held its convention in Cleveland in May 1942, national membership had topped 700,000. It was at this time that it changed its name to the United Steelworkers of America (USWA). By the end of the war, USWA represented about 93 percent of the men working in American mills.[7] Inevitably, its 1937 settlement with the SWOC did not mark the end of disputes between US Steel and its workers. During World War II the Corporation cooperated better with both the federal government and the union than some other leading companies, but as in the industry as a whole, mutual suspicion, conflict, and strikes continued throughout the years. During the war, at an annual stockholders meeting, J. Newcombe Blackman, a stockholder frequently given to wordy statements, revealed the depth of suspicion and dislike of organized labor that lurked not far below the surface. That his remarks were recorded in the minutes suggests they were not regarded as wholly beyond the pale, though in choosing his target he seems to have failed to recognize that for a year, Philip Murray had formally led the steel workers: "We have confidence in our officials, and as far as exorbitant profits are concerned, I won't get my figures from Mr. John L. Lewis, because he is out for himself. Look at him in the picture alongside the President of the United States, and if you don't see a character or lack of it in that face, I will give up."[8] There was a serious strike in 1946, when steel companies, including and led by US Steel, refused to accept a proposal by Pres. Harry Truman for very substantial wage rises. It ended after the companies were allowed to advance steel prices by five dollars a ton. In settling the strike, Benjamin Fairless and Philip Murray were photographed shaking hands across a table, one of many pictures of the two men presenting an apparent harmony between capital and labor, usually on visits to mills. However, over the three years 1946–1948, US Steel claimed that it lost 7.2 million tons of production due to work stoppages in steel, coal, and other areas. Addressing stockholders less than two months before the outbreak of the Korean War and when their operating rate was already over 100 percent, Irving Olds reckoned that since World War II the nation as a whole had lost the production of around 29 million tons of steel because of strikes.[9] When Murray died in 1952, he was succeeded by David McDonald, a close associate for thirty years. Murray came from mine labor; McDonald had been university educated and was accordingly viewed by many in union ranks as out of touch with their everyday circumstances.

Notwithstanding major progress in labor relations, some other conditions of labor were less easily improved. The mill towns remained appallingly drab.[10] Life in these working-class communities was still dominated by the needs of the major steel companies. This was vividly illustrated during the war years. Well before Pearl Harbor, the Defense Plant Corporation announced plans for the building of eleven new open-hearth furnaces, a slabbing mill, and a sheared-plate mill at Homestead. To make room for this very large extension, 1,225 houses had to be demolished and approximately 1,850 families, or 10,000 people in all, were cleared from one of the town's worst areas. There was an increase in the population of nearby Munhall in the 1940s, but in that decade the numbers living in Homestead fell from 19,041 to 10,046.[11]

Technological unemployment became a critical problem. Loss of jobs due to the introduction of better plant and its impact on local economies and society had been a feature of the industry from its early days. Fifty years before this, changes in the rail mill at Edgar Thomson had been highly productive for the Carnegie interests in terms of economy in labor, but there had been little enough in the way of sympathy or practical help for the men whose livelihood disappeared—for instance, in 1885 new machinery displaced 108 out of 132 men on the heating furnaces and rail-mill train.[12] Creation of the trusts at the end of the 1890s had been followed by plant closures in which large workforces, and sometimes the economic base of whole communities, could be destroyed. Now the problem loomed even larger, in part as a result of revolutionary changes in the industry, alternative jobs being notoriously difficult to find, and a new concern (from the Roosevelt administration downward) about problems of regional and local unemployment. In particular, the advance of the hot-strip mill and its coincidence with the Great Depression caused huge lay-offs as many old-style sheet and tinplate mills were closed to make way for the new process. The nub of the problem was identified by *Fortune* in its spring 1936 analyses of US Steel. It pointed out that with a labor force of no more than 125 men, a strip mill could average an output of 2,000 tons of thin flat-rolled steel a day; to make the same tonnage, old-style mills required about 4,500 workers. Roughly, the new technology reduced the labor input per unit of output by over 97percent—or if looked at not so much in terms of productivity but from the employee's perspective, it made 97percent of the workers redundant. At the end of the decade, in testimony given to the Temporary National Economic Committee, it was reckoned that throughout the industry, strip mills had already displaced 38,470 workers; changes in prospects might bring the total to 84,770.[13] When the new Irvin works came into production in 1938, its management had the opportunity of picking a select workforce from the huge number of men who had been laid-off from older plants, whereas some small mill-towns were stranded with few if any alternative lines of work. Places such as Vandergrift and Farrell, Pennsylvania, or Cambridge, Ohio, which had grown substantially in the early decades of the century, now fell away. Though the problem was most pronounced in the sheet, strip, and tinplate sections of the industry, it was by no means confined to them. For example, early in 1936 Carnegie-Illinois closed its small steel, structurals, and billet works at Pencoyd in Philadelphia, which it decided could not compete with bigger, integrated, and more modern operations; from 850 to 900 men were thrown out of work.[14] Some of the smaller integrated works were also falling by the wayside. In summer 1945 McDonald drew attention to the plight of Mingo Junction, Ohio, where US Steel had decided to close a small integrated mill. Although it was eventually sold to Wheeling Steel, in the interim only 60 men out of a population of 5,300 still had jobs.[15] By 1950 the town of Mingo Junction had lost over 14 percent of its 1940 population. In other cases, as at New Kensington in the Lower Allegheny Valley, other industries came in to give a new boost to an otherwise afflicted local economy.

New Regions

US Steel and the Changing Geography
of the National Market

Through the early decades of the twentieth century, it was clear that demand for steel was not only growing and changing in composition but also undergoing major alterations in distribution. Some parts of the industrial core of the nation—now for the first time referred to as the Manufacturing Belt—were, relatively speaking, in decline, while others were improving their standing. New England and, to a lesser degree, Pennsylvania were examples of areas declining, while the Midwest and especially Michigan represented the surging regions. The West, and particularly the Far West, was steadily increasing its share along with the South. The Middle Atlantic seaboard seemed to offer opportunities not recognized at the beginning of the century. As general economic growth spread in these directions, the steel industry and its leading company had to consider appropriate responses.

The Detroit region became a major and rapidly growing center of consumption in which US Steel gained no production foothold. The automobile industry, already heavily localized in southern Michigan, grew largely in the 1920s to become a prime consumer for the nation's steel. By 1926 it was reckoned that one in every thirteen tons of steel produced was already used in Lower Michigan; one 1929 estimate—there were others different from this—was that over the previous five years, motor industry annual steel consumption had risen from 3.5 million to 6 million tons. This demand was exceeded only by that of greater Chicago.[1]

Although large tonnages of steel were now used in automobile manufacturing, the cost of this raw material was only a very small part of that industry's overall costs of production. As a result, although the motor firms were

TABLE 13.1

**Shares of national value added in manufacturing in selected areas,
1899, 1909, 1919, 1929, and 1939**

Year	Pennsylvania	Michigan	Southeast[a]	Pacific Coast
1899	14.88%	3.09%	5.78%	3.03%
1909	12.80%	3.88%	6.75%	4.29%
1919	12.97%	6.48%	7.17%	5.30%
1929	11.21%	6.76%	7.68%	6.28%
1939	10.11%	7.33%	8.67%	6.32%

Based on: U.S. Bureau of Census, *1954 Census of Manufactures: General Summary* (Washington, D.C.: GPO, 1957).

[a] Virginia, North Carolina, South Carolina, Georgia, Florida, Alabama, Mississippi, and Tennessee.

concerned to secure material of good quality at as low a cost as possible, for them any expected saving from entering the steel trade was small, especially as this would require a large capital outlay in a field where they had no expertise. Given his unusual enthusiasm for full integration, Henry Ford's company was the natural exception to this general situation. In 1920 the Ford Motor Company began to produce pig iron; shortly afterward it announced it would build open-hearth furnaces, bar, and strip capacity. By 1929 raw-steel capacity at the Rouge plant in Dearborn, Michigan, was almost 600,000 tons, but Michigan, though now ranking fourth among the states in consumption of steel, was seventeenth in production.[2] At this time Newton Steel at Monroe, Michigan, and Michigan Steel in Detroit were each reported as likely to build raw-steel capacities of about 140,000 tons; much more important for other steel makers was the decision of a new company, Great Lakes Steel, in associ-

TABLE 13.2

**Consumption of steel by railroads and the automobile industry,
1920, 1925, 1929, 1932, and 1937 (thousand net tons)**

Year	Railroads	Automobile industry
1920	8,333	3,623
1925	9,349	4,861
1929	7,171	7,592
1932	1,404	1,990
1937	4,941	7,165

Source: Iron Age, 4 January 1940, 72.

ation with the existing Hanna Furnace Company, to build an integrated operation at Ecorse, Michigan, having a steel capacity of 450,000 tons, all of which would be finished in a hot-strip mill as material for the motor plants. The growth of this capacity was primarily a response to an extraordinary market situation, but the Detroit area was also favorably placed in terms of iron- and steel-making costs. All of the huge flow of water-borne ore from the Upper Great Lakes bound for the ore docks of Lake Erie passed along the St. Clair and Detroit Rivers. Smaller but still very large tonnages of coal, including coking coal, passed in the reverse direction from Lake Erie ports westward. Near the shores of the western part of Lake Erie and the Lake Huron edges of Upper Michigan were suitable limestone deposits for furnace flux. Equally important, the industrial areas of southern Michigan produced very large tonnages of industrial scrap so that the low price for this material made it more than usually competitive with molten iron for use in the open-hearth furnace. To complete the area's attractions, it was protected by distance from existing major centers of steel production, Pittsburgh being some 260 miles away, Gary and Youngstown 240 miles, and Cleveland 180 miles away. By the end of the 1920s, it was estimated that the freight charge on finished steel delivered from Pittsburgh was four dollars a ton above the switching charges that Detroit plants paid in getting steel to the automobile plants. Throughout most of the next decade, local plants chose to set their prices for flat-rolled steel at

TABLE 13.3

Distribution of capacity for the production of steel sheets, 1920 and 1929, and shipments received, 1919–1921, and 1929–1931 (percentage of national total)

Region	Capacity 1920	Capacity 1929	Shipments received 1919–1921	Shipments received 1929–1931
Pennsylvania	23.1	17.0	12.0	8.7
Ohio	52.0	46.2	17.4	22.3
Ill./Ind.	17.7	7.8	13.4	12.7
West Virginia	0.0	5.3	1.4	0.4
Kentucky	0.0	10.2	0.4	0.8
N.Y./N.J.	3.7	2.5	11.6	7.5
New England	0.0	0.0	3.7	2.0
Wisc./Minn./Miss.	1.1	1.4	10.9	7.7
Maryland	2.3	2.0	1.1	0.4
Michigan	0.0	7.5	16.2	28.6
All other states	0.0	0.0	11.9	8.9

Source: C. R. Daugherty, M. G. de Chazeau, and S. S. Stratton, *Economics of the Iron and Steel Industry* (New York: McGraw-Hill, 1937), 72.

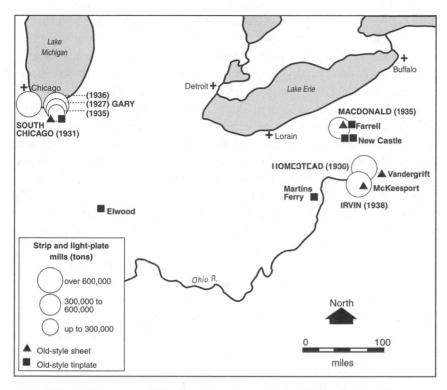

US Steel Strip Mills, Continuous Light-plate Mills, and Old-style Sheet and Tinplate Works in the Northeast, 1940

four dollars a ton above the Pittsburgh base price; the regional customer paid the same, and the Detroit steel producer gained some extra income on every ton shipped.[3]

As the automobile market became more important, distant suppliers continued to compete in it with the Detroit-area firms. For instance, in a three-month period of spring–early summer 1934, Pittsburgh-district mills made total deliveries of National Recovery Administration steel-code products amounting to 1,533,540 net tons. Of these deliveries, 28.4 percent went to the Pittsburgh district, but Detroit was the second-most important destination with 8.39 percent of the total, well in excess of deliveries to Chicago and Cleveland combined.[4] US Steel lagged in its response to this market. In 1934, 21 percent of sales by all steel plants were made to the automobile industry but only 9 percent of those made by US Steel. Within this overall figure, Illinois Steel played a larger role with about 12 percent of its deliveries going to Detroit. In 1934 no more than about 17 percent of American Sheet and Tin

Plate's sheet output went to the automobile industry, but in 1935 this increased to some 28 percent.[5] Not until US Steel built more wide-strip mill capacity could it play a bigger part in supplying high-grade autobody sheet. As it did so the Corporation decided in every case to locate these units in order to take advantage of existing iron- and steel-making capacity and therefore located them at Gary, Youngstown, along the Monongahela River, and at Fairfield. The prime locations of southern Michigan were left to those companies that were forcing their way into the industry for the first time. The aggressive thinking that had planned for Conneaut or had conceived and built Gary had now been contained by a rational response to the economics of massive plant-in-being. Without doubt, Myron Taylor achieved a great deal at US Steel, but in the case of Detroit and its long-term prospects, at least, his administration of the Corporation's affairs was limited by his very expertise in the consideration of overhead costs.

The purchase of Tennessee Coal, Iron, and Railroad (TCI) in 1907 gave US Steel a dominance in the South that it never lost. In 1920 the Ensley plant had an ingot capacity of 1.25 million net tons. The Gadsden works of Gulf States Steel, just over one-quarter as large, was the region's only other integrated plant. A second US Steel plant was begun at Fairfield just before World War I as a rolling-mill operation. In the 1920s Fairfield was gradually extended and rounded out. Merchant-bar mills were installed there by 1923, at the end of 1925 it had four open-hearth furnaces, and by 1928 two blast furnaces were running. Fairfield was equipped with sheet, sheet-bar, billet, cotton-tie, and hoop mills. By 1930 it was capable of 840,000 tons raw steel. Through most of the depressed years of the early 1930s, Fairfield ran at almost full capacity.[6] Over the years it was to become the main US Steel center for growth in the region.

Regional dominance by no means implied a lack of problems for the Corporation in the South. In 1922 over half of its mill capacity in the Birmingham district was in rails. Moreover, the industry there was penalized by the fact that it had been corporate policy from the start not to allow the local management a free hand in pricing. In 1909 US Steel had introduced its new pricing policy for the region, the so-called "Birmingham Differential." For many years in the 1920s, this imposed an extra five dollars a ton above the Pittsburgh base price; in the 1930s it was three dollars. The differential was removed in 1938. In evidence to the Temporary National Economic Committee enquiry in fall the following year, the president of TCI, Robert Gregg, attempted to provide a rational explanation of the differential based on the overall costs of steel making in the Southeast. Orders were smaller, which demanded more roll changes, which in turn increased unit costs. There were

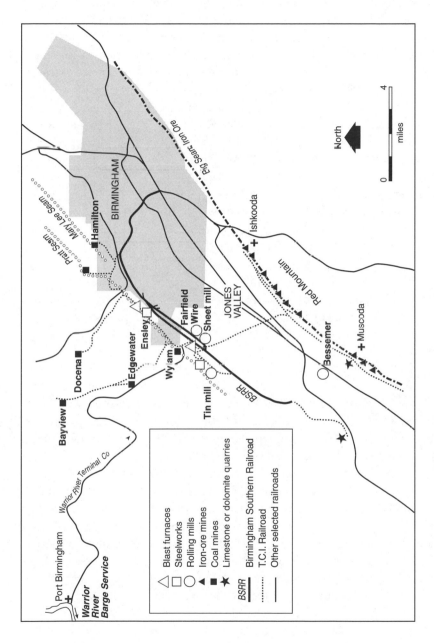

Legend:

△	Blast furnaces
☐	Steelworks
○	Rolling mills
◀	Iron-ore mines
■	Coal mines
★	Limestone or dolomite quarries
BSRR	Birmingham Southern Railroad
┈┈┈	T.C.I. Railroad
────	Other selected railroads

Locations: Port Birmingham, Warrior River Barge Service, Warrior River Terminal Co, Bayview, Docena, Edgewater, Wyam, Tin mill, BSRR, Ensley, Fairfield, Wire, Sheet mill, JONES VALLEY, Pratt Seam, Mary Lee Seam, Hamilton, BIRMINGHAM, Big Seam Iron Ore, Ishkooda, Red Mountain, Muscoda, Bessemer, North, miles, 0 4

Tennessee Coal, Iron, and Railroad Company Works, 1937

further extras: "the cost of doing business brings into the account the selling and distribution of your products. You must bear in mind that the Tennessee Company . . . covers a very large area of the Southern States where you do not have the concentrated buying power that you have in those areas served by Carnegie-Illinois, at Pittsburgh, and Chicago, and as a consequence of traveling expense, of selling expense, our general expense of doing business is substantially higher than is true in the Pittsburgh and Chicago areas." He even maintained the differential had been of service to the South in so far as it had helped build up TCI.[7]

There is some evidence that US Steel did not allow TCI its "fair" share of business. For instance, during the 1921 depression, Illinois Steel was allowed to increase its sales of heavy rails to the natural market areas of Birmingham in the South and Southwest. More significantly, the Ford, Bacon, and Davis analysis of US Steel operations showed that in 1935, tinplate from Birmingham could have been delivered and warehoused along the Atlantic seaboard at an overall cost of $2.48 less per ton than for what the same class of material was delivered from Pittsburgh mills. It also suggested that the Corporation was losing $1 million a year by filling West Coast orders from Carnegie-Illinois mills rather than from Birmingham. Some time later George Stocking, professor of economics at Vanderbilt, pointed out that, though the South and Southwest contained major markets for pipes and tubes, TCI had never been permitted to produce these lines. (In the years 1919–1921, the eleven states from the northern boundaries of North Carolina and Tennessee to Oklahoma and Texas plus California received average annual shipments of 819,000 tons of tubes out of a national total of 1,903,000 tons; in 1920 they contained none of the 2,705,000-ton-capacity for tubes; in 1937 their share of national consumption was only marginally less at 1,023,907 tons out of 2,439,975 tons produced, but the region still produced no tubes.[8] Such circumstances seemed to justify the report's conclusion: "It seems reasonably clear that, if the Tennessee Coal and Iron Company had been independently owned, it would have pursued a more vigorous expansion policy than the Steel Corporation has permitted it to follow. . . . [I]t seems clear that vested interests within the Corporation have blocked or retarded expansion of low-cost facilities within the Corporation's own domain."[9] However, it is well to remember that it is by no means certain that an independent company would have been able to invest as much in southern steel making as US Steel. The independent Gulf States Steel at Gadsden, Alabama, provided mixed evidence for the thesis that independence could have meant more expansion. In 1920, raw-steel capacity there was 25.7 percent of that at TCI. By 1930, with production underway at Fairfield as well as Ensley, Gadsden was only 15.8 percent as big. Capacity in

1938 was 28.6 percent as large as that of the two US Steel plants. In 1937 Gadsden was bought by Republic Steel. Meanwhile, Alabama continued to make progress within the national industry. In 1920 its shipments of rolled iron and steel were 2.77 percent of the national production; in 1930 it's share was 3.16 percent, though this represented a slight drop compared with the mid-1920s.

Throughout almost a quarter-century after it acquired TCI, US Steel played no part as a producer in the West; for even longer it had no significant East Coast operations. In both areas the interwar years saw stirrings that heralded change in this geographical pattern; the changes went furthest in the West. Even there, US Steel's approach to the opportunities was cautious, but it may be wrong to regard this as yet more evidence of entrepreneurial weakness, for as well as considerable attractions to expansion, conditions in the region posed many difficulties. As early as 1896, Henry H. Campbell, who worked for Pennsylvania Steel, summed up the problem presented by the West

TABLE 13.4
Shipments of heavy rails, structurals, and plates to selected market areas in 1920 and 1921 (net tons)

Product	Date and source	To South and Southwest	To West and Pacific
Heavy Rails			
	1920		
	TCI	260,390	25,914
	Carnegie Steel	18,030	4,909
	Illinois Steel	22,048	512,852
	1921		
	TCI	233,645	21,978
	Carnegie Steel	10,021	4,411
	Illinois Steel	48,681	652,471
Heavy Structurals and Plates			
	1920		
	TCI	118,462	1,336
	Carnegie Steel	62,934	77,093
	Illinois Steel	32,154	935,671
	1921		
	TCI	40,593	1,437
	Carnegie Steel	32,580	42,759
	Illinois Steel	12,401	396,923

Source: Iron Age, 19 February 1925, 540–45.

to any who might entertain hopes of building there an industry similar to that in the northeastern states: "There are places in the uninhabited valleys of the northern Rocky Mountains where coal fields and ore beds lie waiting for the coming of the metallurgist; but in such districts the installation to-day of a plant with its dependent community would be done at an enormous financial sacrifice, the labor which could be obtained at exorbitant rates would be unreliable and inefficient, and the cost of transporting the product to market would be prohibitory. The financial equation of a great industry is a combination of social and metallurgical factors."[10]

A few years after Campbell's comments, eastern entrepreneurs and western railroads carried through fuller assessments of the minerals available for any future western iron and steel developments. Joseph Sellwood of the Great Lakes ore shipping firm Pickands, Mather, and Company, acting as consultant for John W. Gates, James J. Hill, and Edward Harriman, reckoned that southwestern Utah contained over 500 million tons of iron ore—possibly twice as much—and that coal was available within about 150 miles, though it was not of good coking quality. US Steel's subsidiary, the Oliver Iron Mining Company, sent out its own engineers to southwestern Utah and recognized that the ideal focus for a future iron and steel industry lay in the north-central area of that state. T. F. Cole, who had been Oliver's president at that time, later told the Stanley Committee their conclusion had been, "when the time comes that Salt Lake City can distribute large quantities of that material, I see no reason why the ores cannot be sent up to the valley of the great Salt Lake and there meet the coking coal from Wyoming and from Utah mines, and be reduced into iron, made into manufactured shapes and sold."[11] Eventually, this predicted form of development was more or less realized, but for the next few decades it was the more populous coastal states that attracted investment—for making steel, not iron.

In population, per-capita wealth, and some lines of economic activity, the western states, and particularly those on the shores of the Pacific Ocean, had long been pacemakers. Many old trades were moving west and new ones were being established there. Some were of little significance to steel makers, but there were important ones that were. In 1904 the amount of oil produced west of the Mississippi River for the first time exceeded that from the East; output of California wells alone amounted to 25 percent of the national total that year. Soon afterward California lost its leadership in petroleum to the Gulf Coast and the mid-continent fields, but the state remained a major factor in production, as late as 1935 for instance generating over one-fifth of a vastly increased American oil output. A second growth sector depended on West Coast agriculture. In 1905 it was noted that whereas twenty-five years before there had not been a dozen canneries west of Ohio, already 80 percent of the plants that "prepare the inconceivable quantity of canned foods marketed annually" were out west, a large proportion of them on the Pacific Coast.[12] In other words, for pipe and tinplate, the West was soon a major outlet. It was important too in consumption of structurals, bars, and wire but less so for sheet or even for rails. When in early 1911 William Corey suggested that for sales, "the time is ripe for further aggressive action," it was the West that he chose to illustrate his theme—but for improved warehousing rather than production. "With the opening of the Panama Canal, it seems probable that by establishing warehouses on the Pacific Coast and possibly maintaining a fleet of steamers designed especially to handle our products, we may be able to recover the valuable trade of this growing section, which has been greatly reduced owing to the operation of the recently enacted tariff law."[13] In fact, the Underwood Tariff, which lowered protection to foreign competition and thereby exposed the Pacific Coast market to more imports, did not become law until 1913. When the opportunity came to use the isthmian water route to the West Coast, US Steel did not have sufficient capacity near the Atlantic ports to take full advantage of it, and Bethlehem Steel became the main shipper by the method Corey had suggested.

In the years immediately after World War I, annual consumption of finished steel west of the Rockies was of the order of 2 million tons, a very minor part of the national total. From the point of view of a major steel producer contemplating prospects for operating successfully other major obstacles were also evident. The coastal market was divided into three quite separate centers of consumption, each with a 1920 population in the range of 670,000–780,000. In the rest of the West, use of steel was scattered over a huge area. As Corey anticipated, the opening of Panama theoretically gave Atlantic-seaboard centers of production marked delivery-cost advantages at Pacific Coast ports. For a time these were not available—landslips along the canal interrupting any assured flow—so that there were practically no shipments there via the canal before 1920. As a result, Chicago mills continued to do well on the West Coast, Illinois Steel shipping 178,000 tons there in 1918. But by 1920 the ships were at last active on the canal route, and by July 1920 the charge for seaborne deliveries from Sparrows Point, Maryland, was only 42 percent of the rail rate from Chicago. During the 1920s Bethlehem Steel actively developed its West Coast business by introducing new ocean carriers and improving warehousing in the main centers of population and industry. Another problem for US Steel was this district's accessibility to imported material, often brought back at very low " ballast" freight rates in boats that had taken agricultural products to Europe. A striking illustration of the effects of this was that, up to World War I, foundries in West Coast states used scarcely any American pig iron. The problem continued after the war, when freight charges from Glasgow to Seattle were frequently as low as $3.65 a ton; the rail charge on finished steel from Pittsburgh to the Pacific was then ten times as high. By 1922 Illinois Steel shipments to the West Coast had dwindled to 1,400 tons.[14]

There were only ten steel producers—other than very small foundries— west of the Rockies in the 1920s. Combined, their 1922–1923 rated capacity was 426,000 tons of ingots. Rolled-steel capacity was 334,000 tons, all but 65,000 tons of which was made up of bars and other categories of semifinished or lower-grade products. A works in Seattle was capable of up to 45,000 tons of structural shapes and 5,000 tons of rails, while another in San Francisco could roll up to 15,000 tons of structurals. By 1928 western works supplied 500,000 tons of the finished steel used west of the Rockies, 1.2 million tons came from eastern works, and 120,000 tons from overseas.[15] Despite the large proportion of steel from eastern works, the small tonnages shipped to the West, as well as the fact that this was made up of a wide range of products, made the area unattractive for important US Steel developments. There was, however, one major exception—the tinplate that had already been singled out for special mention twenty years before. In 1920 the three Pacific Coast states took over one-fifth of all American Sheet and Tin Plate (AS&T) domestic tinplate deliveries. AS&T was not well located in relation to these outlets. At the end of 1922 the company had just over one million tons of capacity in black plate for tinning. Of this, 33 percent was in its Valleys plants, the Wheeling and Pittsburgh areas each had 19 percent, and its most westerly operations— in Gary and Elwood, Indiana, some 1,900 miles from California —produced

TABLE 13.5

Shipments of rolled finished steel by six US Steel divisions to states of the Pacific Coast, western interior, and the United States generally, 1920 (thousand gross tons)

Region	Carnegie	Illinois	TCI	Minnesota	AS&T	AS&W	Total
Pacific Coast	42.6	135.7	2.9	n.a.	139.4	87.0	407.6
Western Interior[a]	1.0	40.8	12.7	0.2	11.6	28.6	94.9
U.S. Total	4,169.9	2,757.3	448.9	96.8	1,388.3	1,694.5	10,555.7
West as %	1	6.4	3.5	0.2	10.9	6.8	5

Source: Iron Age, 19 February 1925, 540–49.

Note: Figures are not available for National Tube.

[a] Idaho, Nevada, Arizona, New Mexico, Utah, Wyoming, Montana, and Colorado.

under 26 percent of the total.[16] In 1929 the Corporation at last acquired subsidiary steel operations in the West. It did so by taking over a small but already relatively complex operation.

The modern western steel industry may be dated from 1909, when the former manager of a small foundry in Portland, Oregon, transferred his attention to California. There, he persuaded a group of local businessmen to finance a company that would be known as Columbia Steel. It began with a single open-hearth furnace and steel-founding operation at a small town forty-five miles northeast of San Francisco, a place that, with remarkable optimism, was later renamed Pittsburg. In 1916 a second open-hearth furnace was installed and in 1918 a third as demand for steel castings increased to supply the region's war-expanded industries. As the outlets for castings fell away after 1918, the Pittsburg plant's first merchant and bar mills were installed in 1920. By 1923 rolling capacity had been expanded by a further

TABLE 13.6

Selected Deliveries by American Sheet and Tin Plate, 1920
(percent of all domestic shipments)

Destination	Sheet	Tinplate
Ohio	15.0	3.3
Michigan	20.1	1.0
California, Oregon and Washington	2.2	22.5

Source: *Iron Age*, 19 February 1925, 546.

TABLE 13.7

Western shipments of rolled steel by six US Steel divisions by type, 1920
(percentage of total deliveries made by the divisions in the USA)

Product	Three Pacific states	Eight Western Interior States
Heavy Rails	1.02	1.96
Heavy Structurals	5.13	0.81
Plates	6.47	0.30
Bars	0.42	0.05
Drawn Wire	3.32	1.56
Sheet	2.19	0.76
Tinplate	22.52	1.04

Source: *Iron Age*, 19 February 1925, 540–49.

100,000 tons and into rod, wire, nails, and sheet. That year a similar small steel and rolling mill operation at Torrance near Los Angeles was brought into Columbia. In 1910 the company's production had been at an annual rate of under 5,000 tons of steel; sixteen years later, it was around 375,000 tons.[17] Torrance was equipped with a sheet mill, and early in 1929 the Pittsburg plant installed the West's first tinplate capacity.

Steel making in the West could draw on big local arisings of scrap as well as supplementary supplies from neighboring areas. In 1927 Pittsburg and Torrance obtained 38 percent of their scrap from within California; the rest was mainly from Hawaii, Mexico, and Central America. But growth in production had already encouraged Columbia into backward integration. In 1922 it acquired ore and coal holdings in Utah and two years later had coke ovens and a blast furnace in operation at Ironton, Utah, near Provo at the foot of the Wasatch Range. The logistics for this small operation were far from easy, for Provo was a 130-mile rail journey from the coking coal of the Columbia mine to the southeast and some 240 miles distant from its ore workings in southwestern Utah. Early in May 1924, Provo shipped its first iron to the Pacific Coast. Three years later it supplied 85 percent of the tonnage used there.[18]

The major growth of the far western markets in the 1920s and Columbia's successes attracted the attention of US Steel, whose outlets there were being squeezed between local and East Coast suppliers. In July 1929 a possible acquisition of Columbia was brought before the finance committee, which decided a thorough investigation should be made as to its advisability.[19] In November the purchase was made, the first acquisition of western capacity by an eastern integrated producer. As well as selling its own output, Columbia became the western sales agent for the Corporation's other divisions. Bethlehem Steel responded at once to the US Steel acquisition by gaining control of the three steelworks of the Pacific Coast Steel Company. Over the next decade, plagued as it was with the difficulties of the greatest of depressions, the necessity to revamp its product lines to meet the changing structure of demand, and the need to invest on a large scale in new rolling-mill technology, US Steel gave to its new western works little of the attention they might have received if the prosperity of the 1920s had continued. Even so, there were discernible if small westward shifts in some of its operations. A notable instance came in August 1931 when, on the recommendation of President Pargny of AS&T, corporate officers decided to close four eastern works, two in sheet and two in tinplate, and to spend $2 million to double tinplate capacity at Pittsburg.[20]

Throughout the 1930s US Steel's experience in the West provided a mixture of positive and negative. In 1936 the operating rate throughout US Steel was 65.6 percent, but Columbia was producing at 92.6 percent capacity; the Corporation's and Columbia's net earnings as a percent of their net sales were respectively 4.5 percent and 12.1 percent. That year the territory served by Columbia consumed only 9 percent of US Steel finished steel but yielded 14 percent of its net profits. Even so, Columbia supplied only 37 percent of the 1,706,000 tons of steel products consumed in the Pacific Coast markets; other divisions of US Steel supplied an additional 4 percent.[21] Plant growth was modest. Between 1925 and 1930, the ingot capacity of the Columbia and Torrance works had increased from 85,000 to 363,000 tons. In the 1930s US Steel spent $20 million on both locations, but their 1938 combined capacity was only 440,000 tons. In important respects the problem of the West's low density of demand had been reemphasized. Demand was large, but it was scattered and insufficient to justify the installation of high-capacity modern technology, and it was not always economical to supply it from within the region. For example, as cold reduction mills became a more economical way of producing the sheets at the big eastern mills, the tinplate operations at Pittsburg became uneconomic and, though relatively recently built, the mills there were dismantled in 1937. In 1933 it was estimated that only one-sixth of steel consumption along the West Coast was met by local production. Four years later, in the relatively good trading conditions of 1937, shipments of steel products by plants in the eleven western states amounted to 623,000 tons, about two-thirds of capacity. Total shipments received in these states were not far short of four times the tonnage shipped by western mills. Bethlehem Steel, shipping from its eastern plants and from operations within the region, was the largest source of supply.[22] At this time the prices of steel products sold on the West Coast were computed on East Coast basing points. Such circumstances meant that at US Steel there was little or no willingness to contemplate going further in the direction of large-scale, fully integrated, western operations. There had already been, though, suggestions of movement in that direction.

As early as 1922, Columbia Steel seems to have had hazy ideas of placing steel capacity as well as a blast furnace in Utah. In 1932 the fifty-six-year-old Ambrose N. Diehl, who had many years of experience at Carnegie Steel and afterward as a US Steel vice president, was chosen to become president of Columbia. After three years' experience with western conditions, he enthusiastically proposed a Utah steel plant. However, US Steel executives would have nothing to do with his suggestion. In spring 1938 a foreign visitor who had spent some weeks in the eastern steel centers thought it sufficient to add only a brief note about the western industry: "A comparatively small steel industry is springing up on the Pacific Coast, based primarily on scrap and some Utah pig, with crude oil as fuel. It seems safe to predict, however, that its influence

TABLE 13.8
**All finished rolled steel and tinplate markets in the West and United States
and US Steel participation, 1936 (in thousands)**

	Tons	Share of US Steel
All finished rolled products in United States	31,474	24.7%
from seven Western states[a]	2,492	32.6%
Tinplate marketed in United States	2,396	27.0%
from seven Western states[a]	526	38.3%

Source: Ford, Bacon, and Davis, "Reports," 200:20.

[a] Pacific Coast states are Washington, Oregon, California, Idaho, Nevada, Utah, and Arizona.

will long remain purely local and it will tend to develop along lines which are complementary to rather than competitive with the major eastern and middle western centers."[23] A little later, in a submission to the TNEC, US Steel made its own attitude clear: "Although it is an important steel consuming area, the West Coast cannot support more than limited steel making capacity due to high assembly costs, particularly in the face of competition from Birmingham and Sparrows Point, both of which can serve this area on a more economical basis." At that time, such conclusions seemed reasonable deductions from normal commercial practice; that they were soon proved wrong was due to the arrival of radically new western market conditions. In any case, if or when the time came to make pig iron on a larger scale in the West, US Steel reckoned it could make it in Utah and transport it to the coast at five to six dollars a ton cheaper than at tidewater.[24]

On the Atlantic seaboard the situation had been very different. The time for large-scale western steel developments had not yet come; over many years at the end of the nineteenth century it had seemed that the day of the East had passed. As the very first moves toward the formation of US Steel were underway, Andrew Carnegie was writing a survey of iron and steel manufacture for the "Review of the Century" issue of the *New York Evening Post*. In it he concluded, "for the making of ordinary steel the East is not a favorable location."[25] Apart from two small steelworks and rolling mills at Pencoyd, in the outskirts of Philadelphia; and at Worcester, Massachusetts; wire mills at Allentown, Pennsylvania; and a small, antiquated, essentially redundant and soon abandoned integrated operation at Troy, New York, in its early years US Steel had no significant production in the East. Over the next two decades the eastern market grew relatively more important and, moving from US Steel to the then

TABLE 13.9

Deliveries to eastern, western, and foreign markets by Carnegie Steel, Illinois Steel, and TCI, 1920 (gross tons)

Division	East[a]	West[b]	Total domestic	Foreign
Carnegie Steel	1,485,709	263,049	4,169,925	490,565
Illinois Steel	12,113	2,356,863	2,757,262	206,641
TCI	728	23,651	448,898	197,841

[a] Includes New England, New York, New Jersey, Maryland, and Washington, D.C.
[b] Includes Michigan, Indiana, Illinois, Iowa, Wisconsin, and Minnesota.

TABLE 13.10

US Steel's and other companies' raw-steel capacity in the East, 1925 (thousand tons)

District	Total district capacity	Of which USS capacity
Eastern (N. Eng., eastern N.Y., N.J.)	629	179 (AS&W, Worcester, Mass.)
Philadelphia (eastern Pa., N.J., Md.)	6,904	240 (A. Bridge, Pencoyd, Pa.)
Buffalo other than Lake Erie shore	103	0

Source: Iron Age, 1 January 1925.

small plant at Bethlehem, Charles M. Schwab built up a major eastern operation. Between 1916 and 1923, enriched by profits made during the war, Bethlehem Steel acquired control of all the other important integrated works in or along the margins of the East—Steelton, Sparrows Point, Lackawanna, and Johnstown. Even before the last of these takeovers was made, US Steel was an almost negligible factor on the East Coast; on the steel it sold into the area from beyond the Alleghenies, the Corporation became progressively more at a disadvantage in the 1920s as rail freight charges increased, "Pittsburgh Plus" pricing was abandoned, and new basing points were established. During these years Bethlehem Steel increased the amount and improved the efficiency of its deliveries of iron ore from Chile. This ore was both richer and its delivered price lower than was the case with Great Lakes ore in the same district. Overall cost considerations at eastern furnaces were improved further by the advance of byproduct coking and by the almost complete replacement of the Bessemer converter with the open-hearth furnace, which could utilize the large eastern supplies of scrap. As was shown spectacularly with heavy structurals and to a lesser extent in other finished products, US Steel was being gradually excluded from more and more of the Atlantic Coast market. The same Panama route that made smelting of Chilean ore commercially attractive also made possible cheaper delivery of steel products to the Pacific Coast from Bethlehem Steel, which in freight-cost terms was closer than any rival.

Now and again during its earliest days, US Steel considered greater involvement in the East only to decide on each occasion not to go ahead. Notwithstanding the scrap available there, the most critical consideration of all throughout most of the Corporation's first half-century was that of ore supply. As early as 1904 the finance committee decided to examine the cost of delivering ore from a large deposit near Acapulco to Atlantic Coast ports, and in 1905, at Corey's request, a special committee was appointed "to consider and report recommendations concerning [the] proposition to establish a plant at some seaboard point." From 1907 through at least 1911, consideration was periodically given to acquiring control of ore deposits in Cuba.[26] There were more discussions about foreign ore after World War I. In 1923, offered an option to buy the iron-ore properties of the Brazilian Iron and Steel Company in Minas Geraes for $5.5 million, after a "full consideration" the finance committee decided to advise the owners that US Steel was "not interested." A year later consideration was given to the purchase of the 10-million-ton El Volcan ore deposit in Sonora, Mexico. In this case executives resolved to spend money to examine the ore. In summer 1926 Elbert Gary mentioned an earlier suggestion that the Corporation look into rich ores near the Chilean coast, and the committee decided to send one or two experts to meet the German who owned the deposits and evaluate the ore. During 1931 Myron Taylor brought to the attention of the finance committee the availability of an option of a deposit of ore in the interior of Brazil amounting to over 470 million tons. Despite seriously depressed trading conditions, the committee decided to take a year's option and to spend $100,000 to determine more fully its quantity

TABLE 13.11
**Carnegie Steel Company's shipments of plates, heavy structurals, and bars, 1920
(thousand tons)**

Product	Total in USA	To East	Overseas
Plates	824.9	131.3	116.2
Heavy structurals	415.5	85.9	101.7
Bars	1,305.7	185.9	128.3

Source: *Iron Age*, 19 February 1925, 540–41.

Note: East is taken here as New England, New York, New Jersey, Delaware, Maryland, and Washington, D.C.

and quality. In each instance no more was heard of these Latin American ore prospects.[27]

In the course of the 1920s as the competitive situation in the East became more difficult for US Steel, Carnegie Steel sold relatively small parts of its products in the region. Then, at the end of the decade, Bethlehem took the offensive with a strong attempt to invade Midwestern markets through a merger with Youngstown Sheet and Tube. This was defeated in the courts, but shortly after that a number of changes began to make the East seem more attractive for steel making. On the one hand, the localities and states from which Great Lakes ore was derived began to impose extra costs. On the other hand, the scrap differential—the extent to which the price for scrap was below that for pig iron—widened substantially during the Depression. Bethlehem benefited from this more than US Steel, much of whose capital was tied up in mineral lands and blast furnaces. In 1928, although having three times the steel capacity, US Steel used only half as much scrap as Bethlehem Steel. Bethlehem was well placed in the eastern centers for bought scrap and now faced no major local rivals in purchasing it.

There were, however, important factors delaying any overt response by US Steel to these conditions. The most important was the prevailing depression. Another was the urgent need during that decade to attend to the revamping of existing capacity. In the second half of the 1930s, the studies undertaken by Ford, Bacon, and Davis not only recognized that US Steel was disadvantaged in the East as compared with some of its competitors, especially Bethlehem, but anticipated that this problem would worsen as the Federal Trade Commission had recently recommended in a report to the president that basing-point pricing should be replaced by quotations on the f.o.b.-mill basis. The Corporation's own foothold in the region in wire and wire products suggested that it might be able to hold its own in this sector, especially after its plants there were overhauled, but in other respects it was undoubtedly at a serious freight-rate disadvantage—in 1937 the Pittsburgh–New York and Pittsburgh–Philadelphia rates on iron and steel products were respectively $0.33 and $0.26 per hundredweight, as compared with $0.175 and $0.105 respectively from Bethlehem. In some products—tinplate was specially mentioned by the consultants—US Steel's Birmingham operations might replace Pittsburgh as a considerably cheaper source of supply, but this of course would worsen the load factor in Pittsburgh mills at a time when the Irvin works was coming into production and looking for markets in tinplate as well as sheet.[28] Apart from the high capital outlay required for a wholly new plant, a continuing and even more important impediment to US Steel expansion in the East remained its lack of assured supplies of ore.

During 1937 Ford, Bacon, and Davis engaged the geologist C. K. Leith to make a study of foreign supplies of iron ore. Leith concluded that probably no single large-scale source could be obtained to support a possible US Steel seaboard plant, although "it appears likely that an aggregate of upwards of 2 million tons per year of standard high grade ore could be secured in peacetimes from scattered sources by a new plant on the Atlantic coast." He reckoned that ore from Venezuela was "the most desirable from the standpoint of proximity to east coast plants," but the ore then known to exist in that country was controlled by Bethlehem or the major Great Lakes ore firm of M. A. Hanna. There were serious dangers in dependence on any outside sources of supply. Inevitably, they were likely to be disrupted or fail completely in time of war, and even if that did not occur, "All of the Latin American sources named will be subject to the rising tide of government taxes, domestic and export, regulation, and nationalization." Great Lakes ore could be a backup for a tidewater plant, though the rail freight on it would be higher than the cost of haulage on sea-borne ore.[29] In 1938, Ford, Bacon, and Davis, in the summary report of their examination of the entirety of US Steel properties and operations, returned to the possibility of an East Coast mill. Venezuela seems to have faded into the background:

> If the Corporation were to build a steel plant on the Atlantic coast, or were to need imported ore for any other reason, it could not count on more than two million tons of standard ore annually . . . from countries which now produce standard commercial ore. These imports could not be depended on in wartime, as they would come principally from Sweden, Norway, West and North Africa, Australia, and Russia, with only a small amount from Brazil. If more reliable sources of ore

TABLE 13.12

Estimated sources of supply for 2 million tons of iron ore annually for a new US Steel East Coast mill, 1937 (thousand tons)

Region	Iron ore	Region	Iron ore
Sweden	500	Australia	250
North Africa	500	Brazil	100
Sierra Leone	500	S. Russia	100
Norway	100		
		Total	2,050

Source: C. K. Leith in Ford, Bacon, and Davis, "Reports," vol. 165.

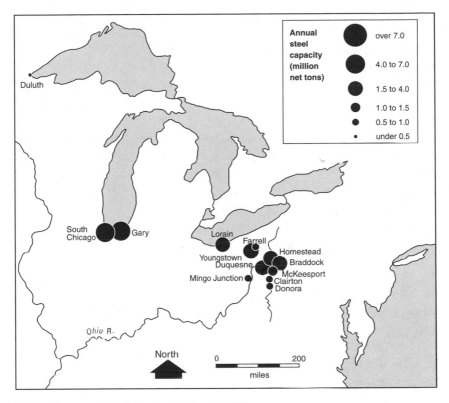

US Steel Integrated Works in the Northeast, 1938

were necessary from abroad, large sums of money would have to be spent on developing known reserves of high-grade ore in South America, principally in Brazil and Chile, or means would have to be discovered for utilizing the lateritic ores of Cuba (where the Corporation owns a reserve of 100 million tons) or of using the high-phosphorus Newfoundland ore, known as the Wabana ore.[30]

By the end of the interwar period, the ore problem still had not been solved, but officials could already foresee that US Steel would eventually find it advantageous to have a much bigger presence in the East. Three years after his visit and commentary on western prospects for iron and steel production, the same English observer, G. A. V. Russell, was back in the United States, this time with a British government team. Again he took the opportunity to visit steel works and talk with those prominent in the industry. He recorded that whereas 51.5 percent Fe non-Bessemer grade Mesabi ore was then priced at about $5.35 a ton, El Tofo ores of about 58 percent Fe were being delivered to Sparrows Point for about $4.90. Spelling out the situation, he deduced some general implications and suggested how US Steel might react: "It is the works on or near the Eastern Seaboard that now have the greatest freedom of choice [of ores] and the handicap under which these operations labored in the palmy days of the exploitation of the Lake deposits, due to their relative inaccessibility to the latter, seems now to be considerably reduced or even obliterated by the cheapening of imported ores consequent upon improvements in transportation and mining methods." He gained the impression that "should major extensions of capacity be more or less forced upon the industry, that any complete works, which might be built, would be likely to be sited in the East or the South, thus halting the almost continuous westwards drift of the center of gravity of the industry. . . . It was learned confidentially that the Steel Corporation have [sic] been actively studying the relative advantages of such points on the Atlantic seaboard as Wilmington and Norfolk and Richmond, Va. as suitable locations for a new integrated steel plant, and it will be highly interesting to see whether one of the sites is chosen for a major new development in lieu of making big extensions at existing works."[31] Six months after that assessment, the United States was plunged into World War II. The necessities of that conflict brought about a massive increase in the nation's steel capacity, an expansion in which US Steel was fully involved. The urgency of the times meant that most of the extensions were at existing plants. There was also a considerable investment in new works. Because of the importance of the Pacific theater much of the extra steel was needed in the West. As a consequence, notwithstanding the Corporation's earlier caution, the West received major extensions of capacity, while in the East construction of a major new plant was delayed for a few more years.

Part Three

Expansion, Prosperity, and Increasing Problems

Government-guided Growth

US Steel in World War II

In summing up the situation of US Steel as he left its chairmanship in spring 1938, Myron Taylor alluded to the Corporation's significance to the nation. His assessment may be difficult to grasp from a modern perspective in which the whole nature of society, economy, and technology has changed almost beyond recognition: "There is a national aspect. The presence of the Corporation gives a measure of protection to the nation, both in peace and war. For it is not without significance for the nation to have always at call the developed facilities of the Corporation and always to be assured of its sympathetic and competent help." Those words were quoted in the annual report for 1941, published after the United States had become a belligerent. In a number of ways, World War II was of great significance in US Steel's history. The conflict provided a definite end to the years of depression and opened up new sections of the nation for corporate expansion, yet at the same time it reinforced both product lines and locations that interwar experience had indicated were becoming progressively less suitable. In these latter respects it helped put off for another generation the wholesale reconstruction that might otherwise have continued the Taylor reforms. Most obvious at the time, the war brought government requirements, expansion policies, and money into the steel industry on an unprecedented scale.

After falling back into severe depression in 1938 the national industry revived in 1939. By 1940, stimulated in large part by the war in Europe, rolled-steel production was at new record levels, exceeding by over 2 million tons the previous high of 1929. Various companies began to expand their capacities. Altogether, about a year and a half after the outbreak of fighting in

Europe, national ingot capacity had been increased by almost 5 million tons. A remarkable feature of this program was that 1.4 million tons of the growth was made up of electric-furnace capacity, representing a more than 80 percent addition for that high-quality steel process. Having achieved this expansion, the industry looked unfavorably on further large increases. In part, its collective skepticism reflected concern about the postwar effect of expansion over recent years in European capacity, especially as it was then expected that the war in Europe might soon be over. The domestic industry, its members concluded, now had enough capacity to face a war.[1] In these circumstances the government began to press for even greater extensions. The industry was suspicious of these plans. To some extent this reflected a reaction against the whole ethos of the New Deal, but there was also a more recent and particular occasion for doubt. At a press conference at the beginning of 1938, Pres. Franklin D. Roosevelt had suggested that the current sharp recession was due to a lack of planning, especially in the steel and automotive industries. He wondered if a scheme of planned production, through which government and business leaders would meet to determine policies, might be possible and prevent overproduction.[2] In 1940, during the TNEC hearings, US Steel was charged with having unneeded steel capacity. Obviously, such attitudes were unlikely to help commend any government ideas of expansion to the industry.

Rated-raw-steel capacity at the beginning of 1940 was 81.6 million net tons. Some Washington analysts previously had described that tonnage as excessive. A year later the Roosevelt administration asked a government engineer, Gano Dunn, to assess the adequacy of the industry for national-defense requirements. In relation to tonnages his report was conservative. Dunn concluded that 1930s experience seemed to show that capacity increases occurred as "a by-product" of cost reduction and that bottlenecks in mineral availability, iron-making facilities, or transport might be as important as steel capacity itself in determining whether the latter was anything more than a nominal figure. The industry's target was 87.6 million tons, only 2.5 million tons higher than the level of early 1941. The steel companies concluded that Dunn broadly supported their contention that no major expansion program was needed. By September of that year, the government had pledged $1.1 billion of public funds to help expansion. The apparent accord between the expectations of government and the plans of industry was broken when, some time after Dunn's assessments, W. A. Hauck submitted a plan to the Office of Production Management for a 10-million-ton expansion program. The firms generally opposed this as excessive. As Irving S. Olds, US Steel chairman since June 1940, put it to the AISI, they did not want "vast additional facilities," especially if built by the government. He defended this opinion with a dogmatic statement rather than argument: "In this country, we cannot dissociate our system of free private enterprise from democracy, because it is the very heart and soul thereof." In fact, between 1941 and 1945 steel capacity within the United States was increased by an estimated 10.3 million net tons to 95 million tons. Though large, this was some 2 million tons below the level envisaged in expansion plans as amended in the first part of 1943.[3] The federal government financed and owned much of the new plant, contributing about half the total outlay; private companies were paid to operate it. Some years after the war, the director of domestic commerce made a rather different summary of the expansion of capacity in 1940–1945 as tabulated. Of the increase in steel, 78 percent was open hearth and 22 percent in electric furnaces, though in 1940 only 2 percent of capacity had been electric. Between 1939 and 1943 alone, national production of alloy steels increased from 3.2 to 13.1 million tons.[4] Ten years after Olds's statement of opposition to government involvement, the author of an officially endorsed history of US Steel had so far forgotten the strength of both industry and Corporation reservations in the euphoria of memories of later achievement that he represented the matter much more positively, writing: "In response to the Government's request in 1941, the steel industry of America pitched in and built new furnaces, rolling mills, and finishing equipment."[5]

US Steel not only played an important part in the wartime expansion program but even proved willing to undertake developments at a loss if Washington wished it. Between June 1940 and the end of 1942, the Corporation spent $282 million of its own money on additions and betterments as well as $436 million of government money in building emergency facilities; in 1943 it spent a further $89 million of its own funds. In total from 1 January 1940 to 15 August 1945, US Steel invested $457 million of its own money and $471 million from the U.S. government. Employment within the various divisions in 1939 had averaged 224,000; in 1943, at what turned out to be its

TABLE 14.1
National Wartime Increases in Capacities, 1940–1945 (in million tons)

Blast furnaces	15.7
Steel plant	15.3
Ore sintering	11.5
Coke ovens	13.4

Source: U.S. Congress, *House Subcommittee on the Study of Monopoly Power* (Washington, D.C.: GPO, 1950), pt. 4a.

wartime peak, 340,000 people worked in US Steel operations.[6] Even so, the Corporation's share of national ingot capacity fell from 35.1 percent in 1941 to 31.8 percent by 1945. In the peak years of World War I, 1916–1918, its shipments had averaged 45.8 percent of the nation's rolled iron and steel; in 1941–1944, they amounted to only 32.4 percent.

By far the largest US Steel extensions—and in fact the biggest at any existing plant in the nation—were made in the Pittsburgh area. This was above all due to the insistence of the U.S. Department of the Navy that expansion in armor-plate capacity should be centered on the expertise at Homestead. Already in 1940 the Charleston Naval Ordnance Plant in West Virginia, an operation built but only briefly used during and immediately after World War I, was started up again under Homestead's management. Then in 1941 the Defense Plant Corporation announced a huge project for Homestead itself. Eleven more open-hearth furnaces with a rated 1.5-million-ton steel capacity, a slabbing mill, 160-inch plate mill, and forging shops were added. To accommodate this plant, 120 more acres of land had to be provided, and Homestead's wartime employment increased by 3,000 to a wartime peak of 15,000. (Two other existing armor mills, at Bethlehem and Midvale, under-

Homestead works, Pennsylvania, at end of World War II. Part of major wartime plant extensions are shown on right. *Courtesy of USX Corporation*

went smaller extensions, their combined steel capacity in 1938–1945 increasing by less than 45 percent as much as at Homestead. In 1944 US Steel also installed an armor plant at Gary.) The need for supplementary sources of steel for finishing at Homestead affected other local plants, especially Edgar Thomson, where new blast furnaces were built. Electric-furnace capacity and a heat-treatment plant were installed at Duquesne.

Throughout the nation, the necessities of war involved a major turn-around in the positions of two market sectors. Demand from the automotive industries had come to be the leading category in the 1930s but now declined massively; shipbuilding, in peacetime a relatively minor activity, as in World War I became feverishly busy in construction of mercantile tonnage. In five wartime years, the United States launched an amazing 53.2 million dead-weight tons. Manufacture of ship plate again became a critical sector of the steel industry. In 1940 Carnegie-Illinois Steel produced 1,360,000 tons of plate, and the following year made 2,000,000 tons. Much of this was consumed in the great yards of the Atlantic seaboard. There, US Steel still operated the Federal Shipbuilding and Dry Dock Company, which it had constructed during World War I. At Federal, the value of ships built in 1941 was already 83 percent above 1940 levels. By mid-1942 plate output at Carnegie-Illinois was running at an annual rate of 3,600,000 tons. Already,

TABLE 14.2

Planned national and regional increases in United States ingot capacity as of mid-1943 and actual US Steel increases, 1938–1945

District	Steel Industry	United States Steel		
	Planned national total increase over 1/1/1940	Capacity (th. tons) 1938	1945	Increase, 1938–1945
Pittsburgh	18.6%[a]	9,041	11,203	23.91%
Youngstown-Wheeling	–	3,315	3,030	- 8.60%
Cleveland-Detroit	1.7%	1,560	1,736	11.28%
Chicago-Duluth	13.4%	9,217	9,691	5.14%
East	15.3%	406	250	- 38.40%
Southeast	30.5%	1,802	2,470	37.07%
West	110.9%	436	539	23.62%
United States	19.6%	25,777	28,919	12.19%

Sources: Iron Age, 3 June 1943, 93; AISI, *Iron and Steel Works Directory of the United States and Canada* (New York: AISI, 1938, 1945).

[a] This figure represents both the Pittsburgh and the Youngstown-Wheeling districts.

many of the strip mills had been adapted to produce heavy ship plate, but in some parts of the country there was an urgent need for the building of additional plate mills. Above all, the necessities of war meant a huge increase in shipbuilding on the Pacific Coast. The nearest major plate mills were in Chicago, but now the needs of these West Coast yards clearly required new western plate and steel capacity.

When the war began there was no large concentration of western steel making, the whole area west of the Mississippi River containing plant capable of only 2.7 million net tons in 1938, under half the size of Gary alone. By 1945 capacity in the same area was 6.2 million tons. In 1938, except for the works at Pueblo, Colorado, which was established almost sixty years before, there were no western integrated operations, and only two other locations had over a quarter-million-ton capacity: the Kansas City, Missouri, works of Sheffield Steel (340,000 net tons) and the 290,000 tons of Columbia Steel's Pittsburg mill northeast of San Francisco Bay. In the mid-1930s, those involved in Columbia Steel interests both at Pittsburg and in Utah had tried but failed to persuade the main decision makers at US Steel to build steel capacity in Utah. Then, before the United States entered the war, Bethlehem Steel put forward a proposal for a fully integrated iron and steel works in the Los Angeles area. This was turned down by the War Department. By the middle of 1941, Hauck had recognized that half the 1.1 million tons of new steel capacity then envisaged as required for the West Coast would have to be in integrated operations. After the attack on Pearl Harbor, the urgency of providing western plate capacity and supporting steel and iron facilities suddenly increased. Altogether, from 1940 to 1945 ship plate demand in the eleven western states increased from an estimated 200,000 to perhaps 3 million tons. For

TABLE 14.3
Western raw-steel capacity (except Colorado), 1938 and 1945
(thousand net tons)

	1938	1945	*Increase*
Total	1,023	3,606	252.49%
Unintegrated works	1,023	1,573	53.76%
Integrated works			
Geneva	0	1,283	n.a.
Fontana	0	750	n.a.

Sources: AISI, *Iron and Steel Works Directory of the United States and Canada* (New York: AISI, 1938, 1945).

Geneva works, Utah, showing the blast furnaces under construction, May 1943. *Courtesy of USX Corporation*

all other categories of finished steel, the rise was much smaller though still impressive—from about 2.2 to 3.7 million tons.[7] Only about one-fifth of the actual increase in crude-steel capacity was obtained by expanding existing or building more unintegrated works; the rest was focused in two fully integrated mills. The Kaiser interests built one of the new mills, that at Fontana east of Los Angeles and especially well placed to serve the large requirements in Southern California. US Steel put up the other in a location far from any one of the main population, industrial, and shipbuilding centers of the West Coast but strategically placed to supply each of them over more or less comparable rail hauls. In short, the Corporation at last took up the suggestion first made at least thirty years before that any western integrated works should be in Utah.

By late October 1941, officials of Columbia Steel were in discussions with the Defense Plant Corporation about the construction and leasing of additional plant facilities at Provo, including an additional blast furnace and a steel plant, the whole costing $90 million.[8] Then the project grew in size and became a separate integrated works. Having agreed to build for the government, in spring 1942 US Steel began work at Geneva on Utah Lake. In August 1943 it came to an agreement with the Defense Plant Corporation under which US Steel would operate the new works for the duration of the war without charge or fee. Geneva produced its first steel on 3 February 1944 and its first plates nineteen days later. Its annual capacity was then 700,000 tons of plate and 250,000 tons of structural shapes. By 1945 its blast furnaces and the nearby smaller plants at Provo and Ironton gave it command of 1.28 million tons of iron making. It was the biggest steel operation west of the Mississippi. Together, Geneva and its other operations meant that US Steel now owned or operated for the government over 52 percent of the steel-making capacity in the region. If it gained ownership of Geneva, the Corporation would become dominant in the West; Bethlehem at the time had only one-third as much capacity there, none of it integrated with iron making.

Filling Out the Production Map

US Steel beyond Pennsylvania and the Great Lakes, 1945–1970

Until after World War II, US Steel had no major facilities outside the two major national ore and fuel production systems—the almost thousand-mile-long movements between the head of the Great Lakes and the Appalachian coal plateaus and the much tighter pattern of supply and production in the Birmingham area. In the East it had locally important manufacturing operations at Worcester, Massachusetts, and Pencoyd, Pennsylvania; fabricating works; and service centers. Over the huge areas of the West, the Corporation controlled a number of small steel operations, above all that at Pittsburg, California, near the mouth of the Sacramento River, and a small iron-making plant in Utah. In neither region, though, had it a single integrated iron and steel operation. Within the first half dozen years of the post–World War II era, it acquired important iron- and steel-making capacity in both regions. The origin of each was connected with national emergencies and the federal government.

By the end of the war, the recently built 1.3 million tons of the Geneva, Utah, plant's capacity had cost $200 million of government money. The urgencies and shortages of the times had meant construction costs had been high, and the resulting heavy overheads would burden commercial operations. More ominous still for peacetime prospects, Geneva and the Fontana, California, works together had a plate capacity five times the consumption of the eleven western states in the boom peacetime year of 1937. These new plants had come into production as western steel consumption was falling from its highest wartime levels. In short, it was obvious that in order to succeed after the war a great deal more would have to be spent to convert them

201

for successful commercial operation, in which they would once more be in competition with well-established flows of material from the East. Geneva's annual finishing capacity was 700,000 tons of plate and 250,000 tons of structural shapes. It would be necessary to modify this heavy emphasis with the addition of lighter and, above all, thin flat-rolled products. There were some compensating advantages. Though dependent on rail movements, the location was a reasonably good one in terms of access to mineral supplies. It was well placed to serve each of the three biggest nodes of coastal population and steel consumption—though in 1944 to reach any one of them it had to bear rail charges of about twelve dollars per net ton.[1]

Speaking in Salt Lake City during February 1945, Benjamin Fairless made clear that the Corporation would be interested in acquiring Geneva when the U.S. government put it up for sale. Within the Roosevelt administration, there was some opposition to US Steel taking over. In reaction, for a time US Steel seemed to withdraw its interest, though in August Fairless wrote to the chairman of the Defense Plant Corporation offering to negotiate to buy a substantial tonnage of coils from Geneva for a new cold-reduction plant at Pittsburg. Meanwhile, strong pressure was mounted by western states and businesses for US Steel to acquire the plant. These took the matter to the subsequent Truman administration, and reportedly at the president's request, Stuart Symington, chairman of the War Production Board, wrote to ask Fairless to reconsider.[2] Late in 1945 sealed bids were invited for Geneva, to be submitted before 1 March 1946. After what Irving Olds described as "a thorough study conducted over recent months," the Corporation put in an offer.[3] Five or six other parties made offers, including Kaiser and Colorado Fuel and Iron, but the War Assets Administration ruled that US Steel was the only bidder that had made "a satisfactory showing of ability to purchase, reconvert, and oper-

TABLE 15.1

Estimated rolled-steel consumption in eleven western states,
1937–1946 (thousand tons)

1937	4,362	1942	8,489
1938	2,670	1943	10,124
1939	3,630	1944	9,587
1940	4,337	1945	7,232
1941	6,008	1946	6,000

Source: Hearings of *United States v. Columbia Steel Company, et al.,* in the U.S. Supreme Court, from R. R. Bowie, *Government Regulation of Business,* Cases from the National Reporter System (N.p., 1949), 1732.

ate." In May 1946 their offer of $47.5 million, although amounting to under one-quarter the cost of construction, was accepted. The Corporation also committed itself to spend another $43.5 million on reconstruction—$18.6 million of which was slated to revamp the finishing capacity at the site, above all converting the plate mill into a plate and strip mill able to produce up to 800,000 tons of hot-rolled coil; the rest of the funds were invested at Pittsburg, involving the completion of a cold-reduction mill to work up to 386,000 tons a year of hot-rolled coil from Geneva. From mid-October 1946, Geneva was also made into a basing point for sheared plate.[4] One other proposed development quickly failed. A year after the Pittsburg mill was announced, executives decided to build a similar mill at the existing Torrance works; the latter was not begun until the early 1950s, was very small, and by the middle of that decade had been discontinued. By 1948 western consumption of steel was more than double the average of the five prewar years, but production in the region was 4.5 times the level of that time. The region's mills accounted for about 56 percent of western steel requirements, while 32 percent came in by rail, and the remainder by sea.[5]

During the early postwar years, the largest independent fabricator of steel in the West was the Consolidated Steel Corporation, which had plate-fabricating yards—for tanks, welded pipe, and other items—in both California and Arizona and structural fabrication near Los Angeles and at Orange, Texas. It decided to sell these operations and talked with both Kaiser and Bethlehem, also contacting US Steel. A committee of the latter reported that the cost of duplicating the Consolidated facilities would be $14 million and that it would take three years to build them. Consolidated's properties had a depreciated value of $9.8 million. In mid-1948 US Steel acquired these works for $8.25 million in order to provide another outlet for Geneva plate and structurals. A government suit against US Steel on the grounds that it would now control 25–30 percent of the fabricating capacity in the eleven western states was dismissed.[6] In the mid-1950s, Consolidated built a large-diameter ERW (electrical resistance weld) pipe mill at Provo, Utah, which worked up more of Geneva's coil and plate.

In summary, the war had given US Steel the opportunity, which the Corporation took, to enter as the major producer in a much-expanded western market area. The much-reduced price at which it bought Geneva in 1946 enabled it, notwithstanding the need to spend liberally to reequip for peacetime operations, to secure this capacity at a price that gave reasonable overheads at a time when the expense of new plant construction was increasing rapidly. Its own initiatives helped it adjust to the new demand situation in the West. But there was no way of overcoming the problem of Geneva's remoteness from its main outlets nor the extra costs involved in the relatively small runs still common in serving a market that, though large in overall tonnage, was inconveniently divided into large numbers of separate lines. Not until May 1948 were steel-product base prices at Geneva cut by three dollars a ton to become identical with those for similar products in the Chicago, Pittsburgh, and Birmingham areas, "the first time . . . a western mill has announced price parity with those for similar products produced by steel mills in the East," Fairless pointed out at the time.[7] It was indicative of the "narrowness" of the western market in this respect that Geneva continued to grow relatively slowly, so that even at its peak, raw-steel capacity was only twice that of 1945. Its share of the total steel capacity in the eleven western states fell slightly, from 28 percent in 1945 to 27 percent in 1953 and 26.6 percent in 1960. A pointer to other than com-

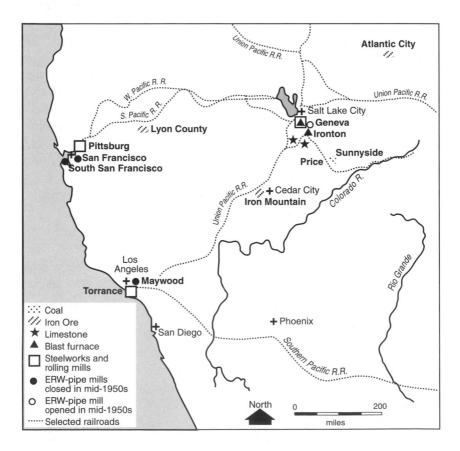

US Steel in the West, 1935–1960

plete satisfaction came in the mid-1950s with rumors that US Steel had thoughts of building an integrated works at Pittsburg and was already buying land for that purpose. It was thought that the expanded works might use Venezuelan ore and that by supplying the California market this would "free" Geneva to supply new fabricating industries in Utah and other mountain states.[8] Soon afterward, a sharp increase in steel imports, which reached levels double those for the nation as a whole, heralded a new phase in western steel making. Not only were no new works built, but existing ones began to experience increasing difficulties.

In the East, US Steel capacity had been only a little more significant than in the prewar West. Its production facilities in the region were dwarfed compared with its main rival, Bethlehem Steel, or even with the nationally insignificant Alan Wood or Phoenix Iron and Steel Companies. Gradual erosion of basing-point pricing between the wars, first with the ending of "Pittsburgh Plus" and then the 1938 removal of price differentials and the establishment of new basing points, had increasingly excluded the Corporation from an important market. The 1938 price changes more or less coincided with the completion of the Irvin works, largely geared toward sales of thin flat-rolled products in the East. Experience during the war and early postwar years provided strong incentives for the development of larger steel operations in that region.

Among the attractions for US Steel was the "pull" factor of growth in East Coast consumption. By 1947 an estimated 11.8 percent of all steel mill products consumed by metal fabricators throughout the nation were used in New England, New York, and New Jersey. Rail-freight rate increases and from 1948 the replacement of all basing-point pricing by the f.o.b. mill system made it even more difficult to supply these growing requirements from distant mills.[9] At the same time as access to the area became more difficult, conditions there for low-cost production were visibly improving. Fuel economy promised at best a marginal gain in the cost situation, but the most dramatic change was in the availability of iron ore. On the one hand it seemed that Upper Great Lakes supplies would either fail or become much more costly; on the other prospects for ore supplies from overseas improved. As early as 1942, with most of the war still to be fought, E. W. Davis of the University of Minnesota Mines Experiment Station had reckoned that the best of the Upper Lake ores would be gone by the first half of the 1950s. In the early postwar years, some expressed an anticipation of unavoidable and almost cataclysmic changes in the distribution of the steel industry. An extreme example was the forecast of the economist Marvin Barloon, who predicted, "much of the industry is going to have to move to other parts of the country" and would have to be built up on the Atlantic and Gulf Coasts. Like existing eastern works, new mills in that region would draw on foreign ore.[10] Other observers, though, had long seen there were possible remedies. As early as 1931, one mining engineer wrote, "Although some believe the future iron ore supply will come from foreign sources and will be brought to interior points via the projected Great Lakes–St. Lawrence Waterway, and others fear a displacement of the interior steel-producing districts to the Atlantic seaboard, the author has faith that American engineering skill will make possible the use of lean ores from the Lake Superior district." Postwar necessities generally justified the faith of the latter. Already by 1945 US Steel was looking into beneficiation of ore in a big way, but even if the research was successful, commercial application might take a long time and in any case would require a very large capital investment, resulting in a more costly lake ore supply. Meanwhile, the grade of the ore was falling; by 1947 that shipped from Mesabi averaged 48 percent iron content.[11]

In contrast to the situation at home, prospects for importation of high-grade foreign ores were improving, though, as considered a few years before by C. K. Leith, many possible sources presented problems. In 1945 *Fortune* magazine reviewed the situation. Some ores, such as Wabana (Newfound-

The Hull-Rust-Mahoning open pit on the Mesabi Iron Range, Minnesota, August 1946.
Courtesy of USX Corporation

land) and Durango (Mexico) were high in phosphorus. Cuban north-coast deposits contained too much nickel and chromium. In Brazil, Matto Grosso ores lay far inland, and Itabira ore, though rich and abundant, would involve huge development costs; Chilean ore was far away and in any case controlled like Las Truchas (Mexico) deposits by Bethlehem Steel. More generally, the possibility of arbitrary taxes, discriminatory laws, stoppage by fighting, and possible confiscation as before remained important disincentives to ore development in Latin America.[12] By the time of Barloon's article, many had already recognized the immense prospects in Labrador, Canada, but development there awaited large-scale investment in railroads and port facilities. Then in 1945 Ben Fairless authorized expenditures for prospecting in the areas south of the Orinoco and west of the Caroni Rivers in Venezuela. By 1947 it had been proved that this area contained large deposits of high-grade ore in the Cerro Bolivar hills near the Orinoco. Cerro Bolivar would become the material base for a US Steel East Coast mill. To work out the practicalities took a few years longer.

In 1944, long before the Venezuelan ores had been proved, US Steel engineers began designing a new integrated plant. By the end of 1946, their plans were already at an advanced stage of formulation and were completed two years later.[13] Earl Hollinshead, director of US Steel management and the real

Cerro Bolivar, Venezuela, ore-mining operations in the 1950s. *Courtesy of USX Corporation*

estate department, examined possible sites for a mill all along the Atlantic seaboard from Norfolk, Virginia, up Chesapeake Bay, along the Delaware River in eastern New Jersey, on New York Bay, up the Hudson River, and as far along the New England shore as Bridgeport, Connecticut. Despite the ground covered in this search, from an early stage the area within the great bend of the Delaware across from Trenton, New Jersey, seems to have been the preferred location. There, over the course of five years beginning in 1944 and with help from the Pennsylvania Railroad, US Steel acquired a considerable number of "worn-out" farms and consolidated them into a 3,842-acre tract. Purchases were completed in 1949, and on 28 December the Corporation announced that the land had been acquired for possible future use of an eastern seaboard mill. On 24 January 1950 in the course of hearings before the Joint Committee on the Economic Report in Washington, Ben Fairless announced his company's decision about East Coast development. Having been pressed by regional interests, US Steel officials had given "thoughtful attention" to the "possibilities" of an integrated works located in New England. Not only had they decided against this, but they also wished to oppose any government help for such a project: "We believe wholeheartedly in competitive, free private enterprise, and for that reason we cannot advocate the building or financing of new steelmaking facilities with Federal or State Government funds, except possibly in an emergency during wartime." It was clear that their own successful acquisition of the Geneva works lay at the back of their minds. The positive side of the statement contained the prospect of the first US Steel greenfield project since Duluth. "After careful consideration of raw materials and the markets in relation to plant locations, we have concluded that if and when United States Steel constructs an integrated steel plant on the Atlantic Seaboard it should be located in the Philadelphia–Trenton–New York area. With this possibility in mind, United States Steel has recently acquired an acreage on the Delaware River near Morrisville, Pa., about 30 miles northeast of Philadelphia."[14] At that time there seemed little likelihood of an early start to construction. For one thing any new capacity might need to be counterbalanced by the retirement of plant elsewhere; more important, ore from the Venezuelan deposits would not be available for some years. To the stockholders' meeting in May, Olds explained that plans were in preparation, "although the exact character of the mill and the time for the commencement of its building are matters which have not yet been determined." Eventually there would be an integrated works, but "it is possible that the mill will be constructed in different stages." A few years later it was revealed that the original proposal had been for sheet- and tinplate-finishing operations only.[15] Eight weeks after Olds's statement, North Korean forces invaded South Korea, and soon the urgent need for war supplies not only hurried along the Morrisville project but ensured that it would be a fully integrated operation from the start. Even so, in August 1950 US Steel seemed still to be thinking of a small plant— an integrated mill of a minimum 700,000-ton ingot capacity.[16]

Within four months of the beginning of the Korean War, the United States government introduced a scheme for accelerated depreciation of capital invested in projects regarded as necessary for defense purposes. The President's Economic Report to Congress, delivered in January 1951 when the war was going badly for United Nations forces (to which the United States made the largest contribution), recommended an expansion of steel capacity from 104 to 120 million tons over three or four years, sooner if possible. Already by rounding out existing plants, US Steel had added 1.8 million tons, or 5.6 percent, to its ingot capacity since the fighting began. It planned to add another 2.5 million tons by the end of 1952.[17] The new plant was now constructed with remarkable speed. Ground was broken at Morrisville on 1 March 1951; production of both iron and steel began on 11 December 1952; the hot-strip mill, the work's key finishing unit, was in production by July 1953.

In providing improved access to the eastern-seaboard markets, the decision to build the new works was unimpeachable. In this respect, at least, the tables were turned on Bethlehem Steel. Late in 1952 officials reckoned that it would cost only 20.7 cents to ship a hundredweight of steel to New York from the new works as compared with 35.7 cents from Sparrows Point, Maryland, and 57.5 cents from Pittsburgh.[18] But the urgency with which the Trenton plant's construction was pushed ahead meant that operations began well before the main supplies of seaborne ore from Venezuela became available or before the Delaware River had been dredged sufficiently to take large ore carriers up to the site. There were other features that pointed to possible future problems. The capital outlay was about twice the original estimate. The works was built with future expansion to 6 or even 8 million tons of steel in the minds of the planners. Initially, it was equipped with a melting shop of nine 290-ton open-hearth furnaces, but there was room for more. Two blast furnaces were built but the layout provided for up to six. Inevitably, in a layout providing for the accommodation of future growth, there was a danger of long internal hauls, though it must be remembered that nearly fifty years beforehand, the engineers of the Gary works had largely succeeded in sidestepping this problem. At Morrisville, as early panoramic views made plain, a comparable success was not achieved: coke, for instance, having to travel about a mile by belt from the ovens to the furnace yards. Moreover, a good deal of the equipment, especially in the rolling mills, had been designed to cope with the anticipated higher levels of production and so was underutilized at lower outputs, resulting in increased unit costs. (An interesting description, recognizing both the advantages and some of the deficiencies was produced as a result of a 1952 British visit led by Charles Goodeve.)[19] By early 1953 the new plant, named in honor of Fairless, who had retired from the US Steel presidency that January, was rated at 1.2 million net tons of raw-steel capacity; a year later it was 2.2 million tons. National Tube installed facilities to produce up to 280,000 tons of tubular products on a site adjacent to the steelworks, from which it drew its slabs.[20] At that time there seemed every expectation of expansion in a setting of continuing regional economic growth. There were projections of 300,000 new jobs in the Delaware valley over the next five years, and the National and Republic Steel companies were said to be considering construction of their own integrated operations in the area. The president of one of the region's chambers of commerce was carried away by the prospects: "The putting into motion of the large Fairless works . . . is likely to go down in history as the most momentous event in our Delaware Valley since the night Washington crossed the Delaware River to deliver his surprise blow to the Hessian mercenaries."[21] In 1956 a third blast furnace increased iron capacity by 50 percent, and the greater supply of hot metal expanded effective steel capacity and therefore made possible much fuller use of finishing facilities already installed. National imports of Venezuelan ore increased from 1.9 million gross tons in 1953 to 13.5 million tons six years later. By that time, with the Delaware deepened, Fairless could make 2.7 million tons of steel. However, rosy expectations of continuing high rates of economic growth and increasing steel consumption were not being as fully realized along the Atlantic seaboard as in some other leading industrial sections of the nation. A "Fairless Industrial District" was laid out on the north side of the works in the hopes of stimulating the development of new manufacturing industries.[22] At the end of the 1950s, although it ranked far ahead of all other plants on the East Coast except for Sparrows Point and Bethlehem, in the latter region, US Steel still controlled less than one-fifth the tonnage of its main rival. In a few more years, Fairless works began to pose serious problems for US Steel executives.

During World War II there was a massive growth of the economy of the southwestern Gulf region. Though increasingly diverse, the basis of this expansion was its massive oil and gas industries. These provided a major market for steel, and by the early postwar years the states of Kansas, Arkansas, Louisiana, Oklahoma, New Mexico, and Texas consumed 70 percent of the "oil country goods" of the United States.[23] Though such a huge consumer, the south-central United States then produced no pipe, supplies coming from northern plants that had been built when the geography of consumption was completely different. However, by the late 1940s, sometimes—as at Lone Star

Fairless works, Pennsylvania. An artist's impression of the proposed plant layout, 1951.
Courtesy of USX Corporation

Steel in northeast Texas—with help from government loans, pipe capacity there was beginning to increase. In 1949 there was talk of a US Steel ERW-pipe mill near the Houston ship canal, but instead the Corporation announced the construction of a 100,000-ton mill one hundred miles away at Orange, Texas. Situated strategically at the intersection of the Sabine River and the Intracoastal Waterway, the mill's products could be distributed by barge, ship, rail, and road over a wide area. US Steel supplied the plant's requirements of steel from northern operations. By 1954, Orange capacity was already 350,000 tons. Birmingham joined Pittsburgh and Chicago as a supplier of its semifinished steel. Then in the mid-1960s US Steel embarked on a process of large-scale backward integration that at last promised a major commitment in Texas. For a time the Corporation seemed to be considering an integrated works for the Houston area.[24] Then in December 1965 it announced that it had spent $20 million to acquire 11,000 acres at Cedar Point near the small coastal town of Baytown. Instead of fully integrated operations, US Steel would build a large electric furnace, continuous casting (concasting), and plate-mill operation to supply both the Orange pipe mills, less than eighty miles to the east, and part of the region's needs for plate for construction and industrial use—in shipbuilding or repair, tank fabrication, pressure vessels, bridge work, and so on—which until now it had supplied from Birmingham or Chicago. When he met the members of the Houston Chamber of Commerce in May 1966, the US Steel president, Leslie Worthington, told them, "We expect this [the Texas works] to become one of the greatest steel producing complexes in the nation, if not in the world."[25] Its 160-inch plate mill was authorized in mid-1967 and was at work in 1970 using slabs brought in from northern plants. By spring 1971 electric furnaces and its own 500,000-ton slab caster were at work.

The economics of the new Baytown operation were complicated. Capital costs were relatively low. Whereas in the early 1970s blast furnaces and BOF (basic oxygen furnace) plant cost an estimated $120 per ton of steel capacity, this could be gained at a cost of only $65 per ton by direct-reduction processes and electrical furnaces. Baytown's electric furnaces used the abundant sup-

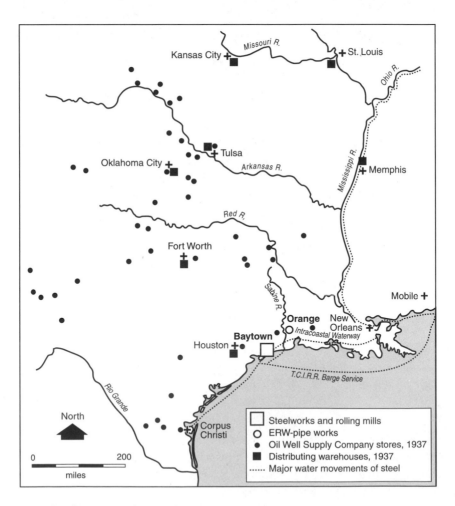

US Steel in the West South Central Region, 1930–1970

plies of local scrap, in a generally scrap-surplus area, and iron briquettes shipped from a US Steel direct-reduction plant in Venezuela.[26] For a time, though, Baytown's plate mill was grossly underused. This facility had an annual capacity of 1 to 1.2 million tons, but through to the mid-1970s the output of slabs from the first melting shop and casting plant was only sufficient for about 385,000 tons of plate a year. Limited availability of slabs from other US Steel works meant that even with materials from these, the Baytown mill could only operate about 40 percent of its rated capacity. At the time, plate consumption in the Houston area was projected to increase by 900,000 tons during the 1980s, and for this reason and in order to get a better loading for the plate mill, in August 1975 officials decided to install two more electric furnaces and two more slab casters, increasing capacity for steel to 1.6 million tons and for slabs to 1.5 million tons. With these extensions it was expected that Baytown would be able to supply 200,000 tons of plate to the Orange pipe works and 880,000 tons for regional plate consumption. The cost of these extensions was put at $129 million. The success of the whole venture depended on the size of future consumption, the share secured by the works, and the price of scrap. It was an uncertain business. Assuming the latter remained at $60 a ton, it was reckoned the outlay for extensions would result in annual savings of $28.9 million, but if its price rose to $80 per ton, the savings would fall to $18.7 million. The first half of the 1970s ended with the probability that a major plant would be built on the 4,000-acre site. The Corporation owned another 10,000 acres nearby that it was reported as planning to use to attract other industries.[27]

Triumph and Marking Time, 1945–1960

By the late 1940s, the American steel industry occupied an exceptional position in the world. The material base of the military victory recently won by the Allies had depended at least as much on the United States's preeminent position in iron and steel as on any other sector of its economy. At the same time, the course of World War II had left much of the capacity in other parts of the globe in ruins. American 1945 output of almost 80 million tons of raw steel amounted to an estimated 62.4 percent of the world total, a share not approached by any single country since the earliest days of bulk steel eighty years previous. As late as 1950, by which time recovery was nearing completion in Western Europe, the United States made 46.6 percent of the world's steel. It was a time for understandable satisfaction. Unfortunately, such a natural human reaction all too easily passes over into complacency. Evidence suggests that for a number of years this was the case both in the industry generally and at US Steel. One instance was provided when, on 24 January 1950, George W. Wolf, president of the US Steel Export Company, testified before the Joint Committee on the Economic Report. His subject was the "General World Steel Situation." In setting out his argument, he took a broad view of the economic history of the previous eighty years. He ended by referring to the "penalty of the lack of integration in the overall economy of Europe" and of the dangers of trade wars and their consequences since "out of such lack of wisdom stem uneconomic practices of every kind and sort." This Wolf contrasted with "our unparalleled American industry, operating under our free individual enterprise system, in a never satisfied mass market. . . . Thus it is that American steel capacity is better placed as to type, variety and quantity of steel products than any other single country." Through the Economic Co-operation Act 1948, which activated Marshall Aid, funds had been made available for putting in facilities "more in accord with the realities of modern steel product demand," but "the European steel industry is still far behind that of the United States, product, quality, and cost wise." In fact, as far as Wolf could tell, the quality gap "in favor of American steel is an ever-widening one . . . [so that] the American seller is justified in a reasonable price premium over that of an inferior European product, as long as he can get it."[1] Within no more than ten years, such breezy optimism was rapidly becoming a thing of the past—"as long as he can get it" by then seemed to have a different meaning. During the late 1950s, the European industry was largely reconstructed, modernized, and rapidly expanding. By this time too, Japan had become a major player in world steel.

It is at first sight surprising that this period of signal achievement, during which the national steel capacity increased as never before (and probably by too much)—from an estimated 95.5 million net tons at the beginning of 1945 to 147.6 million tons within fourteen years (an addition far greater than the *total* capacity in any other nation except the Soviet Union)—should be seen in retrospect to have heralded its decline in standing on the world stage and the nearing approach of a period of unprecedented troubles. In fact, not only did material circumstances change, but leadership and self-satisfaction also fostered a degree of complacency. Two writers have described these years as the "dodo period" of the American steel industry, referring to the extinct flightless bird. Perhaps a more appropriate avian image would be that of the ostrich, a creature still very much alive but nonetheless having a life-threatening propensity to bury its head in the sand and consequently failing to realize that pursuers are catching up with it. In the mid-1950s, there seemed to be an American presumption that not much of significance was happening in the rest of the world. This attitude would soon bring a dreadful outcome to U.S. steel producers. Even when this occurred, there were so many scapegoats that it was easy to ignore the fact that a good deal of the cause of the difficulties was to be found in overweening self-confidence combined with an incapacity to recognize that there were innovators, good practical steelmen, and expansion-minded companies elsewhere in the world. By 1960 the United States produced only 26 percent of the world's steel, a lower proportion than at any time since the early 1870s. All in all it seems that, throughout the first fifteen years of the postwar period, the United States Steel Corporation was no better off, though little if any more culpable, than the rest of the American industry.

A huge expansion of United States steel capacity was carried through in the 1950s, first in response to the pressing demands of the Korean War, sec-

TABLE 16.1
Capacities in the American Iron and Steel Industry, 1946, 1955, 1959
(thousand net tons)

Product	1946	1955	1959
Coke	60,421	72,685	73,098
Iron Ore (home shipments and net imports)[a]	118,192	151,440	170,672
Pig iron	67,340	83,971	94,635
Raw steel	91,890	125,828	147,633
Open hearth	81,236	110,234	126,528
Bessemer	5,154	4,787	3,577
Electrical	5,500	10,807	13,495
Oxygen converter	n.a.	n.a.	4,033

Sources: AISI annual statistical reports.
[a] Estimate based on requirements of the industry at full production. The figure for 1959 may be inaccurate.

ond to support continuing high levels of spending on armaments through the years of the Cold War, and third (and above all), to provide the basic materials for a civilian economy geared to an apparently never-ending increase in consumption. For much of this period, the federal government was pressing the industry to extend. Notwithstanding the huge outlays being made by US Steel and other major companies, there was a strong movement in Washington for the building of major new plants, if necessary helped by government funding or even built by the government itself. Most of the 52.1-million-ton expansion during 1945–1959 was secured by "rounding out" existing plants. Some of this was at unintegrated works, but most in operations combining iron and steel making. A small part of the increase was at new and as yet small operations based on electrical steel furnaces, facilities that in the near future would receive separate recognition as "mini mills." Only 2.4 million tons, or 4.6 percent of the total, was in new greenfield integrated works. The disadvantage of expanding existing works was that it preserved old patterns of production and distribution, though by varying rates of expansion geographically, a multi-plant operator might respond positively to changes in raw-material supplies and markets. The rationale for extending works rather than building new ones was simple: it was much cheaper. A rule-of-thumb estimate was that the cost of a fully integrated works on a greenfield site had increased from about $100 a ton of finished-steel capacity before World War II to about $300 per ton by the mid-1905s. However, where nothing more than enlargement of existing facilities and improvement in production techniques was needed, additional capacity might be obtained for about $100 a ton; though if new open-hearth melting shops or major finishing installations were involved, these might push the cost of rounding out existing works up to about $200 a ton. Even in the latter case, there would be a saving of something like $100 million for every million tons of new capacity as compared with a wholly new plant.[2]

In this period there were important changes in the nature and still more in the distribution of the markets for steel. Decline continued in some heavy steel lines, but there were large increases in construction. Above all, the shift to light flat-rolled steel products, so marked in the 1930s but briefly interrupted by the war, resumed. Responding to this shift, the composition of national finished rolled-steel capacities was changing. Despite its efforts in the 1930s, US Steel remained strongest in lines that were growing less rapidly. The geography of consumption was changing significantly. In a sharp acceleration of trends only faintly visible before the war, markets in the South and West were growing more rapidly than those in the Northeast. The industrial districts within the manufacturing belt showed varying tendencies; generally, its western parts grew most.

One change of this period required initiative, sustained effort, and vast outlays of capital: securing adequate raw-material supply. In fuel, this was much less important than in iron ore. New coking coalfields were opened in the West, and economy in coke consumption continued. Long before, the coalfield-located beehive coke industry had become a miniscule supplement at

TABLE 16.2

Changes in the finished-steel capacity of the U.S. steel industry, 1938–1951, and US Steel finished-steel capacity as a proportion of the national total, 1948

Product	United States change 1938–1951	US Steel share of all capacity 1948
Rails	−19.2%	54.5%
Light structural shapes	−41.7%	32.2%
Bars	+11.0%	25.4% (hot-rolled)
Hot rolled strip	+29.5%	24.4%
Wire rods	+30.3%	38.5%
Plates	+34.0%	56.0%
Skelp and pierced billets	+38.3%	19.4% (skelp)
Heavy structural shapes	+42.5%	47.9%
Hot rolled sheets	+102.9%	21.9%

Sources: G. S. Armstrong, et al., *The Iron and Steel Industry,* vol. 10 of *An Engineering Interpretation of the Economic and Financial Aspects of American Industry* (N.p., 1952), 93; G. G. Schroeder, *The Growth of Major Steel Companies, 1900<N>1950* (Baltimore: Johns Hopkins University Press, 1953), 62–74.

TABLE 16.3

Deliveries of iron and steel products by major geographical regions, 1929–1950
(proportion of total deliveries)

Year	Northeast and Midwest	West and Southwest	South and Southeast
1929	77.1%	16.4%	6.5%
1935	77.0%	16.6%	6.4%
1940	76.9%	14.7%	8.4%
1945	67.8%	23.1%	9.1%
1950	63.9%	25.2%	10.9%

Source: G. S. Armstrong, et al., *The Iron and Steel Industry*, vol. 10 of *An Engineering Interpretation of the Economic and Financial Aspects of American Industry* (N.p., 1952).
Note: The figures are based on the terminations of freight by class-1 railroads.

times of peak demand for the byproduct-coke batteries at the iron and steel plants. In iron ore, though, there was an urgent need to find an alternative to the direct shipment of ores from the Upper Great Lakes. For nearly three-quarters of a century, this region had dominated the national ore supply, and in doing so it had helped check any further large-scale western movement from the Pittsburgh-Chicago-Buffalo core of national production, dominant since the late nineteenth century. Until after World War II, the old ore classifications still existed—standard ore was all direct-shipping ore of 49 percent or more Fe, Bessemer grades had less than 0.045 percent phosphorus content, lean ores contained over 35 percent Fe but were inferior in grade to standard ores.[3] However, following the unprecedented demands made on these deposits during the war—in 1941–1943 US Steel ore railroads moved an average annual tonnage twice that of the average of the previous twenty-five years—it was painfully obvious that this supply system could not long continue. For some time there were suggestions that much of the iron capacity would have to begin using imported ore, and as late as 1950 the U.S. Bureau of Mines went so far as to forecast that within twenty years shortages of ore might limit the industry to 85 percent of capacity.[4] In fact, within a relatively short period of time, a solution to the domestic iron-ore problem was found. While largely perpetuating the old-source region, it radically changed the economics of ore supply.

As early as their final report of 1938, Ford, Bacon, and Davis had pointed to the solution. They anticipated that the proportion of high-grade ores would begin to decline in about fifteen years, but its "place will have to be taken by the concentration of leaner ores from the same district."[5] As this suggested, it

TABLE 16.4
Estimated consumption of finished steel by states, 1940, 1947, 1952, and 1963
(proportion of national total)

State	1940[a]	1947[a]	1952[b]	1963[c]
Michigan	25.9%	21.0%	15.2%	19.5%
Ohio	13.6%	13.8%	14.8%	14.2%
Pennsylvania	11.4%	12.7%	13.6%	9.0%
Illinois	10.0%	10.1%	12.8%	11.0%
New York	5.4%	5.5%	5.6%	6.5%
Indiana	4.6%	5.1%	6.1%	5.7%
Wisconsin	3.9%	3.3%	4.9%	4.4%
New Jersey	2.3%	2.2%	2.6%	2.4%
California	3.4%	3.5%	3.8%	4.4%
Texas	3.7%	4.3%	2.3%	1.4%
All other states	15.8%	18.5%	18.3%	21.5%

Sources: Iron Age, various issues, 1940, 1947, 1952, and 1963; *Steel,* 17 January 1955, 125–26.

[a] All finished steel.

[b] Carbon steel.

[c] Carbon plate, hot-rolled bars, hot-rolled and cold-rolled sheet, and strip only.

had been realized that the country rock containing the workable ores of the iron ranges was itself a low-grade source of iron, though hard, intractable, and in its existing form unusable. During the first fifteen years after World War II, research and development work gradually perfected the techniques for mining, crushing, and concentrating this "taconite" rock so as to produce a high-grade blast-furnace burden. The resulting changes to ore production were on a heroic scale. Shipments of material made from taconite were 62,000 tons in 1950 but 1.45 million tons five years later. As early as 1952 taconite-processing plants with an annual capacity of 15 million tons were under construction; within about fifteen years national "pellet" capacity was projected to be 50 million tons. Eventually, further development of the taconite industry was encouraged by the more liberal tax policy introduced following Minnesota's 1964 "taconite amendment" vote.[6] US Steel was prominent in these developments. By early 1951 its Oliver Iron Mining division had begun to build an experimental 500,000-ton sintering and nodulizing plant at Virginia on the Mesabi Range. Planning for an experimental taconite plant was underway, the report pointedly adding: "A major cost in iron ore mining operations over which US Steel has no control is the taxes levied on ownership and mining of ore deposits. Onerous taxes of this kind can have a crippling effect

TABLE 16.5

**United States and US Steel production and imports of iron ore,
1948, 1955, 1960 (thousand gross tons)**

	1948	1955	1960
USA production	101,003	102,999	88,784
of which Lake Superior	82,277	83,255	70,565
USA iron ore imports	6,086	23,443	34,591
USS total ores mined	43,684	46,551	44,821

Sources: Annual reports of AISI, American Iron Ore Association, and US Steel.

on new projects." Within two years the experimental taconite plant was near-ing completion at Iron Mountain, Minnesota, and with the Virginia unit was to be a first step to a series of US Steel taconite operations.[7]

Coinciding with these developments in Lakes ore was a large increase in the use of foreign ore. Again, US Steel played a major part. With one major ex-ception, expectations that ore-supply changes might cause major locational shifts in the steel industry went unfulfilled. The great importance of market access, the vast capital tied up in existing plants, production in the Upper Great Lakes of a higher-grade and more transportable form of iron-bearing material, and development of new transport routes for imported ore—most notably the St. Lawrence Seaway—were important causes of this inertia. One effect of the richer ore burden, whether from home or overseas, was that few new blast furnaces were required to support increases in steel capacity. US Steel's expansion reflected the changes in patterns of ore supply more dra-matically than in the industry generally.

Between 1945 and 1960, the Corporation's steel-making capacity in-creased by 9.6 million tons, or 30 percent. Yet, compared with the entire industry, it remained overwhelmingly a bulk-steel producer. Its 1945–1946 share of national electrical steel capacity had been 8.3 percent; but by 1959 it was only 3.4 percent. At the latter date US Steel still had no oxygen-con-verter plant, though this already made up 2.7 percent of national capacity. As with the rest of the industry, it had found expansion much costlier than in the past. In 1950, addressing a government committee, US Steel's vice president for engineering put the average cost of replacing facilities at almost double the 1939 level and six times that of 1901.[8] Over the six months following the North Korean invasion of South Korea, US Steel extended its annual capacity by 1.8 million tons to a total of 33.9 million tons. A year later, capacity was

TABLE 16.6

US Steel shares of the production, shipments, and net income of the American industry, 1945–1960 (proportion of industry total)

Year	Iron prod.	Raw steel prod.	Steel product shipments	Net income
1945	36.3%	33.2%	32.2%	n.a
1950	36.0%	32.5%	31.3%	28.1%
1955	33.4%	30.2%	30.1%	33.7%
1960	31.5%	27.5%	26.3%	37.5%

Sources: US Steel annual reports for 1945–1960; AISI statistical reports for 1945–1960.
Note: The net income figures do not include all companies.

TABLE 16.7

United States and US Steel crude-steel production by process, 1959 (proportion of totals for each)

	Open hearth	Bessemer	Electrical	Oxygen converter
United States	85.70%	2.42%	9.14%	2.73%
US Steel	94.70%	4.19%	1.10%	0%

Sources: AISI, *Statistical Report* (New York and Washington, D.C.: AISI, 1959); US Steel records, USX.

34.6 million tons, 5.8 million tons more than at the start of 1946. Work underway by May 1952 would add another 2.3 million tons to potential output. In this instance, more than three-quarters of the increase would be at a new works, for the United States Steel Corporation built the only greenfield-site plant during this period. Fairless works made up almost exactly one-quarter of the Corporation's 1945–1960 increase; Geneva, bought from the government in 1946 and requiring large-scale reconstruction to fit it for a role in the peacetime economy of the West, accounted for another quarter. Over the same period, a few small, less suitably located or otherwise unfavorably endowed operations were closed, sold, or run down, primarily in Pennsylvania. For instance, the 1-million-ton Farrell integrated works at the southern edge of the town of Sharon, having been retained so long only because of a pressing need for maximum production, was now sold to the Sharon Steel Company. A decade later, steel making, though not mill operations, was abandoned at Vandergrift. At the end of the 1950s, expansion at Fairless made it possible to close the open-hearth shop at the Worcester, Massachusetts, wire operations.

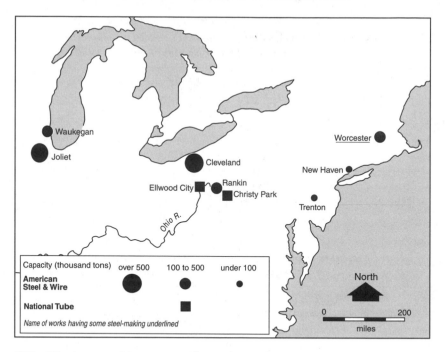

US Steel Nonintegrated Operations in the Northeast, 1957

During this decade it sometimes seemed that there would be no end for expansion, and in 1956 US Steel announced it was planning a long-term program involving an average annual increase of 1 million tons over the next ten years at a cost of $1.5 billion. In fact, in the late 1950s a sharp recession cut into this program and marked a transition to more uncertain times.

As late as the 1950s, some were still suggesting that Birmingham-area expansion would have been greater if Tennessee Coal, Iron, and Railroad (TCI) had been independently owned. There was growth in Alabama during this time. Ensley was still the bigger of US Steel's operations there in 1945. By 1950 ingot capacity in the district had increased only slightly from the tonnage five years before, but at this time a 500,000-ton expansion was announced.[9] Most of this and all subsequent extensions were concentrated at Fairfield, which by the end of the decade was considerably the larger of the Birmingham plants. Except for rails it had all the finishing capacity, in terms of hot mill almost 91 percent of the area's combined tonnage. Even more than Gary, it had been picked out as its district's growth point.

In terms of output and income, the fifteen years following the war was for US Steel as for the rest of the industry a generally good period. Profits were at record levels. The Corporation remained a giant in the metallurgical world. Indeed, reviewing its 1954 performance, the *Economist* was almost bowled over: "Its steel making capacity falls little short of the combined steel production of Britain and Germany or that of Soviet Russia"; its annual report was "not merely impressive—it is staggering." Others concluded that at last the giant had thrown off the characteristics of a "laggard." By 1955, at a return rate of over 9 percent, its profits were as high as those of most other steel companies. *Fortune* hailed "the transformation of US Steel."[10] This high tide of prosperity was marked by the retirement, after a long and distinguished career, of Ben Fairless. Despite some positive interpretations, other aspects of corporate performance during this decade indicated that US Steel was in fact falling further from its once-lofty preeminence. Costs, especially for employment, were rising faster than prices. Output of steel was increasing more rapidly in other parts of the industrialized world, and by the end of the decade it was obvious that, through imports, expansion in foreign production would

TABLE 16.8

Raw-steel capacity of US Steel integrated works in the Pittsburgh, Chicago, and Birmingham areas, 1945 and 1959 (thousand net tons)

Works		1945	1959
Pittsburgh			
	Homestead	4,732	4,426
	Duquesne	2,147	1,741
	Edgar Thomson	2,297	2,529
	McKeesport	1,200	1,392
	Clairton	805	1,064
	Donora	842	1,015
		12,023	12,167
Chicago			
	South Works	4,525	5,589
	Gary	5,719	7,999
		10,244	13,588
Birmingham			
	Ensley	1,568	1,770
	Fairfield	1,092	2,227
		2,660	3,997

Sources: AISI, *Iron and Steel Works Directory of the United States and Canada* (New York: AISI, 1945, 1959).

Ben Fairless and David McDonald making one of their celebrated joint visits to a US Steel works, 1953. *Courtesy of USX Corporation*

seriously affect outlets for domestic producers. Some of the uncertainties implicit in the situation were already being recognized in the final years of the Ben Fairless chairmanship. Speaking to stockholders in May 1953, Fairless outlined a situation that at first sight seemed enviable. Their new East Coast works had opened the previous December and was working its way up toward full operations. The Corporation currently had underway 1,100 separate plant improvement and replacement programs each costing at least $25,000. In the first quarter of that year their operations had been at 102 percent of rated capacity. Yet the way ahead was unclear. If there was to be a new world war, corporate operations would need much more expansion—already three times that century major world conflicts forced up US Steel's capital spending and decreased its profits—if, instead of war, there was disarmament, they might have too much capacity. In any case, whatever the size and state of physical plant, things were not going well enough on the commercial side: "Our facilities are modern, efficient, and strategically located. . . . [B]ut from a purely financial point of view I do not believe that United States Steel is equally well-equipped to meet the extreme demands which may be made upon it by our nation in the future." Fairless alluded to a cost-price squeeze at the heart of this financial problem. "Since 1940, our employment costs have risen 155 percent. Our payments for purchased goods and services have increased by 138 percent. But the price of the steel we sell has gone up only 87 percent." This theme would recur time and again over the years. Looking back on this period of the early 1950s, two later top executives recognized opportunities had been lost. Leslie Worthington had been at Carnegie-Illinois, US Steel Supply, and the Columbia-Geneva divisions before he became president of the Corporation in 1959. Five years later he acknowledged they should have made bigger changes in their product range: "I don't like to look back over my shoulder at what someone else has done, but I do wish we had put more emphasis on lighter steels." Edwin Gott, general superintendent at the Youngstown district works until 1956, when he became vice president in charge of operations, was more outspoken in his reflections on the first half of the decade. "A period like this is not good for a company. Everyone becomes complacent. Our mills were jam full and we had to fight the customers off. When you get used to working that way, orders don't go out on time and quality isn't always right."[11]

At the beginning of May 1955, Fairless was on the eve of his sixty-fifth birthday and retirement from the chairmanship when he addressed the stockholders. As Myron Taylor had done in 1938, Fairless took this opportunity to trace the changes since he had first occupied a leading position in US Steel. Over the eighteen years to 1954, corporate steel capacity had increased 30 percent, shipments by 70 percent, and the workforce by 20 percent. The value of products and services sold had gone up 311.3 percent and employment costs by 309.1 percent. But the price of steel had increased only about 125 percent. Hourly wage costs in 1936 had been seventy cents, but by the end of 1954 these had reached $2.65. Five times as many people were engaged in research work as in 1936. Fairless reckoned the prospect ahead was for continuing growth: "our experts now estimate that if our Corporation is to continue to hold its present position of leadership in the industry at the peak of the prospective demand, we shall have to create the equivalent of one new Fairless Works every two years throughout the next two decades." (Growth at such a rate would have amounted to about 22 million tons—much more than half their current capacity.) All this was positive, but as he left the prime position, Fairless's remarks on top management may have been a perfectly innocent generalization, but a consideration of the age structure of the board on which he now remained as an ordinary director, in addition to subsequent events, give Fairless's words added force: "I am firmly convinced that there must always be room at the top of our management team for young men with young ideas, and a fresh, new, outlook upon the problems that our company must face if it is to keep pace with its growing responsibilities toward a young and growing nation." If the Corporation failed to provide these opportunities, the young would go elsewhere, and "thus our entire organization would soon become stagnant, set in its ways, and dangerously unprogressive in its views, and that, of course, must never, never happen at US Steel."[12]

The honors and responsibilities as chairman and chief executive officer passed in 1955 to fifty-one-year-old attorney Roger Blough—"the first of the so-called 'managers,'" as one of Fairless's obituary writers rather unkindly put it seven years later.[13] Since 1942 Blough had been in charge of US Steel legal affairs. After carrying through major changes to operating structures in the search for increased efficiency, he became a director in 1951 and vice chairman in May 1952. He took over the chair at an auspicious time, 1955 steel shipments being at a level never previously matched—afterward they fell away. Now, in one public statement after another, the new chairman and his senior colleagues emphasized the seriousness of the industry's cost-price squeeze and the threatening tide of foreign competition. In 1957 C. F. Hood, the president, Robert Tyson, chairman of the finance committee, and R. Conrad Cooper, at that time vice president of administrative planning, gave evidence to the Subcommittee on Anti-Trust and Monopoly. Hood pointed out that for many years US Steel employment costs had been rising by over 8 percent a year. In negotiating for a wage settlement from 1 August 1956, the Corporation had offered its employees a five-year agreement with wages advancing 4 percent each year. However, the men had struck and the outcome was a three-year agreement under which employment costs were rising 6 percent annually, compounded. Tyson put it more colorfully. Since shortly before World War II, the Corporation's wage costs had been "experiencing a continuous high speed inflation." The average rate of increase had been over 8 percent, but it was at 10 percent in 1956. Cooper indicated to the subcommittee that between 1940 and 1956, their total employment costs per man-hour had increased three times as much as product per man-hour, and taking into account the investment in machinery and plant, at more than three times productivity.[14] In sharp contrast, United Steelworkers of America (USWA) figures (based on US Steel annual reports) claimed to show how considerably profit had increased in relation to both sales and employment costs since the end of World War II. In any case, by the late 1950s there had come into both industry and US Steel statements a new note of unease, uncertainty, and even doubt about future prospects. Expansion continued—the whole process seeming to have a momentum of its own—as existing operations were rounded out, but although output was still increasing it was not keeping pace with the expenses, and operating rates were generally drifting downward.[15]

In 1958 bad trading conditions brought these problems still more into the forefront. The Corporation's steel tonnage was the lowest since 1946; its annual operating rate, 85.2 percent the previous year, had slipped to 59.2 percent. Shipments of steel products were down by 6.4 million tons, or 27.45 percent from 1957. Despite these conditions, in September the USWA convention unanimously called for "substantial" improvements in wages, hours, and working conditions for 1959. Next month, speaking in Dallas, Robert Tyson again targeted the "cost-push" inflation burdening his company. He believed it "originates in the power of our labor unions to enforce inflationary increases in employment costs, calling in turn for cost-covering increases in prices as an alternative to bankruptcy."[16] Management claimed such increases were particularly dangerous as international competition rose. In an address delivered in April 1958 to the Cleveland Chamber of Commerce, Blough pointed out that wire from Dusseldorf was being sold to a jobbers

TABLE 16.9
US Steel profits in relation to sales and to employment costs

Year	Net profit as % sales	Net profit as % employment costs
1946	5.92	12.6
1947	5.99	14.1
1948	5.22	12.5
1949	7.21	17.5
1950	7.29	18.3
1951	5.23	13.4
1952	4.58	10.9
1953	5.75	14.2
1954	6.01	14.1
1955	9.03	22.9
1956	8.23	20.7
1957	9.50	22.5
1958	8.68	20.3

Source: USWA, *Foreign Competition and Steel Wages* (N.p., 1959).

warehouse in Cleveland for as much as $40 a ton less than by American Steel and Wire Division, whose local Cuyahoga works was a major producer. Next year the St. Lawrence Seaway would be open and foreign steel would come in still more cheaply. He adopted a rather patronizing tone in relation to competition from another quarter: "Out on the West Coast, the industrious Japanese have bought scrap metal at premium prices, hauled it back to their mills in the Orient, manufactured it into finished products, shipped them back across the wide Pacific, and still undersold American producers by such substantial margins as $29 a ton on reinforcing bars." Like his colleagues, Blough attributed the problem above all to higher wages in the American industry. At the same time he noted that American technology had spread worldwide. Its introduction overseas was sometimes helped by United States financial aid, and the nation's technical assistance programs meant that American works were helping to train those who were becoming ever-keener rivals for business. As a result, "we are rapidly losing the technological margin that we have had over other nations, and that has thus far supported American wages at levels high above those prevailing elsewhere in the world." Overall, "America, as a nation, is *costing* itself out of the market."[17]

In 1958 the tonnage of imported steel products was less than 3.1 percent of home production, but shortly before collective bargaining began for a new settlement in mid-1959, the American Iron and Steel Institute began a campaign against imports. USWA alleged that it did this to highlight the lower wage cost of foreign industries and thereby arouse public sentiment against increased wages at home. If so, the move was rather inept. Its advertisement pointed out that American steelworkers earned four times as much as European counterparts and seven times the Japanese rate. There was a picture of a Russian worker named Vladimir Pekov, and to it was appended in bold type the question: "Will he take away an American steelworker's job?" As the USWA International Affairs Department triumphantly commented: "The Russian steelworker has as much to do with the case as the flowers that bloom in the spring. Not a pound of Russian steel products has been imported for as far back as any Madison Avenue idea man can remember."[18] A national strike began on 15 July 1959, and 500,000 workers withdrew their labor, shutting down 85 percent of steel capacity. After 116 days, in which ancillary activities such as railroads and major steel consuming industries, notably automobile manufacture, were seriously affected, the Eisenhower administration brought in a Taft-Hartley injunction that sent the men back to work so that further negotiations could be carried out. At last, after two days of continuous bargaining in Washington, the dispute was settled on 4 January 1960.

As soon as the strike was over, in the course of a television interview

Blough made some interesting comments on the situation. He pointed out that over the last twenty years employment costs had on average increased by 8 percent a year and shipments per man-hour by only 2 percent. The average annual steel price rise of 5.5 percent was attributed to these factors and encouraged US Steel's customers to turn to substitute materials or foreign steel. Next he reaffirmed the importance of high labor costs before moving on to another field in which circumstances had changed: "But how can these products—shipped thousands of miles across the seas—undersell our own right here in the shadow of America's steel mills? Basically the answer lies in the fact that the wages of American steelworkers are from three to seven times as much as steelworkers get abroad. . . . Now until a few years ago, American producers were able to meet this handicap because our modern mills and methods gave us a great technological advantage. But today, many of the worn-out—and bombed out—plants abroad have been replaced with facilities just as efficient as our own. So foreign steel producers have been gaining steadily on us in the competitive race for markets both here and abroad—

TABLE 16.10
Steel capacity, production, and operating rates in the United States, US Steel, Western Europe, and Japan, 1953, 1958, and 1962

	1953	1958	1962
Steel capacity (million net tons)			
USA	120.9	144.2	154.0
USS	32.1	40.2	n.a.
Europe[a]	51.0	74.6	91.7
Japan	n.a.	20.0	36.8
Steel production (million net tons)			
USA	111.6	85.3	98.3
USS	35.8	23.8	25.4
Europe[a]	43.6	63.9	80.4
Japan	8.4	13.4	30.4
Operating rate			
USA	92.3%	59.2%	63.8%
USS	98.4%	59.2%	n.a.
Europe[a]	80.7%	85.7%	87.7%
Japan	n.a.	67.0%	82.6%

Sources: U.S. Congress, *Hearings before the Joint Economic Committee on Steel Prices, Unit Costs, Profits, and Foreign Competition* (Washington, D.C.: GPO, 1963), 339; US Steel annual reports for 1953, 1958, and 1962.
[a] The six original members of the European Coal and Steel Community: Belgium, the Netherlands, Luxembourg, France, West Germany, and Italy.

especially the past two years." According to his reckoning, the decline in exports and increasing imports had caused a loss of production and around 52,000 steel jobs.[19]

Between the early 1950s and the beginning of the 1960s, the steel industry generally and US Steel specifically made huge investments and very large extensions to capacity. As gradually became obvious, there was not the same sustained growth in domestic demand to maintain this expansion, for the economic structure of the nation was already changing in directions that meant increasing prosperity did not require a commensurate growth in the use of steel. Moreover, the home industry was falling behind both the expansion and the performance of operators in other major industrial regions of the world. These changes were now little more than irritants, puzzles, or potential problems. Eventually, though, they would bring a bitter harvest for the American industry and US Steel. Already, in important respects other leading companies seemed to be outpacing US Steel.

17

A Time of Transition, the 1960s

The seeds of future problems for the American steel industry were sown in the choices made during the expansion of the 1950s; the harvest was reaped in operating losses, plant closures, and the lay-off of tens of thousands of workers through the 1970s and 1980s. Between these two periods were approximately ten years of fairly high-level production in the midst of increasing uncertainty. From 1950 to 1959, national steel capacity increased by a remarkable 46 million tons. In tonnage terms, the United States industry still dominated the world of steel at the end of the decade. However, as may be seen in retrospect, in technology and its competitive advantage, as well as in the balance of trade in steel, the United States was already losing out to other developed nations, though its steel companies were generally not willing to admit this. A man who had been employed in US Steel subsidiaries for a number of years and was later to occupy very senior positions was working for the Corporation in Europe in the early 1960s. His position gave him the opportunity to visit many works of the major steel companies. This experience made him realize that not only the lower labor costs of these foreign producers but also in many cases their superior equipment meant that the American industry and his own company would prove uncompetitive in many lines. A sector like the rod-and-wire business of American Steel and Wire (AS&W), which added relatively little value in manufacture and was recognized as inadequately integrated and poorly located, was especially vulnerable, indeed, in many ways indefensible. Those who were then in command at US Steel, though, would not listen. The Corporation would have to struggle on for many more years before all the operations of AS&W were eventually abandoned.

231

During this decade the difficulties of the American industry gradually became more obvious. US Steel output held up well, but its profitability shrank. During the two halves of the 1950s, annual shipments averaged 23.3 and 20.2 million net tons respectively and yearly income as a percentage of sales were 6.12 percent and 8.32 percent. Average shipments during the two halves of the 1960s were 19.5 and 21.5 million tons respectively and income as a percentage of sales amounted to 5.58 percent and 4.56 percent.

The threat from overseas, particularly from Japan, increased in 1960–1964. In the other advanced economies of the world, producers were carried along by rapidly increasing home demand, the result of a later start than that of the United States, into the age of mass consumption. These firms were often helped by their governments, whereas the American industry was frequently harassed by federal administrations anxious to ensure that steel prices were kept as low as possible. By 1960 the Japanese were poised for an expansion that took their production from 24 million to over 100 million net tons within the decade; in the same period United States output grew slower, from 99 million tons in 1960 to a high of 141 million tons in 1969. Much of this increase was by fuller utilization of existing plant; it took seventeen years for American capacity to increase 10 million tons. In such circumstances new technologies could only come in to the American industry as replacement for old plant when the latter was scrapped, not as part of a massive overall expansion such as happened often elsewhere in the world. Steel consumption was still rising steeply, but an increasing part of it was supplied by imports. From 1964 to 1968 alone, the foreign share of total steel supply went up from 8 percent to 17 percent. After 1968 even consumption ceased to rise. In short, the large expansion of the postwar period was coming to an end, international competition was keener, and the domestic industry seemed incapable of stemming the tide of foreign material. Such conditions led to other problems: low levels of plant utilization and the conundrum that, though new lower-cost technologies were essential to enable the industry to compete on equal terms, it was difficult to finance them either from earnings or by attracting new capital. The full consequences of inadequate modernization would be exposed during the 1970s.

There was much complaint of the "unfairness" of foreign competition, but in fact, for the first time for many decades, foreign companies were now operating plant that was equal and increasingly superior to that of the U.S. industry not only in technology but also in scale, organizational efficiency, and location in relation to the contemporary world logistics of mineral supply and steel marketing—therefore also efficient in overall costs. In 1950, 8 percent of the world's output of steel products was shipped across national boundaries, by 1960 this was up to 11 percent, and seven years later it had grown to 17 percent. Between 1960 and 1969, the average freight charge for iron ore delivered from all over the world to Japan fell from $5.90 a ton to $3.65 per ton, even though the length of haul increased from 4,900 to 6,240 miles.[1] Magnificent and indeed quite exceptional though it had been in its heyday, the unpleasant fact was that the American model of supply and production, predominantly between the Great Lakes ore ranges and the Appalachian coal plateau, was being superseded by world-scale raw-material procurement, iron and steel manufacturing, and global marketing. When, year-by-year, adverse balances of trade in steel made the painful recognition of this unavoidable, industry spokesmen commonly fell for the easy option of explaining the situation as due to the fact that foreign steel operations were sometimes owned by, and even more often subsidized by, government. Much play continued to be made of the argument that United States help in rehabilitating European and Far Eastern industry after World War II in addition to the government's later technical assistance programs had played a large part in further building up this newly effective competition. There was, of course, a good deal of cogency in these points, though those who put them forward often seemed to have forgotten that their own industry's history and past huge successes had been helped by generations of tariff protection. It now also became clear that in the early post–World War II years, American steel capacity

TABLE 17.1

US Steel shipments of steel, income, and the distribution of income as dividends or reinvestment, 1957 and 1961–1970

Year	Shipments (million net tons)	Income as portion of sales	Distribution of income	
			Dividends	Reinvested in business
1957	23.4	9.5%	44.5%	55.5%
1961	16.8	5.7%	98.6%	1.4%
1962	17.8	4.7%	98.0%	1.9%
1963	18.9	5.6%	65.5%	34.4%
1964	21.2	5.7%	56.4%	43.6%
1965	22.5	6.2%	48.5%	51.5%
1966	21.6	5.6%	47.8%	52.2%
1967	19.8	4.2%	75.3%	24.7%
1968	22.5	5.5%	51.2%	48.8%
1969	22.4	4.5%	59.8%	40.2%
1970	21.0	3.0%	88.1%	11.9%

Sources: US Steel annual reports for 1957 and 1961–1970.

had been expanded, partly in response to government urgings, far beyond the levels required in the lower peacetime demand at the end of the 1950s and in the uncertain trading conditions of the 1960s. More unfortunate still, those extensions had been mainly in a steel-making technology that was being proved obsolescent but had not yet been amortized.

There was a serious development conundrum: the domestic industry needed higher prices in order to finance modernization, but if prices were raised foreign steel would be attracted at rates that, though aligned on U.S. levels, would naturally be set a little lower. There were also the dangers that price rises would alienate consumers and that, as previously in times of lowered consumption, some producers would choose to cut their prices and go for the business, knowing this would give them high plant-utilization rates and lower unit costs. An important constraint was that exercised by government. While still very much concerned about "administered prices" and restraint of trade, many officials apparently failed to recognize that, in the new globalized production and trading conditions, a considerable degree of cooperative rather than confrontational involvement of national government and industry was essential if domestic producers were to meet challenges from overseas competitors with any real hope of success. The most publicized example of industry-government confrontation occurred in 1962 when, led as so often in the past by US Steel, steel companies tried to raise prices. The Kennedy administration was aggressively critical. Having held the line for a time, one or two companies lowered prices, US Steel itself gave way within a week, and the rest of the industry followed.

Gradually, the foreign-trade figures became more unfavorable. During the 1959 strike, overseas producers had not only made up the shortfall in domestic supply, but they also naturally had done so only through business contracts that gave them more than merely stop-gap status with consumers. Afterward the net inward flow of foreign steel occasionally decreased but never disappeared. Consumption followed an uncertain course, but steel imports rose each year from 1961 to 1968; by the latter they were not far short of six times the level of 1961. Over the same period American production increased 34 percent. The consequences many be illustrated by the cases of tubular products and wire rods. Large supplies of German butt-weld pipe were now undercutting National Tube delivered prices in Houston by 25 percent. Around Los Angeles, foreign pipe had taken as much as 75 percent of the local market. If there had been no imports, it was reckoned the operating rate at the National Tube butt-weld pipe mills would have been 30 percent higher. Rod imports in 1964 reached 39 percent of domestic supply, and US Steel, followed by others, had to cut home prices by twenty dollars a ton to compete.[2] When it made huge extensions to capacity in the 1950s, the industry's long-term forward projections had anticipated that this expansion would continue. In fact, in reaction to changes in demand, capacity, having first increased more than was projected, now increased much less, and production too lagged behind projected levels. This was scarcely a favorable setting for the large-scale introduction of new processes to the industry; yet they were essential to enable steel makers to meet the rising tide of imports.

It is true that during the 1960s the industry still made large investments, but except for the Burns Harbor plant of Bethlehem Steel and a few major new finishing mills such as National Steel's Portage works or Jones and Laughlin at Hennepin, most expenditures by integrated companies went into existing plants. Considerable operating economies resulted but the cost was high in

TABLE 17.2

Projections of national raw-steel capacity and production made in 1956 and actual tonnages (million net tons)

Year	Capacity		Production	
	Forecast in 1956	Actual	Forecast in 1956	Actual
1956	128	128	115	115
1960	144	148[a]	130	99
1965	162	n.a.	146	131
1970	176	n.a.	158	131
1975	186	153	167	117

Sources: *Iron Age*, 29 Nov 1956, 77; AISI statistical reports.
[a] Figure for 1959.

TABLE 17.3

United States consumption, shipments, and net imports of steel-mill products, 1955–1975 (thousand net tons)

Year	Consumption	Shipments	Net imports	Net imports as proportion of shipments
1955	81,629	84,717	−3,088	n.a.
1960	71,531	71,149	382	0.54%
1965	100,553	92,666	7,887	8.51%
1970	97,100	90,798	6,302	6.94%
1975	89,016	79,957	11,631	14.54%

Source: U.S. Federal Trade Commission, *The United States Steel Industry and Its International Rivals* (Washington, D.C: GPO, 1977), 70.

capital terms. In Japan, and to a considerable extent in Europe too, a much higher proportion of the expansion was in new "greenfield" plants using state of the art technology and located at tidewater, where they could benefit from the lowest cost access possible to the world's highest-grade mineral deposits. Mineral-producing countries were often in keen competition with each other, which kept raw-material prices down. By contrast, the "captive" minerals of the Great Lakes and the Appalachian plateau, once such a source of strength, now began to seem more a burden to domestic steel makers. As one later critic summed up, "expenditures went to replace obsolete facilities, to tack advanced rolling and finishing facilities onto old plants, and, especially toward the decade's end, to install costly environmental protection equipment." Within a few years there were unfortunate consequences: "[T]he integrated U.S. producers suffer from significantly outmoded capacity, poor location and layout, mismatched products and markets, and significant technological backwardness in energy productivity, resource utilization, and finishing capability."[3]

Once again, US Steel generally performed less well than its leading rivals. In 1950, as the Korean War–inspired expansion of capacity began, the Corporation controlled 31.7 percent of the nation's raw-steel capacity. Between that date and 1959, though it built the only new integrated works of the decade, capacity had increased only 8.2 million net tons, or by 17.4 percent of the national total. In the early 1960s, the dispute over pricing with the Kennedy administration caused the industry, in a rather petty reaction, to give up publication of capacity figures, but it seems that the relative decline at US Steel continued. Certainly in terms of production, it fell behind almost, if not quite, annually. In the good years of 1955–1957, US Steel had averaged 29.6 percent of national steel output. During the 1960s, despite short-term positive fluctuations, its share was considerably below this level. The years 1965–1969 were again good ones, but the Corporation's raw-steel output averaged only 24.5 percent of the total. It also epitomized the problem of not earning sufficient profits to meet both the need for new investments and other calls on revenues. Understandably, a good deal of what was earned continued to be distributed to stockholders, but at US Steel the proportion rose over the years as the ratio of income to sales became less favorable. This was particularly striking in the first two years of this decade. Income during 1962 was 39 percent of the record level of 1957, but the amount distributed as dividends was 86 percent as large.

Why did US Steel perform worse than the industry as a whole? One factor was undoubtedly that, even after all the corporate reforms, it remained too big and rambling for the most effective management. Another was the continu-

TABLE 17.4

US Steel and some leading rival companies, 1957 and 1963

Company	Sales ($millions)			Operating income ($millions)			Earnings per share ($)			1963 return on
	1963	1957	change	1963	1957	change	1963	1957	change	book value
US Steel	3,599	4,379	− 17.8%	674	1,073	− 37.0%	$3.30	$7.33	− 55.0%	5.7%
Bethlehem	2,096	2,604	− 19.5%	343	463	− 26.0%	$2.11	$4.13	− 48.9%	5.8%
Republic	1,114	1,227	− 9.2%	154	210	− 26.6%	$3.52	$5.45	− 35.4%	7.3%
Armco	933	1,074	− 13.1%	150	165	− 9.6%	$4.12	$4.71	− 12.5%	8.0%
National	846	641	+ 32.1%	184	133	+ 38.7%	$4.12	$3.07	+ 34.2%	11.0%
J&L	836	838	− 0.02%	145	131	+ 11.2%	$5.45	$5.65	− 3.5%	7.6%
Inland	808	764	+ 5.8%	165	142	+ 16.1%	$3.11	$3.45	− 9.9%	9.6%
YS&T	627	680	− 7.8%	104	117	- 11.1%	$3.70	$4.12	− 10.2%	6.9%

Based on: Forbes, 1 August 1964.

ing problem with its product line. Despite the efforts under Myron Taylor three decades earlier, once more the Corporation had to face the fact that too much of its capability remained in heavy-steel products, whereas growth was going ahead more rapidly in lighter lines. In 1962, 20 percent of capacity was in thin flat-rolled products as compared with 60 percent at Inland Steel.[4] Five years later it was pointed out that about 60 percent of US Steel capacity was in products that had accounted for only about 40 percent of 1964–1966 domestic steel shipments. Slightly less than 37 percent was in sheet and strip, which made up at least 40 percent of the industry's total. However, the new eighty-four-inch strip mill then being installed at Gary was expected to increase US Steel's share in these grades to 45 percent of its total.[5]

Corporate-development planners obviously assessed the prospects of the districts within which US Steel operated quite variously, but investments made in oxygen capacity indicated the favored areas. Of the first six installations, two were in Pittsburgh, three near Chicago, and one along Lake Erie. Arguably the recent construction of the Fairless works, including its open-hearth department, ruled it out as a priority for the new technology, but the slower than expected growth of metal-using industries in its natural market area and a generally weak performance by the eastern seaboard in the national economy must also have been important considerations. By 1969 Fairless's annual capacity was 4 million tons. In the early and mid-1950s, western markets had grown much more quickly than the national average, and US Steel had responded with expansion at Geneva. However, before that decade ended, Geneva was surpassed as the region's largest plant by Fontana,

TABLE 17.5
**United States and US Steel production of raw steel,
1955–1957 and 1960–1970 (million net tons)**

Year	USA	USS	USS share of USA
1955–57	avr. 115.0	34.1	29.6%
1960	99.3	27.3	27.5%
1961	98.0	25.2	25.7%
1962	98.3	25.4	25.8%
1963	109.3	27.6	25.3%
1964	127.1	32.4	25.5%
1965	131.5	32.6	24.8%
1966	134.1	32.8	24.5%
1967	127.2	30.9	24.3%
1968	131.5	32.4	24.6%
1969	141.3	34.7	24.6%
1970	131.5	31.4	23.9%

Sources: AISI annual statistical reports; US Steel annual reports.

and it remained wholly dependent on open-hearth furnaces. In the 1960s, as imports into the West Coast increased even more rapidly than in the nation as a whole, US Steel tried to improve Geneva along well-established lines. The three blast furnaces were rebuilt with larger hearths, and between 1957 and 1964 its ten open hearths were upgraded from an average capacity of 265 to 340 net tons per heat. Additions made in 1962 enabled it to produce coils of heavier gauge and greater width. In 1965 Geneva reached what proved to be its highest employment with 6,100 men. After that, major improvements stopped. For a time there was vague talk of a new US Steel integrated works at tidewater near San Francisco Bay, but no action followed. Early in 1964 the US Steel Executive Committee resolved to abandon the open-hearth and primary-mill operations at Pittsburg, California, making the rolling mills there completely dependent on Geneva.

Southern states received relatively little expansion at this time, though the region's consumption of steel was growing rapidly as it went through a rapid industrialization. There was a strong move to improve the quality of the local steel product, and this brought an end to its traditional advantage of low-cost local ore. Even after preparation, the local ore contained only 37 percent iron, was high in phosphorus, and variable in quality. By 1958 officials decided to replace a mixed local red ore–imported ore burden for Birmingham-area blast furnaces with the exclusive use of imported ore. In June 1962 the Red Mountain ore mines were closed. Imports of ore into the Mobile, Alabama, customs district were 2.4 million tons in 1960 and twice that by 1970. Low levels of demand for rails meant that little growth occurred at Ensley, and its mill was sometimes idle. At the end of the decade money was spent on rail-mill improvement. Fairfield, under whose name both plants were managed since the early 1960s, retained much old plant; its plate mill had been built to meet the pressing needs of World War I. But the existence of the hot-strip mill installed in the late 1930s brought expansion of capacity. By the late 1960s it was equipped with a new continuous galvanizing line. The decision to build a wide-flange beam mill and to replace open hearths with oxygen converters proved this plant had been picked out as a future growth point.

At this time the South Chicago operations provided an interesting instance of the way in which political power, even at the local level, could have an important influence on company development policy. Sixty years before, Illinois reluctance and therefore delay in allowing US Steel to reclaim land from Lake Michigan seems to have been one factor inducing the Corporation to start work instead on the wholly new Gary site, which was subject to Indiana regulations.[6] Now there was something of a repeat performance. South works was by 1960 a giant of 5.6-million-tons steel capacity, the fifth largest in the nation, but its site was crowded and the plant layout inconvenient. Already it was occasionally referred to as a marginal operation. In spring 1963 an application was made and approval given to reclaim 194 acres of land from Lake Michigan, providing an addition of over one-third to the plant's area. But by 1965 the Illinois Supreme Court had ruled it was unconstitutional for US Steel to acquire the area for reclamation at a mere $100 an acre. The following January, when the Corporation announced it would build three oxygen converters on the site of the Number 2 open-hearth plant, it had still not received the go-ahead for the site extension. In March 1966 the state supreme court at last gave its approval, and in summer 1967 US Steel confirmed it would build a new thirty-two-foot-diameter blast furnace and was building the BOP (basic oxygen processing) shop. Shortly afterward, a very different attitude from local authorities was shown in the construction of an oxygen steel plant at Lorain, Ohio. The County of Lorain issued an industrial-revenue bond so as to provide the $80 million needed to build this plant, US Steel leasing the facilities and at the end of the lease term to acquire them.[7] Comparison of South works, Gary, and Lorain in the early 1970s seems at first sight to indicate that the first had grown the most. But whereas the last two were now wholly oxygen steel plants, 54 percent of South Works capacity remained open hearth. In 1960, when all three were open-hearth dependent, South works had 34 percent of their joint capacity; thirteen years later its oxygen and electric steel capacity was only 23 percent of their combined totals in these processes; its surviving 4 million tons of open-hearth capacity was now considered marginal plant. It would not be many more years before the whole future of this works would be a matter of growing concern. Three years' delay by Illinois at a critical time in plant development can only have represented an extra adverse factor in South works's long-term competitiveness.

Blough would be sixty-five years old at the beginning of 1969, and in the middle years of the decade speculation began as to who his successor would be. In 1966 fifty-two-year-old R. Heath Larry was made executive vice president, which was taken as an indication that he would succeed to the chairmanship. Like Blough, Larry was a lawyer by training and had been long involved in the Corporation's labor negotiations—in which to later leading officers he seemed to have given too much away in cost-of-living bonuses. Blough and the surviving Morgan influence in US Steel favored Larry as chairman, but the board of directors chose to turn in a different direction. In 1967 Edwin H. Gott was appointed US Steel president, and when Blough retired, he rather than Larry was elected as chairman and chief executive officer. More completely than Fairless—actually more so than any effective head of US Steel since the days of Schwab and Corey—Gott had come up through the steel rather than the law or financial side of the business. Trained as an industrial engineer, from the late 1930s he had experience at the Clairton, Gary, South Chicago, and Youngstown works. From 1956 he occupied a leading position in production control over the whole of US Steel operations. His appointment seemed to promise something rather more vigorous than the policies of Blough's chairmanship, which were rather harshly summed up as the time for the transfer of power approached: "Blough sought to improve US Steel's fortunes largely by paring its work force, consolidating its sprawling divisions, and ending a costly overlap of sales offices. More recently he has loosened the purse strings in a somewhat belated effort to renew the company's plants." Unfortunately, whereas Blough had occupied the chief post for fourteen years, Gott was already almost sixty-one when elected and would have only four years to make his mark on US Steel. He managed to make some important changes, nonetheless. Until this time the chief executive officer had operated from New York headquarters. One condition insisted upon by Gott before he would accept the top executive duties was that he should be based in Pittsburgh, where a major new US Steel headquarters building had been opened the previous year. (Even so, regular meetings of the board of directors were still held in New York, even for a number of years after 71 Broadway was sold in the mid-1970s.) His other condition was that the chairman of the finance committee, which met a week before the directors, should report directly to

Gott and not through the board as in the past. Robert Tyson, who had already chaired the finance committee for thirteen years, had sometimes worked on lines not fully coordinated with his chairman, including a reluctance to spend on major capital projects. Tyson had taken advantage of the old reporting arrangements to influence the board with his own ideas and in this way to some extent undermining the position of Blough. Gott as vice president in charge of production and later as president had experienced necessary plant modernization being impaired or delayed, and since his own election to the board in August 1966 had seen how this sidestepping of the chairman had happened. Now that Gott was chairman and chief executive, Tyson, during his own last year in office, was forced to make regular journeys to communicate with him in Pittsburgh.

In the 1960s and the first years of the 1970s, the American steel industry lost its world leadership, and US Steel began to seem a secondary factor in the U.S. economy. Even in 1960 America's 26-percent share of world steel production had been lower than at any time since the early 1870s. By 1970 it was down to 20.1 percent. Over the decade, US Steel's shipments were up by 12.3 percent and sales by 32 percent, but income was only 47.5 percent as large and income as a percent of sales had fallen from 8.2 percent to 3 percent. In 1970 a merger of the Yawata and Fuji companies produced Nippon Steel, and after sixty-nine years the United States Steel Corporation was displaced from the top rank of the world's steel companies. In terms of assets, it was eleventh in size among all American companies with a total less than one-third that of Standard Oil of New Jersey. Such changes in rank were symbolic. Though still a major factor in world steel and unrivalled in size in the domestic U.S. industry, now more clearly than ever US Steel was an ailing giant leading a sickening industry. The 1970s were to bring crises of unprecedented severity.

Part Four

Decline, Reconstruction, and Prospects

18.

Response to a Technological Revolution

The Large Blast Furnace, Oxygen Steel Making, and Continuous Casting

During the course of the 1950s, the United States began to fall behind the best world steel manufacturing practices in a number of respects. Much of this was for very practical reasons, but the long-term implications were serious. Into the middle of that decade, the United States had the biggest blast furnaces, but after that others pushed ahead. In steel technology and the early stages of further processing, the nation's steel makers were also falling behind the improved, indeed ultimately revolutionary, processes first introduced elsewhere. America had led the noncommunist world in expansion of capacity for many decades. Now the industry outside the United States would grow much more than that within it. In such circumstances it was natural that the overseas industry would be equipped with the best contemporary plant. If this happened, the apparent technological lead of the United States would narrow and might be lost altogether. This was one major reason for the increasing technical lag of the American industry; the other was the legacy of complacency from years of unrivalled leadership. During that time, which with the benefits of hindsight may be seen as critical, parochialism was so complete that even a fairly assiduous reader in the United States of the commercial iron and steel press might well have been excused for not realizing that there were important producers elsewhere and that steel men in those other parts of the world were not complacent in nor content with their secondary roles in the industry. US Steel, as the unquestioned leader of the industry, was especially delinquent and lax in its approach toward technological innovation.

In the postwar decades, technological changes were particularly important in two vital fields, in which subsequent events proved the American ap-

TABLE 18.1

Raw-steel production in the United States, by US Steel, and in Japan, the EEC, and the world, 1929, 1955, and 1975 (million net tons)

Producer	1929	1955	1975
United States	61.3	117	116.6
US Steel	24.5	35.3	26.4
Japan	0.7	10.4	113.8
EEC[a]	51	80.2	137.4
World	125	297.2	712

[a] EEC figure comprises production of West Germany, France, Italy, United Kingdom, Belgium, Luxemburg, and the Netherlands, though the 1929 figure includes the whole of Germany and excludes the Netherlands.

pearance of invulnerability had been most deceiving, iron making and bulk-steel production. In 1954–1955, excluding those few units that concentrated on ferromanganese, the seventy-nine blast furnaces at US Steel plants were estimated to have a combined iron-making capacity of 27.7 million net tons, a tonnage itself far greater than that of any state in the world except the Soviet Union. The average annual capacity per US Steel blast furnace was about 350,000 tons. Its most recently built works contained larger units: 400,000 tons per furnace year at Geneva and only a little short of 570,000 tons at Fairless. Although for so long it had seemed inconsequential to the iron-making industry, Japan within only a few years made this furnace record look antiquated.

In 1954–1955 Fuji and Yawata Steel, which fifteen years later recombined as Nippon Steel, together had four integrated works and twenty-two blast furnaces. Jointly, they could produce 5.88 million net tons of pig iron a year, 21 percent of US Steel tonnage; average furnace capacity was a little over three-quarters that of Corporation furnaces. At the largest works, Yawata, the average size was only 179,000 tons; the largest furnace, at Hirohata, was of 358,000 tons. In the course of nearly twenty years, US Steel closed three iron-making plants—Donora, Duluth, and the small operation at Ironton, Utah. It reduced the number of stacks at almost all other works and built a number of big, new blast furnaces, notably Furnace 13 at Gary, a large unit at South Chicago, and the "Dorothy" Furnace at Duquesne, Pennsylvania. By 1974 US Steel controlled sixty-seven furnaces with a total annual capacity of about 31 million net tons and a furnace average of about 465,000 tons. Over the same period, furnace numbers at Nippon Steel increased only from twenty-two to twenty-five, but by the mid-1970s their annual iron capacity exceeded 45 million tons. Average capacity per furnace year was now 1,814,000 tons, almost four times the US Steel figure.

This extraordinary achievement was in large measure a result of the wholesale replacement of old plant in long-established works by new units—for instance, the six furnaces at Yawata now averaged 1,583,000 net tons annual capacity. Another favorable factor was the construction of greenfield works, providing an opportunity for implementation of new designs and construction. Since the commissioning of Fairless, only one other greenfield works had been constructed in the United States, the Burns Harbor plant of Bethlehem Steel; during that same time the Japanese built at least nine—Chiba, Mizushima, Fukayama, Kashima, Kakagowa, and four additional Nippon Steel works at Nagoya, Sakai, Kimitsu, and Oita. The Japanese built new works of necessity to meet rapidly increasing demand. It was economic to do so because the investment required for construction on a virgin site in Japan was considerably less than in America. (In 1967 Edwin Gott reckoned capital outlay on a new integrated works in the United States would range upward from $300 per ton capacity, whereas in Japan a similar works would cost only $115–127 a ton. Unfortunately, he did not provide evidence to back this cost comparison.)[1] Japanese use of the highest-grade furnace burdens, procured from ore and coking coal deposits spread widely throughout the world, was another strong asset for their furnace operators. Perhaps the most important factor was that the great surge in levels of steel production justified the construction of very large units of plant containing the best furnace technology. National contrasts in hearth diameter, a traditional measure of furnace size, are striking. By the mid-1970s, one of the furnaces at Fairfield had a hearth diameter of 345 inches, the largest in the Pittsburgh area was then one of the Carrie furnaces at 354 inches, Fairless had two 370-inch stacks, and South Chicago one of 384 inches; the only US Steel giant furnace was Gary Number 13 with a hearth diameter of 480 inches. In contrast, seven furnaces at five Nippon works averaged 429 inches in diameter. Kimitsu had one blast furnace of 527 inches and the old, but greatly extended and thoroughly overhauled, Yawata works complex contained a 531-inch-diameter furnace. More important even than size was superior operating experience.

In Western Europe too the 1950s and 1960s were decades both of great expansion and of building greenfield works at tidewater locations providing easy access to the world's richest mineral resources. However, as in America, the European industry was mature, with established patterns of training and procedure and therefore not so readily open to new lines of thought as in Japan. There were new coastal works at Cornigliano and Taranto in Italy, at Bremen in West Germany, in France at Dunkerque and at the end of the period at Fos, at Selzate in Belgium, and at Newport in South Wales. Many older works carried through major reconstructions. For example, in its Rhine-front complex of Ruhr district works, Thyssen had by 1974 an iron-making capacity of 15.2 million net tons. Most units there were of modest size—the seven furnaces at Hamborn averaged only 325 inches in diameter—but the largest furnace, the 551-inch diameter Schwelgern furnace, whose rated annual capacity was 3.76 million net tons, was far bigger than anything at US Steel.[2]

For long an apparent reflection of the continuing influence of US Steel's past preeminence was its reluctance to take up foreign technology, especially from Japan. This proved a costly prejudice, for those who did not share it gained from their open-mindedness. For instance, in 1979, when Inland Steel completed a new blast furnace, it was reckoned that this unit would cut iron costs by at least three dollars a ton. Inland drew on Nippon Kokan (NKK) for the technology. The reason for the superiority of the Japanese design was by no means obscure. It was largely attributable to that nation's greater experience with very large modern furnaces. Inland had built only one new furnace since World War II, but NKK, expanding from a much lower base, had put up thirteen, each bigger and more productive than the last. By 1977 over 51 percent of Japanese blast furnaces had an inner volume in excess of 2,000 cubic meters as compared with only 2.6 percent of American furnaces. For some time US Steel ignored the significance of this greater experience, but eventually it too decided to draw on the abilities of Japanese engineers. Having fortunately moved on from its 1950s assumption that the Japanese could not know anything that American engineers did not, when difficulties arose with the forty-foot Gary furnace, officials decided to consult the Japanese. This particular case of technology transfer left the Americans amazed by how much their leading international rivals now knew. However, it must be stressed that size and tonnage, for so many decades taken as leading indices of American success, were as always no adequate measure of its extent. As an industry expert who experienced the whole of this learning period much later put it, "The large blast furnace has been touted as a visible symbol of Japanese supremacy, but steel companies concerned with profitability rather than symbolic supremacy have devised much more cost effective ways to make iron." More recent American practice with much smaller furnaces than the Japanese giants, or even the United States' own biggest stacks such as those at Gary and Indiana Harbor, has proved that iron can be made with very little higher processing costs, much lower capital outlay, and more flexibility with less in the way of hostages to fortune.[3]

In iron making, the United States industry was overtaken by overseas producers who, for one reason or another, were installing bigger, more productive blast furnaces. Their technology was essentially the same in type but just better and bigger. In steel making there was a much more dramatic change in direction, and the American industry as a whole, and certainly US Steel, lingered too long with older open-hearth-processing methods before moving over to what was by then unquestionably recognized as a superior process for bulk grades, the oxygen converter.

The basic oxygen top-blown converter was introduced in Austria in 1952 as part of an attempt to deal with a difficult raw-material supply situation. In 1953 two steel works in that country, equipped with this new process, produced 370,000 net tons of steel—a mere 0.33 percent of the record tonnage made that year in the United States. Next year the technical director of Oster-reichisch-Alpine Montangesellschaft, the firm that had pioneered the new process, visited the United States. In an address to the Metals Branch of the American Institute of Mining Engineers, he expressed the opinion that "The oxygen steelmaking process will be applied chiefly in the production of mass steels" and suggested his company's product was particularly well suited to the needs of that key sector of the consumer durable trades, the automotive industry: "Because of the favorable conditions under which rimming steels of particularly low carbon contents can be produced by the oxygen steelmaking process, the opportunity they offer for production of quality steel sheets and strip is obvious, especially for deep drawing applications."[4] Over a very few years his expectations were fully realized. In fact, it can now be seen that, almost a hundred years after Bessemer's announcement of the process that transformed the nineteenth century metallurgical world, here was another revolutionary innovation. Bessemer's process had been central to the transfer of leadership from the Old World to the New; the technology now introduced was to do more than anything else to undermine the apparently secure pre-eminence of the American industry. It did so, though, only after being converted into a tonnage process. In that achievement the American industry had an important part, but US Steel was not an important player.

For some years after its introduction, contemporary observers considered

TABLE 18.2
The basic oxygen process in the United States, Japan, and the EEC
(nine members), 1955–1975 (share of regional production)

Year	USA	Japan	EEC
1955	0.3%	n.a.	n.a.
1960	3.4%	11.9%	1.6%
1965	17.4%	55.0%	19.4%
1970	48.1%	79.1%	42.9%
1975	61.6%	82.5%	63.3%

that the new process was unlikely to upset the great structure of big, well-established integrated producers using open-hearth furnaces. Through the mid-1950s large extensions of open-hearth capacity were underway.[5] Retrospectively, this can be seen as a period of unique but lost opportunity for American producers to get established early in the new technology. Between 1950 and 1959, national raw-steel capacity increased 46 million tons: expansion over the following seventeen years would be no more than 10 million tons.[6] Yet in the 1950s, after Fairless works was completed, American companies built no new integrated works, and most of their increased capacity came from extending existing melting shops. Some of these investment programs were legacies of the Korean War, but it is anachronistic to claim, as historian Paul Tiffany has done, that government pressure to expand capacity quickly to meet the needs of this war partly explained why companies in the United States lagged with oxygen steel. On the other hand, a perfectly understandable desire to avoid disruption of firmly established and profitable routines is always a strong deterrent for those who might be tempted to embark on uncharted ways. More generally, the range of explanations of the scant attention paid to the new process has been wide, ranging from fears about pollution to a lack of incentive to experiment as a result of the comfortable administered-prices regime under which the industry operated.[7] By contrast, other nations' industries expanding at this time, particularly that of Japan, having relatively much less capital tied up in old technologies, were ready to move forward with new ones. In 1958 basic oxygen steel made up only 1.55 percent of United States output. *Fortune* magazine was perhaps rather too sweeping when, looking back, it wrote that in the 1950s the United States steel industry installed approximately 40 million tons of melting capacity that was "obsolete when it was built," but there was a strong thread of truth in its hyperbole. No new open-hearth furnaces were built in the United States after 1958, and by 1960 only 3.4 percent of U.S. production was in oxygen converters as compared with 11.55 percent in Japan; five years later the respective proportions were 17.4 percent and 55 percent. By that time the cost advantage of the new process over the old was reckoned to range from two to ten dollars a ton.[8]

In the early years, spokesmen for the industry denied that the new process was advantageous or, even more, perversely claimed they were leading in it. A prime example of the latter position, and one that owed nothing to company loyalty, involved Meyer Bernstein of the United Steelworkers of America (USWA). Referring to oxygen steel making in spring 1963 before the Douglas Committee on Steel Prices, Costs, and Competition, Bernstein claimed, "American engineers and the American steel industry have shown a great deal more imagination than the Europeans in developing and perfecting this process."

He seems to have been thinking in the main of Austria as representative of Europe.[9] Further attempts have been made to exonerate the industry from any charge of culpable delay in adopting the new process. For instance, early in 1968 the *US Steel Quarterly* published a short article whose purpose was clear from its title, "Delinquency? Hardly: America Steps Up Oxygen Steelmaking Sharply." In its attempt to refute the hypothesis that the industry had been a slow adopter, the writers were extremely careful in their choice of dates and of those regions included among "other" producers—including growing communist output as well as that of the Third World—and carefully avoiding comparisons with leading Western producers. As industrial propaganda, it was a well-crafted presentation of carefully selected facts. "The American steel industry, beginning in 1964, has turned out a substantially larger part of its production by the rapid and relatively new basic oxygen process (BOP) than has the balance of the steelmaking world, and during 1967 continued its rapid installation of BOP facilities. Conversely, production of steel by the older, slower open hearth method has declined almost 20 percent since 1955 in the United States, while elsewhere in the world it has increased by more than 105 percent."[10] In fact, the oxygen steel process was already sixteen years old and the United States industry had held on to an "older, slower . . . method" longer than many other leading steel makers. Almost a decade later, by which time the desirability of the oxygen process had long been universally acknowledged, a Federal Trade Commission study of the American industry in its world context again claimed that from 1956 to 1964, and over the next decade as well, the rate of adoption of the basic oxygen furnace in the United States had been more rapid than in any other steel-making country.[11] This conclusion was dangerously deceptive, for its basis was the tonnage of new oxygen steel capacity divided by the change in total steel capacity. By this time steel expansion in the United States was slower than in many other countries. Above all, in the share of oxygen steel in total production, it was indisputable that the United States was no longer a pacemaker.

As with the hot-strip mill, it was not the largest companies that led the way into the new steel process. In 1954, helped financially by General Motors, the then unintegrated McLouth Steel became the first company to install oxygen converters. McLouth's location and markets were eminently suitable for expansion, but though it was located in an area of large-scale availability and low prices for scrap, it had to acquire blast-furnace capacity for the first time in order to supply the molten iron needs of its new steel plant. Much more important in the development of the new industry in the United States was the second unit, built by Jones and Laughlin (J&L) at Aliquippa, Pennsylvania, in 1957. Here, a skilled operator proved that it was possible to obtain as many as thirty-three heats a day from the eighty-ton vessels. In short, this marked the point at which the oxygen process was conclusively proved capable of supporting large-scale steel operations. Four years later the experience gained at Aliquippa was applied in the new 2x230-ton oxygen steel shop installed at the J&L Cleveland works. Steel makers from throughout the nation and indeed from the world came to Cleveland to learn how to make large tonnages of oxygen steel. There too it was necessary to build a new blast furnace to permit full-capacity operation of the converters. But not only had J&L scaled up the process, they also had saved on capital outlay. Kaiser Steel, US Steel's major rival within the West, made the only other oxygen steel investment before the 1960s. Having only one plant and facing keen competition from much bigger companies including those shipping from the Atlantic Coast by sea, Kaiser had to be willing to take risks. Elsewhere opportunist entrepreneurship was important. By 1962 Armco, Colorado Fuel and Iron, Great Lakes, and Sharon Steel had also installed the new process.[12]

US Steel was a latecomer to the oxygen process. Looking back in 1967, the chairman of its finance committee, accountant Robert Tyson, claimed the quality of their open-hearth plant as a legitimate excuse: "A good open hearth shop with half its economic life can't be scrapped. You wait until you cross the economic line."[13] Though some were later to suggest that US Steel made a mistake in the 1950s to install so much new capacity in the older technology, capital-cost comparisons were by no means straightforward. At Cleveland, J&L reportedly projected that increased steel capacity could be obtained through the oxygen process for a capital outlay of only $15 per annual ingot ton as compared with at least $40 a ton for additional open-hearth tonnage. During the late 1960s at one of Bethlehem Steel's plants, two oxygen converters were installed to replace nineteen open-hearth furnaces. The labor force was reduced from 800 to 300, and analysis determined that the cost of a ton of steel was cut by $4 to $5; but the capital outlay on the new plant was $110 million.[14] For its part, the Corporation took many years to edge up to the point of decision. When in 1967 its board chairman refuted allegations that American steel companies had been "delinquent" in not adopting oxygen steel-making earlier, he was speaking of the whole industry, but his statement was also an attempt to justify US Steel's past policies. Much of his case depended on other factors, which may also be identified as cases of entrepreneurial failure, such as the suggestion that greater speed with the new process would have required more bulk supply of oxygen than was then available or that there were no refractory bricks that were sufficiently durable. Certainly, difficulties comparable with the issue of brick quality were not of a magnitude that would have been allowed to hold up previous major advances in the industry's technology.

This executive also pointed out that early oxygen converters were too small to be used to optimum effect by American producers, produced only a limited range of steels, and that since the converters needed more molten iron than open hearths, a bigger expansion program would have required costly new blast-furnace capacity, coke ovens, and iron ore. In fact, J&L had made the new process into a bulk producer almost ten years before, and as US Steel had a higher than average availability of this backup capacity, it might have been expected to be at least a national pioneer of the new technology. Perhaps the strongest argument against quicker adoption of the new process in the United States was its abundant supply of and well-organized market in scrap. The fall in scrap prices as the industry turned to oxygen steel helped foster the growth of an electric steel industry, in which again US Steel played only a small part. Conversely, the switch to oxygen steel could make fuller use of existing blast-furnace capacity or even justify the building of new high-capacity furnaces. Roger Blough pointed out that early oxygen steel operations had created serious pollution problems, but this failed to acknowledge that the open-hearth practice also often involved serious environmental degradation, though it took a different form. His conclusion was similar to that of Tyson, "to have scrapped hundreds of millions of dollars worth of good and serviceable open hearth furnaces in order to replace them with a process that was then in its relatively primitive stages, would have been actually wasteful." (It should be noted that by 1966, an authoritative source reckoned one-third of the nation's existing steel capacity was in urgent need of replacement.) In further mitigation of slow adoption of the new process, Blough pointed out that the application of oxygen to open-hearth furnaces had greatly increased the productivity of that process. Clearly he had forgotten, or chose to overlook, the lessons learned nearly forty years before about the economics of the strip mill versus what proved to be a dead end in the improvement of old pull-over or even automatic mills.[15]

Why had US Steel not learned from its earlier, painful experiences with the strip mill, or with ERW pipe, or with the Grey mill? Why had it not at least experimented with oxygen steel or even installed a pilot plant? After all, compared with any other company, the Corporation had a large number of mills, and even if the argument about the continuing viability of good-quality open-hearth furnaces was valid, the closure of works such as Donora and Clairton only a few years later indicated that US Steel already had operations marginal enough to be used to evaluate a new technology without serious interruption of the overall production program. Within another decade US Steel would be closing many open-hearth melting shops and commenting as it did so on the great age of some. Could not at least one high-cost melting shop in one of the

TABLE 18.3

United States production of steel and pig iron consumption by various processes, 1970

Process	Steel produced (th. tons)	Pig iron consumed (th. tons)	Pig/steel ratio
Oxygen process	63,330	51,730	81.7%
Open hearth	48,022	32,204	67.1%
Electric furnaces	20,162	453	2.2%

Sources: AISI annual statistical reports.

bigger works have been shut down in the mid-1950s to make way for a large-scale experiment in oxygen steel making? Certainly this had been a favorable time for US Steel financially so that the cost of such enquiries would have had minimal impact on overall performance. In 1955–1957, average annual income of $379 million was just over twice the average of the three previous years, during which $633 million had been reinvested in the business. Would it not have been wise stewardship of future prospects to put some of that into evaluation of technologies that well-proved companies in other parts of the world, and soon in the United States itself, thought significant enough to employ large scale? In short, making due allowance for the claim that open-hearth technology and the state of most of the US Steel melting shops meant this was not a time for a wholesale transfer to the new process, there seems much to suggest that the corporate attitude to oxygen steel making in the 1950s was an example of entrepreneurial failure.

Belatedly, US Steel edged toward commitment to the new process. In his chairman's address to the annual meeting of stockholders in May 1959, Blough recognized the Corporation faced "two related challenges" in the fields of competition and "research and technology." He pointed out that low wages overseas now presented an unprecedented threat to the American industry: "In the past we could meet such disparities in employment costs because of abundant capital, better tools, and methods. Today, however, foreign producers are adopting the latest and very best of equipment and methods. . . . To compete successfully with other nations in the years ahead will depend primarily upon our capacity as a nation to acquire and pay for new, more efficient tools of production."[16] The major steel strike that year was accompanied by a surge in steel imports, and the inflow did not cease when the dispute was settled. This provided a spur to innovation.

At the spring 1962 annual meeting there were at last signs of movement.

In reply to one stockholder, Blough stated, "We have plans to install the oxygen steelmaking method."[17] That August, the first US Steel commercial-scale venture into the new process was announced: two 150-ton vessels at Duquesne to replace open-hearth furnaces built in 1907. In August 1963, the executive committee decided to build an oxygen steel shop at Gary as step one of a two-stage process of replacing its Number 1 and 2 melting shops at an overall cost of over $40 million.[18] When unveiled to the public six weeks later as a replacement for steel capacity originally installed in 1917, officials stressed the new plant could make a heat of steel eight times faster than an open-hearth furnace and that it would be "fully equipped with dust recovery units operating with gas cleaning equipment, so that any air pollution problem will be precluded." By now the size of the US Steel technological lag was considerable. Oxygen steel capacity totaling 11.04 million tons had been installed in eight plants. Of an additional 21.6 million tons of scheduled capacity, the two US Steel operations made up only 4.5 million tons.[19] Next year, by which time the oxygen process accounted for 12.2 percent of national production and 43 percent at Kaiser Steel, US Steel commissioned its first unit. By 1965, 32.89 million tons of BOF capacity already existed in the United States or was expected to be completed that year. Duquesne, with 1.5 million tons capacity, was then the only operating US Steel unit, though the 3-million-ton Gary unit was being completed.[20] In short, controlling almost one-quarter of national raw-steel production, US Steel would have only 13.7 percent of the nation's oxygen steel capacity. By mid-1967, 25 percent of US Steel capacity was basic oxygen steel as compared with 38 percent for J&L, 40 percent for Republic, and 33 percent for the American industry overall.[21] In 1968 officials decided to install converters at Lorain; soon a unit was under construction at South Chicago, the next year one was scheduled for Edgar Thomson, and by 1971 a second unit was being built at Gary. When completed, the latter was expected to enable the Corporation to produce over 50 percent of its steel from oxygen converters.

The first important steps in research into continuous casting, the process of pouring molten steel directly into molds to produce a continuously moving shape that goes directly to the finishing mills, dated from the late 1940s and early 1950s. In contrast to the situation in oxygen steel making, the United States undertook much of the early work. Such a process replaces costly, space-consuming primary rolling mills, decreases the amount of scrap reduced, increases yields and productivity, and cuts energy costs. Yet, having been one of the pioneers in research, the United States was relatively slow in application. By 1969 only 2.9 percent of US crude steel was continuously cast

TABLE 18.4

Distribution of oxygen steel-making capacity in the United States, 1965, 1969, and planned by 1970 (million net tons)

Area	1965 (Sept.–Oct.)		1969 (Jan.)		Planned by 1970	
	steel	portion	steel	portion	steel	portion
Detroit	6.7	26.1%	8.0	14.8%	10.8	14.6%
Pittsburgh-Wheeling	6.0	23.2%	14.6	26.9%	14.6	19.7%
Chicago	1.9	7.3%	8.6	16.0%	17.8	24.2%
Cleveland	1.9	7.3%	4.7	8.6%	6.9	9.4%
Cincinnati	1.4	5.4%	1.4	2.6%	3.4	4.6%
Youngstown	1.5	5.8%	1.6	3.0%	1.6	2.2%
Buffalo	2.5	9.6%	4.7	8.7%	5.7	7.7%
St. Louis	n.a.	n.a.	2.2	4.1%	2.2	3.0%
Northeast Coast	n.a.	n.a.	4.3	7.9%	6.8	9.2%
South	1.5	5.8%	1.5	2.8%	1.5	2.0%
West	2.4	9.4%	2.5	4.7%	2.5	3.4%
USA	25.9		54.0		73.8	

Sources: Steel, 11 October 1965, 65; Iron and Steel Engineer; Federal Reserve Bank of Cleveland Economic Review, October 1969, 7.

as compared with 4 percent in Japan and 7.3 percent in West Germany. After that, the United States rapidly fell further behind, respective proportions in 1975 being 9.1 percent, 31.1 percent, and 24.3 percent.[22]

Again, it was smaller companies that spearheaded the use of the new process so that by 1968 firms with only about 3 percent of national steel capacity had 90 percent of continuous-casting, or concast, production. Sometimes overlooked is that these firms had most of their finishing capacity in smaller products for which concasting was easier than for heavier lines. For larger firms, the challenge was to convert the process over to large tonnages, above all to slabs. US Steel carried out a good deal of research on concasting and by 1968 had installed 2.25 million tons of it in its plants—2 million tons at Gary and a 0.25-million-ton unit at Torrance. Edwin Gott later claimed that experience with the Gary unit was encouraging and that the Corporation was ahead of the world in "knowing about" slab casting for sheet and strip. However, great secrecy surrounded these installations, and one conclusion drawn from this was that larger concasting was by no means wholly successful. US Steel's own calculation was that the concast unit at Gary might reduce costs of carbon steels by as much as twelve or thirteen dollars a ton.[23] By 1974 the Corporation had 4.8 million tons of concast capacity. Of this almost two-thirds was at the two Chicago works. US Steel was by no means unique in its slowness, though some other leading companies were far ahead. Bethlehem then had no concast capacity; in proportion to its size, Inland had about the same commitment to the process as US Steel. At more or less the same time, Kawasaki Steel, only one-third the size of US Steel, could make 5.14 million tons of concast steel.[24] It was to be more than a decade before, though then very rapidly, US Steel closed the gap. As late as 1988, only 36 percent of its raw steel could be continuously cast; in 1990, 53.1 percent of output; and by 1994, 99.4 percent.

In relation to the development and introduction of new technologies, US Steel was eager to use its own research and development facilities rather than learn from or hire technology from others. Having installed BOF converters, it was a few years later diverted into a modification of the process that eventually proved a dead end. In 1972, with Maxhutte of Germany, it formed US Steel Maxhutte to market the process known in Europe as OBM and in US Steel as Q-BOP. In this, the hot metal in the converter was bottom rather than top blown. It claimed that this might produce higher-quality steel in a shorter period of time with less pollution. Developers thought that it might allow use of as much as 20 percent more scrap than in the BOF process, but this proved not to be the case.[25] At Gary three BOF were converted to the Q-BOP operation; Fairfield was equipped with two 200-ton BOP vessels. However, they were disappointed in the hope that the process would be widely adopted by others. The process seems to have been more complicated and slightly more expensive to install than the ordinary oxygen steel process. At the end of the 1960s, a new corporate subsidiary, US Steel Engineers and Consultants (UEC), was set up to offer technical advice worldwide. UEC gave technical advice to Kobe Steel's new Kakogawa works and on butt-weld pipe to Kawasaki's longer-established Chiba works. Within a very few years, the flow of know-how was to be strongly established in the opposite direction.

In conclusion, it is important to stress once more that in any consideration of technological innovation one must remember that by the 1960s steel production in the United States was growing more slowly than in other leading economic regions of the world. Between 1960 and 1970, output of hot-rolled steel in America increased from 68.3 to 76.8 million tons, while that of the original six European Economic Community (EEC) countries and of the United Kingdom grew from 68.1 million to 100.5 million and in Japan from 17.1 to 75.8 million. Technologically, in comparison not only with the Japanese—who tended to receive most of the public attention—but also when considered alongside leading European producers, US Steel was by the latter date obviously lagging. This may be summed up in a comparison from the early

1970s of the bulk-steel equipment at US Steel and in the plants of five European companies that together made almost the same raw-steel tonnage—31.7 million tons for US Steel; 31.8 million tons for the Europeans. In the case of US Steel, eleven works were involved, in Europe, nine. In addition to the oxygen and open-hearth capacity, there were also electrical and, for the European companies, some basic Bessemer (Thomas) plant. Inevitably, this to some extent reduces the direct comparability of the two experiences. It seems generally incontrovertible that for one reason or another, US Steel—like most other leading American firms—was slow to adopt the new steel-making technologies that, as subsequent events proved, would dominate world production. Similarly, it is difficult to avoid the conclusion that, whatever the mitigating circumstances, the fundamental cause was entrepreneurial failure. In any event, a high price was paid in reduced international competitiveness.

TABLE 18.5

Open-hearth and oxygen steel plant at US Steel and in leading European steel plants, c. 1972–1973

	Open hearth		Basic Oxygen	
	USS	Europe	USS	Europe
Number of plants	7	6	6	8
Number of furnaces/converters	112	36	14	31
Tot. cap. per heat/blow (tons)	25,611	6,225	2,980	5,450
Avr. size per furnace/converter (tons)	228.7	172.9	213	176

Based on: R. Cordero, *Iron and Steel Works of the World,* 6th ed. (London: Metal Bulletin, 1974).

[a] The European operations involved were those of Thyssen, Krupp, Hoesch, Sollac, and Usinor.

Long-term Changes in Corporate Organization and Location

From the beginning, the operations of the United States Steel Corporation were expected to benefit from a more efficient coordination of operations in which there would be economies of scale, a removal or at least decrease in duplication of production facilities, and reduced cross-hauling. Given the size, complexity, and structure of what took shape in spring 1901, the realization of these sensible aims understandably proved difficult. US Steel was a holding rather than operating corporation, and the early challenge was to strike a balance between the continuing ambitions of those who controlled its individual companies and a logical new strategy that could take fuller account of the width of its combined resources in minerals, plants, and key personnel. Even when, bit-by-bit, better coordination was secured, there remained such open questions as the relationship between the corporate headquarters and the divisions and plants, and whether product group or area was a better basis for organization. In some respects these problems were not solved until the ninth decade of US Steel's history.

In the earliest years the struggle for preeminent power between the chairman and president was conclusively resolved first in the triumph of Chairman Elbert Gary over Pres. Charles M. Schwab and then in the consolidation of this victory through the former's dominance of subsequent presidents William Corey and James Farrell. The Executive Committee, which had represented the importance of the premerger companies or combinations, was dissolved on the day when Schwab resigned the presidency, and afterward the US Steel Finance Committee was the essential powerhouse of central control. For the last twenty years of his life, Gary was chairman of both the board and the finance committee. In this period there was also division between the power of central control, as represented above all by the control of the purse strings in 71 Broadway, and division or plant initiative. The self-regarding ambitions of the premerger companies surfaced time and again. An outstanding instance is revealed in the minutes of the Carnegie Steel Company in the form of its reluctance to let business pass over to Illinois Steel. For the latter, the construction of the Gary works was a triumph over its Pittsburgh-centered rival. Over the years, "Carnegie power" in the affairs of the US Steel center waned with the retirement and resignations not only of Schwab and Corey but also of James Gayley, W. B. Dickson, and Alva Dinkey.

During the early years there was some simplification of the corporate structure. National Steel was merged into Carnegie Steel, American Sheet Steel and American Tin Plate became one company, Shelby Steel Tube was incorporated into National Tube, which also took over the Lorain works. Even after this there remained a legacy of the days before 1901 in the form of two fully integrated firms—Carnegie and Illinois—relatively light on finishing capacity and the descendants of the major finishing groups whose iron- and steel-making capability was slight compared with their rolling-mill capacity—like American Sheet and Tin Plate and American Steel and Wire. Gradually, greater coordination between divisions and plants began to appear. Early in 1911, in his retirement speech, Corey recognized headway had been made in the direction of corporate planning: "Another development which seems worth mentioning has been the adoption of definite plans for extensions, with a view to the logical segregation of products which have involved not only the building of new plants and extensions to old ones, but also the abandonment of isolated plants containing obsolete equipment, the location of which did not seem to justify rebuilding or further expenditure." There was "still room for improvement," but "if every proposed expenditure in the future is studied, not only from the aspect of the particular plant in question but also in its relation to the Corporation as a whole, I feel that much will be gained, both in the reduction of costs and in improved service to customers."[1] Even so, a generation later the reports of Ford, Bacon, and Davis emphasized the need for US Steel to have what they called a planning board.

After Gary's death and the brief interregnum under J. P. Morgan Jr., Myron Taylor took the process of reorganization much further. Major steps were taken toward the concentration of production through the creation of Carnegie-Illinois and the merging into this new group of American Sheet and Tin Plate. Formation of US Steel of Delaware brought control of operations to Pittsburgh, and this process of coordination reduced duplication of facilities. Those in middle management were very conscious of the resultant tightening of the organization. A few years after World War II, the Taylor reforms were taken still further under the guidance of Roger Blough. In 1951 there began a greater coordination of policymaking through the operating policy committee (OPC), described by Blough as "the active decision-maker." The OPC was made up of the chairman, president, chairman of the finance committee, general counsel, and the six corporate vice presidents. In the same year control over production was reorganized. A new organization, Central Operations, with its head offices in Pittsburgh, controlled the integrated works in Pittsburgh (except for McKeesport and Donora) and the mills in Youngstown, the Chicago area, and Fairless, about 65 percent of the Corporation's steel capacity. Five other former steel and processing companies were also reshaped as divisions: American Steel and Wire, consisting of eight works—only two of them fully integrated—had its headquarters in Cleveland; Tennessee Coal, Iron, and Railroad (TCI) controlled from Birmingham; Columbia-Geneva with its offices in San Francisco; National Tube, comprising six works or parts of complexes of operations, centered on Cleveland; and American Bridge.

Although superior to what had gone before, this left the divisions each with its own sales force and to some extent in competition with one another. In the 1960s another arrangement was tried, operations being largely organized by type of product: Heavy Products, Sheet and Tin Products, Tubular Products, and Wire Products. Again there were untidy overlaps or local exceptions. Sheet and Tin included the whole of the Geneva plant, plate and structurals as well as sheet and strip. Irvin works was included in the Sheet and Tin Products division, but the Edgar Thomson furnaces and slabbing mill that supplied its raw materials were in the Heavy Products Division. The 80-inch hot-strip mill at Gary and its 38-inch continuous strip mill both had been in Central Operations previously, now the former was included in Sheet and Tin and the latter under Heavy Products. Clearly this was not due to oversight; it was rather a reflection of the impossibility of making a clear-cut division of assets originally assembled from a varied inheritance that had subsequently grown over another sixty years in a rather piecemeal fashion. The fact of the matter was that there was no single *correct* way of organizing US Steel operations.

In 1970 corporate operations were again revamped, returning to a geographical basis, though different from any before. There were now two groups, Eastern and Western Operations. On 1 January 1974 this in turn was replaced by four areal divisions, each with a distinct geographical area within which production and sales could be coordinated: Eastern Steel Division centered on Pittsburgh and including Youngstown, Lorain, and Fairless; Central Division based in Chicago; Western Division, with San Francisco again in control; and the Birmingham-centered Southern Division, which also covered US Steel Texas operations. By the mid-1980s, as capacity was sharply reduced, division of operations either by product or by area became much less important, and the 1974 structure was abandoned. With the creation of USX, broad product groups took precedence in organizational considerations. In 1984, operations were divided into Steel and Related Resources; Oil and Gas; Chemicals; Manufacturing, Engineering, and Other; and Domestic Transportation and Utility Subsidiaries. Even now, the structure of the steel and related operations was so complicated that it was difficult to get at various real costs of operation, a situation that made it difficult to produce a rational, overall framework for planning necessary cutbacks. At this time the old plea for planning at last received effective implementation, the "long-term planning" of the 1970s giving way to a new "strategic planning." By this time too there was a corporate policy committee whose duties were to assess proposals for expenditure or closure, balancing one-off costs against annual savings on the corporate profit and loss account. In the 1990s a hybrid form of organization was adopted in the US Steel Group.

In its earliest years there was no doubt that the operating core of US Steel lay in the headwaters region of the Ohio River, above all in the mill towns along the Monongahela River but also in the Mahoning and Shenango Valleys and in a number of works along the Ohio centered on Wheeling, West Virginia. Movements of the market, byproduct coking, new conditions of transport, and the abolition of the "Pittsburgh Plus" pricing system progressively lessened the dominance of this area during the Corporation's first twenty years, an erosion most spectacularly marked by the building and expansion of the Gary works in Indiana. By 1920 Gary was the Corporation's largest plant, and South Chicago works also exceeded any other operation. But the combined crude-steel capacity of Pittsburgh-area mills was 25 percent more than that around Chicago. Additionally, the combined capacity in the Valleys and in the neighborhood of Wheeling was not far short of 70 percent as large as that at the head of Lake Michigan. The two works then operating on the Lake Erie shore ranked that district a poor fourth. Over the next half century there was a strong lakeward shift so that from 41 percent of the total capacity in the western part of the belt the lakeshore plants increased to 58.4 percent. There was an even stronger westward component, a result of which being the Chicago share of total capacity rose from 30 percent to 48.9 percent. Such long-term trends resulted from more varied shorter movements. During World War II, for instance, the Pittsburgh area recovered a good deal of the ground it had lost over the previous few years before again falling away during peacetime.

As before World War I, a leading theme in US Steel through the 1920s was

TABLE 19.1

US Steel raw-steel capacity in the main centers of the
western manufacturing belt, 1920–1973 (thousand net tons)

Location	1920	1938	1945	1960	1973
Pittsburgh[a]	8,565	10,354	12,523	12,339	10,500[b]
Chicago	6,856	9,987	10,244	13,588	15,350
Valleys/Wheeling	4,667	3,713	3,394	2,712	2,500
Lorain	1,450	1,747	1,944	2,678	3,000
Cleveland	1,086	n.a.	n.a.	n.a.	n.a.
Other centers in region	251	21	24	25	60
Total	22,875	25,822	28,129	31,342	31,410
USS nationwide	25,047	28,915	32,291	n.a.	38,000
Proportion of USS capacity in region	91.3%	89.3%	87.1%	n.a.	82.6%

Sources: US Steel, various records, USX; AISI, *Iron and Steel Works Directory,* various issues.
[a] Includes Vandergrift.
[b] Excludes "reserve" capacity at McKeesport.

The Gary works' hot-strip mill, 1950. *Courtesy of USX Corporation*

the relatively greater growth of capacity outside, rather than within, the Pittsburgh district. For the Corporation and independents alike, Chicago was now the prime growth area. In the "Pittsburgh Plus" hearings, Gary revealed that manufacturing costs in US Steel's Chicago-area mills were about 80 percent those in its Pittsburgh operations.[2] Chicago commanded a much bigger local market—by 1929 a six-county area there had 142 percent more wage earners and 176 percent greater value of production than the four counties centered on Pittsburgh: Allegheny, Beaver, Westmoreland, and Washington.[3] Additionally, not only were mills on Lake Michigan, as always, better placed to serve the extended markets of the northwestern interior, but they also had superior access to the concentrated and now rapidly growing outlets in southern Michigan. (Even so, due to competition from other companies between 1920 and 1932, 75 percent of the former business done in the Detroit market was lost by Carnegie and Illinois Steel combined.)[4]

As well as leading in the expansion of growth both in capacity and in production, Chicago steel output fluctuated less than that of Ohio or Pennsylvania. Even the depression of 1921 was less severe there.[5] As in earlier periods, a good indicator of competitiveness was the rail trade. In the early 1920s, Gary, South works, and the much smaller Milwaukee mill had a rated annual capacity of 0.6 million tons of rails; Edgar Thomson (ET) was rated as able to make up to 1.5 million tons. Yet in the three years 1919–1921, deliveries to the home market from the three Chicago-area mills were 1.9 million tons and

TABLE 19.2

**Estimated production and consumption of rolled steel
by sections of the United States, 1920 (thousand tons)**

Section	Production	Consumption
East[a]	16,000	10,025
West[b]	4,700	8,975
South[c]	750	1,630
Pacific[d]	200	1,020
Total	21,650	21,650

Source: Iron Age, 26 February 1925, 610.
[a] Includes Ohio, West Virginia, and Maryland.
[b] All states west of Ohio except those along the Pacific Coast.
[c] All states south of Nevada, Utah, Colorado, Kansas, Missouri, Illinois, and Indiana.
[d] The coastal states: California, Oregon, and Washington.

from ET only 0.68 million tons. Conversely, Homestead plate capacity was then more fully utilized than that of US Steel Chicago plants.[6] Overall, Pittsburgh retained the largest concentration of capacity. Abolition of "Pittsburgh Plus" was not followed by an end to investment around the city and also served as a spur to efforts to improve distribution from its plants. Over the next four years, Carnegie Steel alone invested $100 million in the area as well as $5 million on a fleet of steamers and barges to deliver its products to southern and western markets accessible from the waterways of the Mississippi basin. The Ohio, as the route of access to this system, carried slightly less than 400,000 tons of iron and steel from all producers in 1923 but 1.3 million tons five years later. By the early 1930s, National Tube was operating a depot at Memphis to which it made deliveries by water, shipping onward to the major tubular-product markets throughout the Southwest by railroad. On lines that Andrew Carnegie had recommended almost a quarter-century before, the Carnegie Steel Company also made shipments to Detroit by rail to Conneaut, Pennsylvania, and from there by water.[7]

During the 1930s US Steel made major capital outlays for modernization, resulting, against the drift of the previous thirty years, in the relative standing of the main works of the Pittsburgh area and increasing within the Corporation. In the extensive review at the end of his chairmanship, Myron Taylor commented on the geographical distribution of US Steel capacity: "The steel industry, although in the popular mind a heavy industry with fixed locations, is, as a matter of fact, in some respects a mobile industry. . . . [I]n a country the size of the United States . . . as consuming centers shift, the factors of location will change." Clearly, as a long-established amalgamation, US Steel was not in a position to choose a plant distribution pattern ideal for the times, but through elimination of old mills and the construction of new ones it could substantially modify the overall pattern. Taylor summed up the guidelines they had adopted: "It was decided, after a great deal of study, that the Corporation could achieve its highest efficiency by grouping its main producing units in the Pittsburgh district, the Chicago district, and the Birmingham district." In June 1936 he had stressed to the finance committee as "imperative" a $110 million expansion and modernization of flat-rolled product manufacture in these three areas and near Cleveland. At the end of the day and with the decisions made, Cleveland saw little of this outlay, though for a time through the Lorain works, it would continue to be an important contender for investment.[8] Over these years substantial shifts in the regional balances were achieved. The Lake Erie shoreline and still more the Valleys and upper Ohio districts declined absolutely. The South improved its standing but remained of secondary significance. Pittsburgh's share had fallen in the 1920s, but during the next decade more than made up the ground it had lost. Chicago advanced greatly overall, though less than in the 1920s, partly because of the 1936 closure of Joliet steelworks. In absolute terms Chicago expansion during 1930–1938 was half that in Pittsburgh.

The overall interwar contrast in development trends between Pittsburgh and Chicago was nothing new. No new integrated iron and steel operation had been set up by US Steel or its subsidiaries in the Pittsburgh area after 1902— or by any steel company later than the Monessen works of Pittsburgh Steel or the Aliquippa operations of Jones and Laughlin in 1907. Within the broad area at the head of Lake Michigan as well as Gary works, important competitors such as Wisconsin Steel, Inland, and the Sheet and Tube Company of America (from 1923 incorporated in Youngstown Sheet and Tube) had also constructed new integrated works. Greater Chicago had a bigger, more highly diversified manufacturing economy than western Pennsylvania, and though steel production was increasing rapidly in the former, it involved a much smaller proportion of the manufacturing economy than in the older areas. Despite Pittsburgh's efforts to organize water transport, Chicago-area mills were better placed to serve the Mississippi lowlands. The use of the byproduct coke oven and general fuel economy favored districts remote from the coalfields, and local steel-using industries and the huge natural market area gave Chicago the important advantage of cheap scrap as the advance of open-hearth steel reduced the relative importance of pig iron. Another important asset, particularly with some of the newer types of plant such as byproduct

TABLE 19.3

Number and raw-steel capacity of US Steel integrated works by district, 1920, 1930, and 1938 (thousand net tons)

	1920		1930		1938	
District	Works	Capacity	Works	Capacity	Works	Capacity
Pittsburgh	7	8,256	6	8,537	6	9,709
Valleys/Up. Ohio/Columbus	7	4,723	4	4,010	3	3,713
Lake Erie shore	2	2,536	2	2,576	1	1,747
Chicago	3	6,856	3	9,303	2	9,987
South	1	1,254	2	1,982	2	2,018
Other (Duluth)	1	605	1	605	1	336
Total	21	24,230	18	27,013	15	27,510

Based on: AISI, *Iron and Steel Works Directory of the United States and Canada* (New York: AISI, 1920, 1930, 1938).

coke ovens, open-hearth melting shops, and continuous-rolling mills covering large acreages, was the availability of very large, flat sites and easier access for transportation facilities as compared with Pittsburgh, and indeed with all the inland districts of the Ohio–Lake Erie belt. Early in the 1920s, a mid-Ohio valley rival gave a vivid but realistic description of site and transportation difficulties at the head of that river: "The near Pittsburgh producing units have long since outgrown their general development limitations, and it is almost an axiom that the further we remove from Pittsburgh both upward on the Allegheny and Monongahela as well as southward on the Ohio, the greater will be found the freedom for balanced development and the clearer will be found the general atmosphere." At the heart of this great producing district, transport facilities were so cramped by topography that, "Several days frequently elapse before car loads find their way into or through the intricacies of the local Pittsburgh labyrinth."[9] In contrast, there remained large, flat, undeveloped, and accessible tracts; potential plant sites; or, alternatively, opportunities to reclaim land from the lake in the Chicago area. Ford, Bacon, and Davis found the South Chicago works, dating from the early 1880s, very cramped, but critics still admired "the supreme logic of the Gary plant—the engineer's dream."[10] The impressive overall advantages of the Chicago area were at this time illustrated by the experiences of a rapidly expanding rival. At Indiana Harbor, Inland Steel, which normally bought scrap for two dollars a ton less than the Pittsburgh price, obtained a site for its new hot-strip mill by reclaiming fifty acres from Lake Michigan and sold 75 percent of its output within 150 miles.[11]

Although locational change continued, the choice of Pittsburgh as the location for a hot-strip mill showed the immense power of inertia represented by the massed capacity of an old industrial area. There were two important aspects to this, the prime one commercial, and the secondary social. A year or two after this period, US Steel documents summed up the first: "such large and

TABLE 19.4

Iron and steel wage-earners as a portion of all manufacturing wage-earners in selected industrial areas, 1929

Pittsburgh	37.7%	Buffalo	12.0%
Youngstown	56.9%	Cleveland	11.8%
Johnstown	71.7%	Baltimore	11.3%
Chicago	8.9%		

Source: G. E. McLaughlin, *Growth of American Manufacturing Areas* (Pittsburgh: University of Pittsburgh Press, 1938), 108.

complex equipment cannot be moved in response to geographical shifts in demand, and only extraordinarily great differential advantages of a new location justify scrapping existing facilities embodying large unamortized investment and long remaining service life." The other, more tenuous factor was summarized by *Fortune* when commenting on the increasing disadvantages of a Valleys district location: "but the Valley mills cannot be dismantled. They have created large communities about them, and one cannot scrap 250,000 people as one can write off 250,000 dollars."[12] In 1935, when it decided to build a hot-strip mill at its McDonald works, US Steel remodeled a blooming and slabbing mill installed in Youngstown works twenty-six years before in order to supply its slab requirements. Unfortunately, time would show that rather than continue such arrangements it was indeed possible to write off even large steel communities, and that by some this would be deemed essential in the wider interests of US Steel. However, that time was a long way off, and experiences in the late 1930s seemed to show there was a third inertial factor, a strange, almost sentimental attachment by top management to preservation of the greatest concentration of inherited capacity.

Together, these various factors temporarily reversed Pittsburgh's relative decline. In doing so, they also influenced decisions that went against the recommendations of Ford, Bacon, and Davis. In August 1934 the US Steel Committee on Flat Rolled Products outlined a program of "rehabilitation" and "improvement" for their sheet and tinplate operations. A first step, carried out by spring 1936, involved expansion for both products in Chicago. Stage two focused on Pittsburgh. The local managements of US Steel wanted to go ahead at once in spending over $60 million on new sheet and tinplate capacity; in an interim report their consultants urged that money should not be spent before the completion of market analyses for eastern areas of the country. For a time Ford, Bacon, and Davis thought the market for sheet in the area might be met from the surplus capacity of the new 100-inch plate mill at Homestead, if it could be made flexible enough to roll either plate or sheet gauges. In September 1936 Robert Gregg submitted a report from US Steel engineers recommending the new capacity be installed at Clairton, where the existing blast furnaces, steel-making plant, and rolling mills, except for a 22-inch structural and 18-inch bar mill, would be demolished. After the site had been cleared, an 80-inch strip mill and a cold-reduction mill should be built there. It would obtain steel from an enlarged open-hearth shop and new slabbing mill at Edgar Thomson.

Ford, Bacon, and Davis felt strongly that Lorain might be a better location than Pittsburgh for the new mill. They drew executives' attention to the fact that the Corporation's own engineering studies had not been accompanied by similar reports from the sales side. A study undertaken jointly by B. H. Lawrence of US Steel and H. E. Whatcher of the consulting firm FB&D was later recognized by the consultants to have made assumptions both about depreciation charges and transport costs that unreasonably favored a Clairton location. The first of these mistakes resulted from failing to recognize that the new development might eventually require all plant to be wholly new; in other words, it might prove to have been unjustified to assume that making use of existing ET facilities made Clairton a cheaper investment option than Lorain, where a greater amount of new plant from coke ovens and iron making through to steel and primary mills would be needed from the start. A second error was in assumptions about applicable rail tariffs.

Lawrence and Whatcher had suggested that capital spending at Lorain would be $72.3 million as compared with $59.6 million at Clairton. They put annual production costs at Lorain $634,000 lower. Further examination produced projected annual savings on operations at Lorain of $828,000. Distribution costs there would be $269,000 less than from Clairton. Ford, Bacon, and Davis made clear that they believed "no adequate market data had been submitted to show that the development should be at Clairton." Yet, despite these strong warnings, early in December 1936 Ben Fairless, as president of Carnegie Illinois, wrote a letter that rushed the decision in favor of Clairton. He pointed out that US Steel sheet capacity as a proportion of the industry's total had fallen from 35 percent in 1915 to only 13 percent and if they failed to act at once would be down to 10 percent in a few more years. Immediate installation of the strip mill and sheet and tinplate capacity at Clairton was preferred over Lorain mainly because of "the important eastern tinplate situation." (In fact, Ford, Bacon, and Davis showed that a Birmingham mill should be able to deliver tinplate in New York for $4.42 per gross ton less than Pittsburgh, $64.92 as compared with $69.34.)[13] The Fairless letter brought an immediate decision for Pittsburgh. Named the Irvin works, the new hot-strip mill was commissioned in December 1937. Regardless of what had so recently been considered a clinching argument in its favor, of its initial 600,000-ton capacity only one-sixth was tinplate. For sheet and coiled strip, the biggest outlets in southern Michigan were almost twice as far from Pittsburgh as from Lorain. Instead of using the Clairton site, Irvin works was built on the plateau above the Monongahela between Clairton and Dravosburg, massive earth-moving operations adding very greatly to site preparation costs. The Clairton works was retained and Irvin provided a continuing raison d'etre for Edgar Thomson.[14] When Carnegie-Illinois forced through this decision, it could not be foreseen that decades later the contentious lifeline now thrown to the Monongahela agglomeration of capacity would be the only justification for any of it to survive.

As in the 1930s, the extensions undertaken by US Steel during World War II, largely at the request of and to a considerable extent financed by the federal government, interfered with and partly reversed the trends of locational change. Nowhere in the manufacturing belt was this more pronounced than in the position of Pittsburgh, due to the prime significance of investments made at Homestead. However, long before the war was won, changes were being contemplated that would reinforce the drift against both Pittsburgh and Youngstown and boost Chicago. Although these recommendations would not actually be followed through in the postwar world, they represented a rational response to the changes in the circumstances for production. They came in the form of a "Comprehensive Plan" for Carnegie-Illinois produced in 1943. This provided—unrealistically as it turned out—for a 13.5 percent reduction in the division's steel capacity from present levels. The Chicago, Pittsburgh, and Youngstown districts would all face cutbacks, but the fall would be smaller in the first than in the second, and largest of all in the last. Among the works whose steel capacity was slated for closure were Farrell, Mingo Junction, and Vandergrift. In structurals, the existing capacity of 1,507,000 tons at Homestead and Clairton would fall to 1,323,000 tons, largely because of the complete closure of the Clairton structural mills. In contrast, at South Chicago capacity for structurals was projected to grow from 680,000 to 977,000 tons. Gary annual rail capacity would increase from 657,000 to 713,000 tons, but all four rail mills at Edgar Thomson would be dismantled, removing 711,000 tons capacity. Again, this plan was never fully carried out.

TABLE 19.5
Regional shares of US Steel ingot capacity in integrated works, 1920, 1930, and 1938 (portion of total)

Region	1920	1930	1938
Pittsburgh	34.1%	31.6%	35.3%
Valleys/Up. Ohio/Columbus	19.5%	14.8%	13.5%
Lake Erie shore	10.5%	9.5%	6.3%
Chicago	28.3%	34.4%	36.3%
South	5.2%	7.3%	7.3%
Other (Duluth)	2.5%	2.2%	1.2%

Based on: AISI, *Iron and Steel Works Directory* (New York: AISI, 1920, 1930, 1938).

TABLE 19.6
**Carnegie-Illinois districts' 1943 and projected postwar
steel capacity (thousand tons)**

District	Present capacity	Projected capacity	Projected as portion of present
Chicago	10,193	9,914	97.3%
Monongahela valley	8,095	7,102	87.7%
Youngstown	4,034	2,284	56.6%
Total	22,322	19,300	86.5%

Source: Carnegie-Illinois Comprehensive Plan, 1943, USX.

After World War II the gradual shift of emphasis from Pittsburgh resumed. The size of the problem of adjustment was increased by the $65 million purchase during summer 1946 of the government-financed wartime extensions at ET, Duquesne, and Homestead. That year the public was treated to different interpretations of the prospects from two successive presidents of Carnegie-Illinois. In spring J. Perry, speaking in Pittsburgh, courageously pointed out that rising freight charges, exhaustion of the best coking coals, and movement away of what he called "the density of consumption" meant the area must diversify industry, for "we must logically expect that when it is necessary for some of the present steel manufacturing facilities in Pittsburgh to be replaced they will be installed at locations nearer to the concentrated consumption of steel, and there will be some reduction in the total ingot capacity in the Pittsburgh area."[15] In July Perry retired to become an assistant to Fairless. A few days later his successor, Charles R. Cox, suggested that despite "tremendous handicaps"—including that "I would assume we can melt steel here (Chicago) cheaper than in Pittsburgh"—because its plants represented an investment of almost a billion dollars, steelmen would do everything possible to protect these assets; "fight for all we are worth to keep Pittsburgh the steel center of the world." As a result Chicago might never take over the leadership in steel.[16] Two years later, when basing-point pricing was abandoned, the f.o.b. mill pricing system that took its place produced added cost advantages for those producers and consumers in close proximity. In the months that followed, some claimed to make out a movement of steel consumers to Pittsburgh, but this was never nearly commensurate with area capacity.[17]

In late winter 1949, when the effects of the new pricing system were still unclear and with more rail-rate increases ahead, Benjamin Fairless spelled out the situation to the Pittsburgh Chamber of Commerce. It was very much what Henry Bope had said forty years before, but Fairless's laborsome detail drove home its unmistakable logic: "In 1901, when US Steel was formed, its steel ingot capacity in the Greater Pittsburgh District, including Youngstown and Johnstown, was about 6.9 million net tons. Today its Pittsburgh district capacity totals roundly 13.5 million tons, a gain over 1901 of 6.6 million tons. In the same period US Steel's total capacity rose from approximately 10.6 million net tons to 31.3 million, an increase of 20.7 million. Our Pittsburgh capacity, it will be noted, has nearly doubled, but US Steel's total capacity has virtually tripled. . . . This means that greater gains have occurred in other districts than in the Pittsburgh area." He went on to the causes of the change and to consider solutions of the problem: "Pittsburgh has lost ground because, with its relatively large capacity it has had to reach farther for tonnage. . . . Regardless of what happens to basing points, Pittsburgh must enlarge its local steel consuming market."[18] When in 1949 the annual postwar rise in steel production was at last interrupted, it was noticeable that Chicago-area output held up better than that of Pittsburgh. Some of the US Steel facilities in the latter district were, as Cox's successor recognized, "becoming marginal or definitely too old for profitable operation." Duquesne especially was mentioned in this respect.[19] Apart from 1949, there was generally a seller's market until 1953, so that the ending of basing-point pricing had little immediate effect. Major growth in the carriage of steel by truck rather than by rail also helped confuse the picture.[20]

Tubular products provided a good illustration of the problems of market access and of the fact that there were now also better locations for assembling steel-making materials. The Pittsburgh-Cleveland-Buffalo industrial triangle produced about three-quarters but consumed only one-tenth of the nation's tonnage of pipes and tubes. As early as 1946 there was a strong move within US Steel for the transfer of more of the National Tube operations to Lorain and Gary, leaving behind only minimal operations at the National works at McKeesport. This did not happen, and indeed an ERW-pipe mill was built at McKeesport in 1950. By 1960, whereas steel capacity at Lorain was almost 38 percent higher than in 1945, at McKeesport the increase was only 16 percent. A second McKeesport ERW-pipe mill was built in 1964, but by this time the main part of the works there was on the edge of a rapid decline.[21] With more than half the national output of sheet and strip, this same industrial core area used less than one-eighth of that produced.[22]

Although the market was moving away, the massed capacity of the Pittsburgh area yielded important operating advantages. There could be efficiencies achieved through common services and flexibility from linked operations. An important instance from the wider Pittsburgh district involved Vandergrift, some twenty-one miles northeast of the main concentration of capacity in the Monongahela valley. Rolling-mill operations there were accompanied by 500,000 tons of cold metal open-hearth capacity. Vandergrift had become an important producer of electrical-grade sheets, which were produced on hand mills until 1947 when a cold-reduction mill was installed to take over the work. It now sent its own steel slabs to Dravosburg to be rolled into coil on the Irvin hot-strip mill and then received it back for further rolling. In 1953 the supply system was changed, the open-hearth furnaces at Clairton from that time producing the necessary silicon steel slabs for Irvin, which then sent the coils to Vandergrift. A year later the Vandergrift melting shop was abandoned, but at the same time this plant became a growth point for stainless-steel sheet manufacture, receiving some operations until then conducted on hand mills at the Wood works at McKeesport.[23] Another example of increasing operational links between local works involved a combined 670,000-ton increase in ingot capacity underway in 1957 at Homestead, Edgar Thomson, and Duquesne, much of it to support expansion of plate, structurals, and forgings at the first. A few years later when the decision was made to build the new ERW capacity at McKeesport, slabs were produced at Edgar Thomson and rolled down at Irvin into coiled strip for the pipe works.[24] Through linkage economies of this kind, the size of new capital outlays could be reduced. In such ways a Monongahela valley complex was coming into being twenty years or so before reference to a "Mon Valley works" became common usage.

In January 1956 the chairman announced that market studies had shown, while there would be expansion in many areas, major increases in capacity would be needed in Chicago and the West. This was a response to general advantages in relation to the wider geography of the national market and to the huge growth of consumption nearer to the steel mills. Between 1946 and 1955, estimates reckoned that 935 new metalworking firms moved into the Chicago metropolitan area. By the end of this period, these newcomers employed 155,000 workers, and Chicago had a total of 587,000 metalworking jobs as compared with 252,000 in the Pittsburgh area. Local mills were also well placed to supply the metal-working plants of Detroit (534,000 workers) and more easily still the requirements of the 129,000 employed in Milwaukee metal working.[25] By early 1957, US Steel was placing contracts for the expansion of Chicago capacity by 1.3 million tons. At more or less the same time, the Corporation planned substantially smaller Pittsburgh increases—a 90,000-ton open-hearth extension at Duquesne, 120,000 tons at Edgar Thomson, and at Homestead 365,000 tons in open-hearth steel and 95,000 in electric-furnace capacity—together half the new tonnage planned for Chicago. Because of closures of plant built in wartime and smaller extensions from 1945 through to the end of the 1950s, the six integrated Pitts-

Duquesne works, Pennsylvania. *Courtesy of USX Corporation*

burgh district works merely maintained their combined capacity; in the same period capacity in Birmingham increased by half and in Chicago by one-third. Gary further extended its lead over South Chicago.

The 1960s brought important changes in the pace of expansion and continuing westward drift within the manufacturing belt. Between 1960 and 1970, Pennsylvania's share of national raw-steel output fell from 23.9 percent to 22.8 percent; over the same years the proportion from Illinois and Indiana went up from 22.3 percent to 23 percent. These were small changes, but they built on decades of earlier movements in the same direction. There were further shifts in the proportions of raw materials used in iron making, but these were minor considerations in comparison to the changes in the geography of consumption. As it had been for many years, Lorain was seen as an attractive proposition for plant modernization; in contrast the Youngstown area was conspicuously absent from almost all lists of large new expenditures. In relation to the long-term viability of Youngstown's considerable capacity this was ominous. In mid-summer 1962, as that district's operating rate fell to between 33 and 35 percent, US Steel announced it would idle both the Ohio steelworks and the finishing mills at McDonald. Nine months later it seemed that the cloud over them had lifted when plans were announced to spend millions improving McDonald capacity for quality bar. The general superintendent greeted this as "another indication of US Steel's confidence in the Youngstown area."[26] But, as was highlighted in a 1966 description of the 43-inch hot strip at McDonald, it had been a tenuous success. This "robust veteran," a "hold-over from the earliest days of continuous strip mills," was now operating with a new control system that enabled it to compete in terms of accuracy of gauge with the most modern strip mills. "It can be done—old mills *can* be updated and kept competitive.... The US Steel plant is boot-strapping—both because it must to survive inside US Steel's internal competition for business and capital money—and because as part of Youngstown it *must* improve to survive in an allegedly unfavorable steel location."[27] Harsher circumstances might mean such a struggle would become insupportable.

TABLE 19.7

Raw materials used per ton of iron, steel consumption, and raw steel production in the United States (net tons)

	1947	1958	1963	1968
Materials per ton of pig iron				
Iron ore, agglomerates, etc.	1.8	1.60	1.53	1.56 (1967)
Coke	1.0	0.80	0.67	0.63 (1967)
Limestone	0.4	0.33	0.28	0.26 (1967)
Steel consumption				
United States total (th. net tons)	39,400	44,800	58,100	n.a.
Regional distribution				
East	26.0%	22.4%	19.7%	n.a.
Central	57.8%	55.5%	58.5%	n.a.
South	10.6%	13.7%	14.1%	n.a.
West	5.6%	8.5%	7.8%	n.a.
Raw-steel production				
United States total (th. net tons)	84,890	85,260	109,260	131,460
Distribution by selected producing district				
Pittsburgh	26.3%	21.5%	19.9%	19.3%
Youngstown	13.2%	8.3%	8.2%	8.2%
Chicago	20.2%	21.5%	21.1%	20.4%
Detroit	3.7%	5.3%	7.7%	7.0%
Southern	4.7%	6.0%	5.7%	6.4%
Western	5.1%	6.7%	6.4%	6.5%
All others	26.7%	30.8%	31.0%	32.2%

Sources: AISI; Federal Reserve Bank of Cleveland Economic Review.

Much more dramatic were the changes of the 1960s in Pittsburgh and Chicago. Large outlays on high-output blast furnaces and oxygen converters could not be replicated in many plants; in other words, as so often before, new technologies meant still more concentration of capacity. US Steel operated two integrated plants in the Chicago area; in Greater Pittsburgh it had six. Obviously not all of the latter could be modernized. As so often in the past, not only technical advance, but changes in raw-material supplies and markets also worked against Pittsburgh. When the expansion of the 1950s passed over into recession and then to slower growth, these changes were reflected in development. The years 1955–1957 brought unprecedented high levels of production. Extensions were made by all companies, national rated capacity increasing by 15 million tons. In 1958 there was an extremely sharp fall, from the average 115 million net tons of the previous three years to 85.2 million. Despite this decline, during 1958 national capacity increased by a further 6.9 million tons. Average annual output of raw steel in the four years 1959–1962 was 97.2 million tons, 8 million tons less than the 1950–1957 average. Imports were increasing. Companies realized that retrenchment was required. The prospects of various areas were reassessed. *Iron Age* estimated that in 1962, 53.8 percent of the hot-rolled, cold-reduced, and galvanized sheet and strip was consumed in Indiana, Illinois, Wisconsin, and Michigan, but these states produced only 30.3 percent of the total.[28] For Pittsburgh, major outlets for strip-mill products had always been distant. The plate market was drifting away—a significant sign was the opening of what was claimed as the world's largest plate mill at Gary in 1962. Gradually, loss of this trade would impinge on Homestead prospects. There was more immediate action at other plants in the area. In June 1960 the open-hearth department at Edgar Thomson was closed, and production did not resume until August 1961—and then at only five of the sixteen furnaces. The slabbing mill started up again at the same time.[29] While ET was down, the Irvin strip mill was supplied from other plants. The district's first important, permanent closures were announced in 1962. The wire works at Rankin was small, but it had been one of the founding operations of American Steel and Wire; in that sense its abandonment was symbolic. More impressive was the end of iron and steel making at Donora and of most Clairton operations. During spring and summer 1964, there were further periods of closure at Edgar Thomson, and the mid-1965 decision to dismantle rail-mill facilities there marked the end of a remarkable ninety-year history. In 1966 rod and wire units at Donora were shut down. Nevertheless, top US Steel executives continued to give generally up-beat assessments of the Mon Valley, and although in the mid-1960s one after another of its major new investments seemed to focus on Chicago, at the end of the decade Pittsburgh-area cold-rolled sheet capacity was almost doubled.[30]

At the US Steel Annual Meeting on 4 May 1964, a stockholder asked if development in Chicago meant Pittsburgh's position would deteriorate. He received an ambiguous answer: "Not so far as US Steel is concerned. We have put many new facilities there, and I'm sure, in the years ahead, that additional ones will be installed there. The Chicago area happens to be growing at a faster rate than the Pittsburgh area from the standpoint of steel consumption."[31] On a more positive note, Duquesne was chosen as the location for the first US Steel oxygen steel plant. Then in fall 1965, US Steel chose to close the blast furnaces and open-hearth capacity at McKeesport, keeping these "in reserve"; the National works in the meantime obtained its steel from Duquesne across the river. This proved to be a permanent arrangement.[32] The adverse balance between closure and new investment brought a further sharp relative fall in district standing.

In summer 1966 a small group of top steel executives from various companies active in the Chicago area spoke to the Investment Analysts Society of Chicago. Arthur V. Wiebel, executive vice president for engineering, represented US Steel, but the most interesting contribution came from Robert E. Williams, president of Youngstown Sheet and Tube. He explained that the area that could profitably be served from Chicago extended from Canada to the

The Ohio works, surrounded by the residential districts of Youngstown. *Courtesy of USX Corporation*

Gulf and from the Rockies to a line drawn between Detroit and Cincinnati. At its core was the seven-state area of Illinois, Indiana, Wisconsin, Minnesota, Iowa, Missouri, and Michigan, which accounted for almost 40 percent of national consumption of steel products. For flat-rolled products, the proportion might be closer to 45–50 percent.[33]

Despite its general growth, there were also declines and some closures in the Chicago area. In its first fifty years Gary produced 21.4 million tons of rails. Peak output of 681,000 tons was reached in 1921. However, by 1958 output was only 112,000 tons. The old 42-inch Gary strip mill was abandoned in the early 1960s, but an 84-inch mill took its place. One of South Chicago's melting shops was idle in 1964. Nevertheless, there was less run down and more investment in Chicago than in Pittsburgh. As a result that Midwestern district increased its lead in production capacity and output. By 1959 raw-steel capacity at the two Chicago plants was already 11.7 percent greater than at the six integrated Pittsburgh operations. Over the 1960s the gap widened, and by 1973 the two works around Chicago had a raw-steel capacity of 15.3 million tons; the three that now survived in Pittsburgh contained only 10.5 million tons. In oxygen steel, Chicago was 34.8 percent ahead.

The National Steel Industry
since 1970

Early in 1970 the former president of Crucible Steel was quoted as looking toward expanding world market prospects for steel with scarcely contained enthusiasm: "Just imagine what's going to happen in a place like Nigeria. Its the ninth most populous country in the world and these people aren't going to be walking around in T-shirts much longer. All over South East Asia people are going from pushing a wagon to getting a bicycle. My god, if every Chinese had a bike." A few months earlier, the International Iron and Steel Institute had held its annual meeting in Tokyo. In his chairman's address, Logan T. Johnson, chairman of Armco, spoke of the new global context in which the industry would now operate: "Vanishing space, global communication, and scientific discovery have turned our planet into a single neighborhood. . . . New technology which moves so quickly from one nation to another provides one of the most easily recognizable benefits of business ties that extend across friendly borders and around the world."[1] The realities behind such world pictures did indeed present great opportunities. Some national industries would do well, but for American steel companies the new circumstances indicated that the days of secure, relatively sheltered possession of the world's biggest market were passing. Harsher commercial conditions might mean a bitter struggle for survival.

Although there had already been serious problems, into the early 1970s expressions of optimism about the prospects of the industry were common. Profits and profit margins had fallen sharply over the last few years, but, as the New York correspondent of the *Financial Times* saw the situation in late fall

1972, "an industry which has had its share of problems in the last decade appears to be on the mend."[2] Events soon proved this a far too comfortable conclusion, and during almost the entire last third of the twentieth century, US Steel and other long-established integrated steel companies—in other old-industrial economies as well as in the United States—would have to struggle through a succession of crises. The American industry lost its world primacy. For eighty years it had out-produced all others; by 1996 the world's leader in terms of tonnage was China, Japan was second. Collectively, the nations of the European Union had long been a larger producer. In an industry that, worldwide, has been conspicuously slow in operating in a multinational framework, much of this decline in status resulted from the inevitable spreading of industrialization. But the problems of the U.S. industry also involved unique circumstances. One of the most important was that, as compared with other major steel industries, American companies were forced to face the challenges of the time with their own strength and resources alone; there has not been any stronger or more active U.S. government participation than a series of updated versions of the time-honored policy of tariffs on imported steel, efforts rewarded with indifferent success. Serious problems of overcapacity and technological backwardness have been tackled in a setting of uncertain but generally lower demand and keener competition both from within and from outside the domestic economy.

In raw-steel tonnage, 1970 was the industry's third best year to date. There were even better performances in the next few years, peaking in 1973 at over 150 million net tons, but already it was clear that all was not well. Expansion in other countries was faster and better maintained. From 1971 the USSR was a bigger producer, though this achievement was at best ambiguous, its output being by no means comparable either in terms of quality or real cost. Much more significantly, Japan and the European Economic Community (EEC) were soon well ahead. For the first time, the developing world increased in relative standing in iron and steel production. Apart from the USSR, EEC, and Japan, combined output in 1970 was no more than 10 million tons, or 8.4 percent, more than that of the United States; twenty-one years later, it was 212 million tons, or 268 percent, ahead. Moreover, some of these growing producers became important suppliers to sections of the American market. For one reason or another, domestic mills proved less able than their main overseas competitors to sell at competitive prices. The situation worsened dramatically in the 1980s; in 1970 American mills produced 20.1 percent of the world's steel, but by 1991 their share was only 10.8 percent.

The internal problems of the industry were complex, but their essence may be summarized briefly enough: steel consumption had not increased as

TABLE 20.1
Steel production in the world and in the United States, 1970 and 1991
(million metric tons)

Area	Crude steel production		1991 as portion of 1970
	1970	1991	
World	594	734	123.6%
USA	119	79	66.4%
USSR/CIS	116	133	114.5%
Japan	93	110	118.3%
EEC (6 countries)	109	105	96.3%
UK	28	16	57.1%
All others	129	291	225.6%

much as was expected in the early 1970s, competition was keener, and for a variety of reasons, many understandable if not excusable, major firms dragged their feet in facing up to them. In fact there seems to have been a general psychological problem, one natural enough in an industry that had led the world for almost a century. As one spokesman put it, "There's an attitude in the steel industry that says, 'if it worked last year, it will work this year.'"[3] If true, such a mind-set, strangely at odds with the traditional view of a dynamic world-leading business, was particularly ill suited for the challenging times lying ahead. Perhaps one of the most significant indicators was that by the late 1980s Nippon Steel was reckoned to spend more on research each year than all of the American companies combined.[4]

This period provided an extreme example of the general proposition that as an advanced society matures, its leading sectors use smaller amounts of steel than in earlier phases of economic growth. Installation of new infrastructure—bridges, pipelines, commercial buildings, and such—continues, and there is always replacement demand in these and other sectors, but the pace of development in these sectors falls sharply. The age of mass consumption had meant a shift to lighter, and preeminently to thin, flat-rolled products, but though demand in this field still rises as affluence widens down through the socioeconomic groups and advertising ensures that replacement demand is kept lively, after many decades there is no longer the same sustained forward thrust. Early in the twentieth century, during the last period preoccupied with infrastructure building, steel was the largest industry and US Steel was the nation's biggest company. This was followed by two or three generations in which the leading sector and companies were in automobiles and petroleum. Now, the vanguard is occupied by consumer and producer services, information technology, and entertainment, in all of which steel and other metals represent only a minor part of the material input. The implication of such changes for some steel-using sectors was vividly illustrated in what, for the industry, were the crisis years of the early 1980s. In the ten years to 1980, the average annual growth rate of the nation's GDP was 3 percent and from 1980 to 1983 1.5 percent. Generally, 1984 manufacturing output was 7 percent higher than that of 1979, but output of tractors was down 65 percent and of railroad equipment by 80 percent while production of computers and electronic components rose 50 percent. Consumption of steel was almost 20 percent below the 1979 level.[5] Peak annual demand for steel in the automotive industries had been 22–23 million tons and in containers some 7.5 million tons; by the mid-1980s these were 12.5 and 4 million tons respectively.[6]

Even during the years of high production in the early 1970s, there were signs that all was not well. For the whole industry, 1972 shipments were 8.1 percent greater than those of 1964, but net income was almost 22 percent less. Profit margins averaged 6 percent in 1964 but were 2.8 percent eight years later. Already the American Iron and Steel Institute was urging the necessity for remedial measures, including limitation of imports and tax relief to encourage investment in new plant and to offset the rising costs of pollution controls.[7] The industry's peak year came in 1973. Raw steel production was 151 million net tons; capacity was 155–160 million tons. It was pointed out that the operating rate was uncomfortably high, for the last 10 percent or so of capacity could only be utilized at much higher than average costs of production. In such circumstances it seemed that large extensions of plant would be necessary. In early 1970, P. D. Block of Inland Steel forecast that within ten years national output could be over 180 million tons—if all went well. Expansionist thinking continued through the next few years. In early autumn 1974, the Munich meeting of the International Iron and Steel Institute was told of plans to add 240 million tons to the world's 750 million metric tons steel capacity. Of this increase, it was anticipated that 28.5 million would be North American. A year later there was dispute among experienced analysts as to the size and cost of the expansion. W. T. Hogan reckoned the United States required 25 million tons more capacity, costing about $486 per ton; D. F. Barnett put the need at no more than 15 million tons and believed it could be obtained for $300 a ton.[8] In fact, instead of increasing, output first fell back slightly and then more sharply so that the annual average of 1979–1981 was only 123 million tons. Capacity also declined, from 1977 to 1983 by about 9 million tons. As this went on global demand was also faltering. By the early 1980s, the former confidence in worldwide expansion had been shattered:

capacity was now reckoned at 200 million tons in excess of consumption. The problem of global overcapacity persisted, in 1987–1988 the US Steel chairman suggesting there were about 75–100 million tons too much.[9] Ten years later the latter figure still applied. Though there was a strong general reaction from the former mood of expansion, it proved selective both at home and overseas. Before 1990, capacity had been trimmed over a number of years in advanced industrial countries, often drastically. Meanwhile, large extensions continued to be made in the Third World, though even these were less than those projected in the mid-1970s. An important result of the shortfall in global demand was an increase in international trade in competitively priced steel products.

By 1990 United States capacity was put at about 100 million tons, roughly one-third less than in the early 1970s, but this striking decline hid the fact that the structure of the industry had also been substantially changed during this period. Since the United States first became a net importer of steel in 1959, foreign companies have supplied a varying but generally growing part of home demand. From the 1960s, various programs for import restraint were agreed between industry and government, but inflows remained at a high level. Denouncing this growing flow from overseas in terms sometimes amounting to moral outrage, company spokesmen, perhaps understandably, failed to show much understanding of the realities and aspirations for economic advance in many other nations. Most of the imported material came from the EEC or Japan, but in spring 1983, in a presentation at Washington, the US Steel chairman also roundly denounced growing imports from the developing world:

[O]ver the past five years Third World nations have been building steel industries, financed not by private capital but by capital borrowed by the governments of those nations from world money markets. To my knowledge, every single steel plant built during this five-year period in emerging countries is either owned by the government or is massively subsidized and protected from fair, worldwide competition. It is an irony that the steel produced has not been for Third World consumption, but primarily for export. From the standpoint of the American steel industry, these exports couldn't be more unfair or damaging: steel coming from new and efficient mills financed by readily available international capital . . . , steel from mills owned or subsidized by foreign governments . . . , steel produced by some of the world's lowest-paid workers . . . and steel unloaded on American shores at prices consistently below the cost of production. Little wonder that the American steel industry cries foul as it watches its domestic markets overrun by cheap, artificially priced imports.[10]

In 1984, at what was then a record 29.5 million tons, foreign mills supplied 26.6 percent of total consumption; four years later, under so-called voluntary restraint agreements, they were at 20.3 percent. Imports were at a low of 17.5 percent in 1990 but afterward climbed irregularly to some 23 percent in 1996. By October 1998 they were running at an extraordinary 35 percent of apparent supply. Early in 1999 this led to new import restrictions. There was, however, an important constraint on government action to reduce the inflow. Domestic steel-consuming industries have argued persuasively for their own need to keep costs down so as to meet overseas competition. They point out that they employ far more workers than the steel mills—one estimate of the early 1980s was twenty times, another a decade later roughly thirty times as many.[11]

The question of lower overseas costs of steel production deserves fuller attention. There seems to be at least some evidence that in the 1970s and into the 1980s such a gap existed, but that well before the end of the century its existence was more doubtful. In the earlier period, many foreign mills were better located for access to the world's best or cheaper raw materials. They were of later construction and therefore incorporated improved technology, including bigger blast furnaces, newer steel-making processes, continuous casting and more efficient rolling mills. As late as 1984, 9 percent of American steel was produced in open-hearth furnaces as compared with none in either the EEC of Japan. By the mid-1980s, EEC countries processed 65.1 percent of steel in continuous casters, Japan concasted 89.1 percent, but the United States produced only 39.6 percent of its steel in this way.[12] As a consequence of superior technology, foreign competitors had taken over America's traditional leader-

TABLE 20.2
Imports of iron and steel products into the United States
(thousand net tons)

1950	2,029	1975	13,919
1955	1,467	1980	17,878
1960	4,088	1985	27,632
1965	11,964	1990	17,169
1970	14,609	1995	24,409

Sources: AISI annual statistical reports.

ship by having work forces that were much smaller, and in claiming that high productivity was a more important asset than low wages. Like the United States, they had a great deal of capacity surplus to the needs of their home markets, and most of them had been traditionally more interested than the Americans in export markets. An interesting pointer to the situation was that in 1986, the ratio of steel production to home consumption was 122 percent for the EEC, 141 percent in Japan, and only 78 percent in the United States.[13] Improved efficiency meant that foreign industries, with large excesses of production over consumption, could penetrate more of the United States market. By 1978 the price for a "basket" of domestically produced steel on the Atlantic Coast was $340 a ton; the Japanese were reckoned able to deliver it there for $320 a ton. There is however an important caveat: at the time at which Japan had such a marked general advantage, it was estimated that production costs at the best and worst plants within the United States varied by as much as $50 a ton. Clearly, not all companies or plants were equally in danger of succumbing to an onslaught from outside. By the late 1990s, many in the American industry argued that for their main overseas competitors, costs of production were not lower than their own, though the prices at which overseas companies sold in the United States certainly were. Some of these sales were made at below costs—that is, the steel was dumped on the market. This could enable the foreign producer to run at full capacity, though not necessarily to make money. Indeed, high levels of production are not always the best indices of commercial success.[14]

Unfortunately, at the same time as they were suffering from foreign competition, there was a sustained and mounting attack on long-established integrated producers from a rapidly expanding and quite different steel sector at home. In this case, there was no possibility of providing any nationally coordinated defense. Ever sharper rivalry has been coming from domestic "mini-mills." Such operations, though in some cases originating decades earlier, first became important competitors for the bigger, fully integrated companies in the 1960s, when they received their name. Such operations short-circuit the traditional production cycle by cutting out blast furnaces and all the manufacturing and mineral operations on which large-scale iron making depends. Instead, they operate large electric furnaces using scrap, employing continuous casting whereas the bigger producers still depended for their semifinished material on primary mills involving both much larger capital outlay and higher process costs, and finishing their steel in recently built rolling mills.[15] In the 1960s, mini-mills were generally small—from 60,000 to 200,000 tons a year—and their products were largely confined to such low grades as reinforcing bar and other bar and channeled products. They have increased in number, size, and range of products annually.

As compared with integrated operations, the biggest economy of a mini-mill is in capital outlay. As long ago as 1969, it was said that a new BOF steel plant with two 300-ton vessels giving a steel capacity in excess of 4.25 million tons of raw-steel a year, would require the iron output of two 6,000-tons-per-day blast furnaces. Such an installation would cost about $400 million. The same tonnage of steel could be derived from eight 150-ton electric-arc furnaces costing $7 million each, a total outlay on the "hot end" of a mini-mill operation of less than $60 million. A decade later the president of a leading mini-mill firm, Nucor, presented an even more favorable comparison: "[T]o put up a new integrated facility costs about $1,000 a ton of annual capacity. We've never put one up for more than $100 a ton."[16] Mini-mills have other advantages. Their recent formation and greater freedom of locational choice, especially from the constraints of mineral supply, enabled them to develop in areas having good access to markets, often bringing with this advantage the additional benefits of abundant local scrap supplies and lower wages than in major metal-producing centers. In fact, their distribution differs sharply from that of the integrated industry. Their new-style technology has often been accompanied by more flexible managements and less of the traditional hierarchical divide between blue- and white-collar workers. They are generally associated with nonunionized labor. Sometimes their labor is cheap—the head of Nucor made much of the attraction of rural areas for their expanding operations, stressing the suitability of ex–farm workers as steel-plant operators—but often the men are paid as well as those in the integrated industry; the attraction of a lack of unionization lies more in the absence of job demarcation and in general flexibility of outlook and practice than in the size of the wage packet. Productivity is consequently high, and the primary stages of traditional processes are sidestepped. In the late 1980s despite recent improvements, many integrated firms still required 5.5–6 man-hours per ton of steel products; the figure for mini-mills was then commonly 1.5–2 man-hours. A few years later the fact that their labor productivity was often 200–300 percent better than at many integrated works was reckoned to give mini-mills up to a $100 a ton cost advantage.[17] Their smaller size also meant quicker response to market opportunities.

The Achilles' heel of mini-mills has always been recognized as lying in the possibility of high or rising charges for electricity and scrap. Year after year, though, this threat failed to materialize. In fact, from the mid-1950s and through the 1960s, the scrap price generally drifted down relative to the cost of pig iron. At the beginning of this period, there was a difference between them of less than $10 a ton. By the 1960s, though, the differential in favor of scrap was well over $30; pig was then $60 a ton and number-I heavy melting

TABLE 20.3

**Estimated distribution of raw-steel capacity in integrated works
and mini-mills, 1993 (portion of national totals)**

Region	Integrated works	Mini-mills
Illinois	4.90%	18.54%
Indiana	33.89%	5.53%
Ohio	19.81%	2.21%
Michigan	12.46%	2.66%
Pennsylvania	8.45%	4.15%
All other states	20.48%	66.90%
Tonnages	69,340	18,070

Based on: Iron and Steel Engineers, *Directory of Iron and Steel Plants* (Pittsburgh: AISE, 1994).

Note: The figures for integrated works are more reliable than those for mini-mills.

scrap, the most attractive material for steel makers, was $28–30 a ton. Later, other factors were involved. The advance of continuous casting meant that much less process scrap was being produced than had been the case with primary rolling mills, but at the same time the collapse of open-hearth steel making removed an important rival for the same material. Generally, the trend of scrap prices provides a check to mini-mills when steel markets are buoyant but gives them a boost in times of lesser demand, for the scrap price rises and falls respectively in these two phases. In contrast, the cost curve for integrated works is flatter. In the 1990s, when scrap prices did at length seem likely to rise to inconveniently high levels, mini-mill companies campaigned to restrict scrap exports—by mid-1999 amounting to as much as 500,000 tons per month—but without much success. Seeing the warning signs, some of them also began to seek out cheaper alternative methods of supplying their iron requirements than the building of conventional blast furnaces. These have included investment in directly reduced iron (DRI). Sometimes, they have ventured into pioneering technologies such as direct steel-making processes based on iron carbide, but these have not been successful so far. Inevitably, investment in such avenues of actual or possible advance increases the mini-mills' capital investment per ton of products shipped.[18] Through the late 1990s, scrap prices continued to be rather volatile. The break point, above which mini-mills were disadvantaged on the material account, was then reckoned at about $130 a ton; any price below that level favored them. In 1997 scrap was $140–150 a ton; by fall 1999 it was down to about $100.

Gradually, the mini-mills have extended their operations into a range of products previously beyond their scope. By the end of the 1990s, no major integrated plant retained important heavy structural capacity; it had all been lost to these smaller rivals. Most threatening of all, as anticipated by some many years ago, the mini-mills began to move into sheet and strip, vital categories for the conventional industry and until this time made only by large and costly hot-strip mills. They did so by introducing a process pioneered by a German plant-equipment maker—Jones and Laughlin engineers seem to have gone far in developing a similar method before giving up the search. It involves the casting of thin slabs that are then rolled down into strip-sheet in a strip mill. Nucor began this process in 1989 in a new works at Crawfordsville, Indiana, and its lead was soon followed by others. Though they cut out expensive reheating and some of the intermediate steps by which quality might suffer, in the earliest days their process only enabled them to make sheet "adequate for some purposes," but they were already expected to follow the same course as in their much earlier invasion of the bar market—gradually moving into higher grades and thereby capturing increasing fractions of the total demand.[19] A key step in this process was the introduction in the mid-1990s of techniques to provide a "calmer" meniscus in the concast mold so as to give better control over the quality of the thin slabs, reducing metallic inclusions and surface blemishes that impaired the quality and limited the use of their sheets for the most exacting purposes.[20] In late 1993 three companies alone—Nucor, the new Steel Dynamics, and the Canadian group Dofasco—were already reported to have or to be building on the order of 6 million tons of mini-mill sheet capacity. As one engineer tellingly put it, this was more or less equivalent to the capacity of the Gary strip mill and would be competing in a market already well served by existing plants.[21] By 1995, $3.5 billion had been invested in flat-rolled product mini-mills in the United States and Canada. That year they produced only 3 million tons of hot-rolled bands and other thin flat-rolled products; by 2000 they were expected to ship 14 million tons.[22]

As with the general situation of the new as compared with the older steelmaking methods, savings on both the capital and production accounts are immense when high-grade thin flat-rolled products are made in mini-mills—by early 1996 the cost-saving potential of new technologies already known were put at 35–40 percent.[23] Already, through the 1990s, competition from the new thin-cast sheet mills caused a huge investment in upgrading existing strip mills, as at Inland Steel, Armco, Gulf States Steel at Gadsden, and US Steel's Gary mill, a development that seems to threaten yet more overcapacity in this vital part of the industry.

Why then in this sector particularly and in the steel industry generally have large companies not responded to the challenge with a wholesale abandonment of their traditional mineral-iron-steel-finishing cycle of production and its replacement with their own electric steel, concast, and mill capacity? The answer is partly found in the existing high investment they would have to write off, in fears of the long-term trend of scrap prices if there is a complete switch to electric steel, a shortage of funds in a setting of increasing uncertainty, and finally the recognition that integrated operations can still offer important advantages of their own. The productivity gap between integrated and mini-mill operations has narrowed. Now, the best-integrated firms can match the profitability of the mini-mills. By the end of the 1990s, it was estimated that in making hot-rolled merchant-grade band, mini-mill productivity was of the order of 0.75 man-hours per ton as compared with 1.5 man-hours at the best integrated works. But the latter improved their standing in being equipped to also turn out the value-added products such as cold-reduced band that mini-mills were only beginning to make at the end of the decade. Over the six years to 1999, operating profit per ton at AK Steel (the former Armco, now joint-owned with Kawasaki) exceeded that at Nucor. In short, the alleged cost advantages of the mini-mill sector are by no means as apparent as often claimed and certainly not sufficient to justify the provocative suggestion by the chief executive officer of Nucor that, if he owned an integrated works, he would swallow his pride and sell pig iron to the mini-mills.[24]

All in all, smaller steel makers have made striking headway. In 1976–1977 they produced no more than about 5 percent of the nation's steel. This increased to 15 percent by 1981, 22.7 percent ten years later, and an estimated almost 50 percent by 1998. In 1997 their leading representative,

TABLE 20.4

**Steel ingot capacity of integrated works and mini-mills in the United States,
1973 and 1993–1994 (thousand tons)**

	Number	Ingot capacity	
		1973	1993–1994
Integrated works abandoned, 1973–94	25	69,150	n.a.
Continuing integrated works	20	84,870	69,340
Mini-mills	31/33	5,005	c18,070

Sources: R. Cordero, *Iron and Steel Works of the World,* 6th ed. (London: Metal Bulletin, 1974); Iron and Steel Engineers, *Directory of Iron and Steel Plants,* (Pittsburgh: AISE, 1973, 1993, 1994).

Note: For an additional three mini-mills in 1973 and one in 1993–1994, no capacity figure was available.

Nucor, was already in tonnage terms the nation's third biggest steel producer. At that time one executive, admittedly a mini-mill operator, forecasted that in ten years time as much as 90 percent of the nation's steel production might come from three integrated works and six or seven big mini-mills.[25] Already, over many years and involving a wide range of finished products, a veritable massacre of integrated works has been occurring. In the twenty-two years to 1995 during which the raw-steel capacity of mini-mills increased from about 5 million to 47 million tons, that of integrated works fell from 154 million to 64 million tons.

Over this long period there have been important changes in the raw-material supply situation for integrated producers. Energy costs have become burdensome, partly because of higher purchase prices, partly because of the increasing obsolescence of the plant in which the energy is converted—old coke ovens and blast furnaces, too many surviving open-hearth furnaces, and primary mills involving reheating costs—at a time when many foreign steel makers as well as domestic mini-mills have turned more rapidly to continuous casting. Regarding iron ore, though, the tables have at last been turned on the traditional practices in the industry. After providing rich, direct shipping ore for more than a century from huge open-pit operations in which the level of production could be fairly quickly and relatively cheaply adjusted up or down, output from Lake Superior now depends on capital-intensive installations that convert 25–30 percent Fe taconite rock into pellets of 65 percent or so iron content. Technically speaking, this process is highly successful, but over the last quarter-century purer, higher-grade ores have been opened in many parts of the world. Again cheaply worked in open pits; often handled by newer, higher capacity, and lower unit-cost plant; and shipped without expensive up-

TABLE 20.5
A comparison of Nucor and US Steel, 1984 and 1997

	Steel production (million tons)	Sales (millions)	Profits (millions)	Net income per dollar of sales (cents)
1984				
US Steel	15.1	$6,488	$154	2.37
Nucor	1.5	$660	$44.5	8.40
1997				
US Steel	12.3	$6,846	$575	6.74
Nucor	9.7	$4,184	$294	7.04

Based on: US Steel annual reports for 1984 and 1997 and Nucor statistical record.

grading, these ores have become available to foreign companies much more readily than either present-day Great Lakes supplies or other foreign ore can be delivered to the generally interior-located American furnaces. Together in 1985, Minnesota and Michigan produced 98 percent of home ore; the average price of overseas ores that year was 40 percent lower than that of Lakes ore of the same iron content.[26]

Labor, of course, has much to lose from rundown in steel, and given the often-confrontational stance of the firms throughout this century, it is understandable that workingmen should oppose rationalization programs that in the relatively short period between 1973 and 1990 reduced the industry's employment from over 500,000 to 170,000. Modernization nearly always sweeps away jobs, and workers will usually resist such a process even when it can be argued persuasively that absence of draconian measures would eventually lead to even greater lay offs. Regardless, over the past two decades, labor numbers have shrunk much more than output, and the men have often felt compelled to accept new, less favorable working and pay arrangements as the precondition for continuing operations and for some, the retention of their jobs.

As society has become more concerned about environmental quality, legislation has imposed higher standards on a basic industry in which, traditionally, pollution was seen as an unavoidable accompaniment of production. The United States has been in the forefront in requiring higher standards, and large old-style steel plants have borne much of the cost. Unfortunately, pressure in this direction grew rapidly at the same time that the industry was moving into its crisis of overcapacity. The firms had to make huge outlays to cut out a great deal of plant, replacing it with units not only more efficient but also environmentally cleaner. A 1975 report from the American Iron and Steel Institute and the Arthur D. Little Group suggested that until 1983, investment expenditures should be $5.5 billion annually. Of this, $2.3 billion would be for plant replacement—largely demolishing open-hearth melting shops and installing oxygen converters or electric furnaces—$1.7 billion to meet an anticipated need for expansion of capacity, and $1.5 billion for the socially desirable but, from a company point of view, unproductive, fields of pollution control and environmental improvement.[27]

All steel producers have been adversely affected by a severe cost-price squeeze. There have also been institutional obstacles to all attempts to break out from these difficulties. Both the squeeze and the slow reaction to it may be combined under the heading of too much capacity and resulting low operating rates, inefficient plant, and high costs for inputs. The tendency for capacity to grow more quickly than production has been a characteristic of the industry for more than a century, but in the past a long-term general upward trend in demand sooner or later justified any expansion. Frequent bonanza years compensated for each run of bad ones. But since the early 1970s, depressed consumption has meant long-term underuse of capacity, and low operating rates resulted year after year in poor earnings in relation to capital. In 1970 the return on net worth in steel was only 4.4 percent, ranking the industry fortieth among forty-one manufacturing groups; by the latter part of that decade, average return on equity was 7 percent, only half the average for all manufacturing. Poor performances, low prices, inadequate returns on investment, and bad prospects meant it was difficult to raise funding to modernize plant. It certainly seemed impossible to hope that major "greenfield" operations could be viable. By the early 1980s the consulting firm Booz Allen reckoned such a project would cost as much as $2,000 per ton of annual capacity. A commercial operation should be looking for a 25–30 percent pretax return on capital outlay, which would necessitate a price of about $500 a ton. Steel was then selling at about $400 a ton.[28]

Such complex conditions remain a burden to a mature industry whose locations were chosen and infrastructure and plant built in a previous age. Although it can and must be modernized, it is difficult to make such an inherited structure fully competitive with new facilities and plants. In short, as the Old World realized too late a century ago, no industry can discount its economic history. Existing companies have to bear the resulting extra costs. The world rankings of leading American companies have slipped. For the bigger firms, an essential consideration in a strategy for survival has been the dispositions of other long-established integrated rivals. Inevitably, some proved better able to compete and survive than others, perhaps because of larger or more modern plant, a more viable range of products, advantages of location, or better direction and management. During this period there have been much reduction of capacity, many plant closures, a number of failures, and a series of mergers or joint ventures. Many companies that failed were reconstructed and rejoined the race, often freed as a result of declared bankruptcy from their burden of debt. In all this ferment the corporate map of the industry has been transformed.

In 1968, LTV acquired the until-then staunchly independent Jones and Laughlin. A year later, Lykes Corporation gained control of Youngstown Sheet and Tube. By 1984 LTV controlled Republic—approval of the United States Department of Justice for this acquisition requiring the sale of the latter's Gadsden and Massillon works. When Lykes and LTV came together, a major new group was produced. Even so, by 1986 this new concern had to file for Chapter 11 bankruptcy protection. Alan Wood was bankrupt by 1977.

Table 20.6
**World rank and raw-steel output in million net tons of American
steel companies, 1972, 1976, and 1997**

Company	1972		1976		1997	
	output	*world rank*	*output*	*world rank*	*output*	*world rank*
US Steel	28.4	2	26.3	2	12.3	8
Bethlehem	17.0	4	17.5	4	9.8	14
Republic	9.6	11	8.9	14		(see LTV)
National	9.1	12	10.0	11	6.0	32
Armco	7.9	15	7.0	18	4.4	45
Inland	7.2	16	7.4	17	5.4	33
J&L	6.7	19	6.4	20		(see LTV)
LTV[a]	n.a.	n.a.	n.a.	n.a.	8.3	21
Nucor	0.1	?	0.4	?	9.7	17

[a] By 1997, includes Republic, J&L, and Youngstown Sheet and Tube.

Kaiser, once full of ambitions for steel in the West, collapsed, and the site of the once great Fontana works was soon afterward partially built-over. McLouth, the nation's pioneer oxygen steel maker, also failed and was no longer an integrated producer when reconstructed. Rather than invest in modernization, International Harvester sold its Wisconsin Steel Division in 1976. Four years later, mismanaged by its new owners, it too closed. Bethlehem withdrew from steel operations at Lackawanna, Johnstown, and in the mid-1990s closed those at Bethlehem itself. In 1971 National Steel acquired Granite City and in the course of the next few years made it the source of supply for its own finishing operations at Portage on Lake Michigan, though it continued for a number of years to refer to plans for full backward integration at the latter. Eventually, National disposed of its Weirton Steel subsidiary by sale to the workers, who seemed for a time to be pulling it back to viability, but by the late 1990s they too were running up quarter after quarter of major losses. A final example may highlight how difficult it was to survive these traumatic times. In May 1968 the Wheeling Steel Company acquired Pittsburgh Steel. During the early 1970s, Wheeling-Pittsburgh disclosed plans for a multimillion dollar development program involving acquisition of new raw materials and plant expansion, and over the five years to 1983 it spent over $500 million on modernizing its integrated works at Steubenville and Monessen. By spring 1985 the company was bankrupt. It was, however, revived and ten years later still owned 2.5 million tons of steel capacity, now confined to Steubenville.

Though their various burdens have been great, and some of them quite beyond their control, it is difficult to escape the conviction that, generally speaking, directors and top management in the industry were slow to react to the challenges of the times. There have been many extenuating circumstances, including the reluctance of the federal government to protect one home industry when others might suffer as a result, and failure or inability to raise prices to levels remunerative enough to permit obtaining money for an unceasing upgrading of plant and reduction of costs with the installation of the best contemporary technology. The industry's leaders sometimes appeared to have forgotten that no commercial supremacy won in the past could ever safely be regarded as part of the natural order of things. Although they have achieved a great deal in rationalizing their industry since the mid-1970s, in 1985 alone fifteen steel companies were removed from the American Iron and Steel Institute's lists. Even in those firms that survived, thirty complete works and 560 sections of works were closed. Between 1982 and 1986, the industry lost $12 billion, shut down over 40 million tons of capacity, managed to improve labor productivity by nearly 40 percent, and reduced costs by 35 percent.[29] Clearly there remains a wealth of expertise, resolution, and consequently competitive power. This provides the context within which US Steel has carried through its own extensive rationalization program.

The Chairmanships of Edwin H. Gott and Edgar B. Speer

In 1969 Roger Blough was succeeded as chairman and chief executive by Edwin H. Gott, who four years later was replaced by Edgar B. Speer, who in turn was forced by illness to retire in spring 1979. In such a large organization as US Steel, it is tempting to assume that a single individual cannot move and give direction to the whole, that what happens is a product of collective thought and decision making. Nevertheless, earlier experience with Elbert Gary, Myron Taylor, or Benjamin Fairless suggests that the times call forth the man responding to opportunities or capable of tackling the problems, and that a particular leader may give a distinctive flavor to policy at any one time. Both of Blough's immediate successors were "steel" men. Just before he took over, Edwin Gott was described as a "production wizard . . . a hard driver who has pressed for plant modernization in the face of stiff foreign competition."[1] His period of office began with thoughts of considerable further increases of steel capacity both for the nation and the company, but it turned out to be far from the most favorable time to inherit the top job. Except for his last full year, annual shipments fell, income as a percent of sales fell, and raw steel output declined from 25.5 percent of the national total the year Gott took charge to 23 percent four years later. In the difficult trading conditions of the mid-1970s US Steel was even harder pressed to retain its position. Speer's incumbency ended with little sign of any root-and-branch reconstruction but rather the oncoming of a far greater crisis than any since the early 1930s. A year later, US Steel accounted for only 20.8 percent of United States steel output.

Publicly, at least, Gott firmly identified himself with expectations of continuing expansion. In 1967, as president, he had spoken of a world steel mar-

TABLE 21.1
United States Steel Corporation, 1969 and 1979

	1969	1979
Raw steel production (million net tons)	34.7	29.7
Steel products shipped (million net tons)	22.4	21.0
Number of integrated works	11	9[a]
Number of employees (thousands)	204.7	171.6
Steel products per employee-year (tons)	109.4	122.4
Income/(loss) (millions)	$217.2	($293.0)

Sources: US Steel annual reports for 1969 and 1979.
[a] By year's end.

ket increasing 40 percent over the next decade and affirmed, "US Steel intends to be ready to meet this demand." Early in his chairmanship, he said: "World wide demand will grow faster than the pessimists dream. Forget that 2.5 percent growth rate everyone predicts. Steel growth could exceed 4 percent to 6 percent worldwide."[2] The Sixty-ninth Annual Meeting, held in Houston on 4 May 1970, gave Gott an opportunity to reaffirm his confessions of faith. Both the times and the place seemed auspicious. A few miles to the east at Baytown, the Corporation had already built a major new plate mill and was then completing an electric-furnace plant and continuous slab caster. He addressed stockholders on the upbeat theme of "a world of opportunities." Gott was optimistic as to US Steel's place in a future economic order: "The decade of the Seventies will be a time of great and exciting opportunities in this country and in the world. This nation is on the verge of becoming the world's first trillion-dollar economy. Before the decade is out, the world will become a billion-ton-a-year consumer of raw steel. These factors forecast a world of opportunity for US Steel. With our resources and capabilities in metals and minerals technology, and our skilled manpower, US Steel will participate profitably in this world of opportunities." It seemed a reasonable proposition. In some directions experience confirmed the cheerful expectations. Within a few years economic growth and inflation had ensured that the U.S. economy had indeed far exceeded the $1 trillion level. However, worldwide growth in steel output did not continue, settling instead at roughly two-thirds of the forecast billion tons. US Steel's own output followed an uneven course but on the whole fell away. In fact, in only two years would US Steel ever produce more raw steel or ship a greater tonnage of products than in 1970. In seven of the next twenty-four years, it incurred multimillion-dollar losses. In late 1972, as part of his final year-end statement, Gott was still projecting a worldwide increase in steel demand of 25 percent by 1980. It grew only little more than half as much. More striking still, United States output of crude steel in 1980 was 16 percent below 1972, and at US Steel the decline was greater than 24 percent.

As from its earliest years, US Steel controlled huge mineral reserves. In 1970 its holdings of bituminous coal were reckoned to be about 3 billion tons and of Lake Superior taconite, capable of being converted by processes already in use, as equivalent to 2.5 billion gross tons of pellets.[3] The former was equal to 153 years' output of coking coal for its own and others' use at that year's level, the latter to 320 years of current shipments (though these were substantially increased in the next few years). Whether or not keeping such massive reserves was good housekeeping seems then not to have been questioned. Indeed, the Corporation was still making large extensions to it mineral holdings. Some fifty areas worldwide were under geological "review," including nonferrous ores, for as Gott stressed, US Steel saw itself as "a large materials company, not just a steel company." Interviewed in February 1973 just before his chairmanship ended, Gott was still looking to strong benefits from mineral exploration, by now active in eighty countries. Seven years ahead, he reckoned, US Steel would "be so large in iron ore it won't even be funny."[4] He could not anticipate the sense in which that would indeed be true.

Despite general expectations of continuing growth, the early part of the 1970s saw the elimination of one integrated works, the 800,000-ton Duluth mill, "one of the oldest and in many respects the most marginal of the Corporation's existing steel plants." In the earliest postwar years, it had sold about a third of its output into what had long been regarded as its natural twelve-state market area and the rest to the works at Joliet, Illinois. From the first years of the 1950s onward, only about one-fifth of its steel had been processed by its own finishing mills. This could no longer be justified, and in spring 1970 the board decided to shut down the two blast furnaces, eight open-hearth furnaces, coke works, and bloom and billet mills. (Nevertheless, the coke ovens remained at work until 1977, when state pollution controls made them uneconomic.) The rod, bar, and wire mills continued, though, using brought-in steel. Sixteen hundred of the 2,300 workers lost their jobs. The Realty Division planned to work with the community in turning the plant site into an industrial park.[5] Within a few years a similar fate would overtake many plants that at the time seemed fully secure.

When Edgar B. Speer became chairman he was fifty-six years old and had the prospect of over eight years at the post in which to make an impact on the course of US Steel's development. He had worked in turn at Youngstown and more briefly at Gary, Duquesne, and Fairless, receiving from Gott the enviable

Edgar B. Speer, chairman of US Steel, 1973–1979. *Courtesy of USX Corporation*

endorsement that no one in the world knew more about steel production. Gott had been able, energetic, and dignified; Speer was forceful but less easy than his predecessor in human relationships. Both men, brought up in the steel world, started with the presumption that things would go on as before, which meant that the United States—and US Steel—would continue to dominate. It is particularly clear that after twenty-three years in the works, Speer brought to his tenure at the top post a "romantic" attachment to this "Big Steel" image of the past. The public pronouncements of his early years were marked by expansionism. In February 1973 he claimed that as compared with most foreign operators, the American industry as a whole, because of cheaper raw materials, was probably more efficient by five dollars a ton. In contrast, some foreign companies had long been convinced that one of their great advantages was in not being tied to a fixed, controlled source of minerals. Speaking to Washington reporters at a dinner a few months later, Speer reckoned US Steel would expand by 14–15 percent over the next five years. As late as 1975 he seemed breezily confident: "There isn't any question the American steel industry is going to meet the demands of the economy. I'm not an optimist. I'm an historian. That's history."[6] His tenure of office began with production at a high level, though with intimations that all was not well. Before it ended six years later, conditions were much worse, and there had been a noticeable reluctance to introduce radical measures to cope with the problem.

Optimism was marked in 1972–1973 by modernization and new capacity in such a relatively low-value and—as was to prove—vulnerable product as wire rods in the South Chicago, Fairless, Joliet, Cuyahoga, and Pittsburg, California, works. Steel shipments were at record levels in the first nine months of 1973. Though claims that large expansion of capacity was needed had not been fully substantiated, US Steel was planning to step up capital spending from the $335 million average of the past three years to over $900 million, so that by late 1976 it would have 5 million tons more capacity. Oxygen steel making was expanded at Gary and 600,000 tons, or 30 percent, was added at Fairfield. Projections then suggested that the southeastern and the southwestern Gulf states would need an additional 7 million tons of steel products from all sources by 1980.[7] At the annual meeting in San Francisco on 5 May 1975, Speer was still reassuring in his outlook: "I have been associated with US Steel for exactly one half of the 74 years that it has been part of American industry, and I can't recall another time when the future has been more favorable." In fact, things were already beginning to go wrong.

In his final report on 1972 business, Gott forecast that long-term national consumption of steel would increase by about 2.5 percent each year. In his own first report, Speer had suggested a compound rate of 3 percent, but by 1975 he had returned to the Gott figure; the Corporation's own market research suggested 2–2.5 percent. Apparent domestic consumption of finished steel in 1973–1974 averaged 121 million tons. Assuming a 2 percent growth rate, this would give a 1979 figure of over 133 million tons. In fact, by that time consumption was down to 115 million tons, and whereas imports contributed only 10.6 million tons to the earlier total, by 1979 these had risen to 14.7 million. At US Steel the decline was a good deal greater: shipments of steel products averaged 25.9 million tons in 1973–1974 and only 21 million tons in 1979.

More immediately embarrassing for those who had so recently made expansive forecasts, US Steel's 1975 shipments were the lowest for fourteen years, though income was below only the record level of 1974. Speer explained the apparent anomaly by pointing out that they had been working for some time to extend nonsteel activities, "so that we can cushion the impact upon US Steel's profits of cyclical trends in the steel markets." Steel products were already declining rapidly as a contributor to group earnings. In 1973, the Corporation made 35.5 million tons of raw-steel and shipped 26 million tons of steel products. To supply this output, it operated ten fully integrated works, two other plants with steel and rolling mills, and at least eight other rolling mills, wire drawing plants, and other production sites. In addition to steel operations it had nine divisions and twenty "principal subsidiaries." Counting only the fully integrated operations, the Corporation's average output of steel per works was 3.5 million tons. By 1972 Nippon Steel had a steel ingot capacity of 50.9 million net tons and produced that year 36.2 million net tons of raw steel—from eight works. Moreover, all of its works were at tidewater—necessarily so given the geography of Japan—whereas, except for Fairless, none of the US Steel integrated operations were. In short, while the American steel producers were struggling with the legacy of a mature industry and contemporary operations in old locations, and US Steel in particular with its inheritance from the original agglomeration of capacity in a trust formed three-quarters of a century earlier, both the national industry and the Corporation were having to face up to ever more acute competition from newer, better located plants in an ever more global competitive setting. One estimate retrospectively made by a very senior official of the time was that in the mid-1970s, from 25–30 percent of US Steel capacity was good, though not fully world-competitive; roughly 40 percent was, by that standard, obsolete; and the remaining 30–35 percent might do well in years of good trade. It was a pity that there were now fewer of the latter.

US Steel plant and equipment expenditures over the following three years, 1975–1977, averaged $870 million a year. Unfortunately, in an apparent attempt to spread satisfaction with the modernization process, the spending was distributed widely rather than concentrated. As a result, there were world-competitive units in a number of the works, but no complete works of that standard. Consequently, over the next decade many of these excellent individual units would have to be abandoned along with the rest of the operations of which they were part.

Speer was guilty of scattering investment funds, but he also considered alternate solutions to their problems. It seemed an attractive proposition to short-circuit some of the labor necessary to make old industrial structures over into something better by choosing instead to build a state-of-the-art, integrated plant on a virgin site in a location that offered the nearest approach now possible to the advantages enjoyed by Japanese or European tidewater mills. Eventually it would be realized that there was not the essential growth in the national market to absorb the output both of new works and of reconstructed old ones. In other words, given financial and other constraints, the only practicable way of advance was to rationalize the old industry, however

TABLE 21.2
Annual indices during E. Speer's chairmanship

Index	1973	1974	1975	1976	1977	1978	1979
Sales and revenues (millions)	$7,031	$9,339	8,380	8,608	$9,610	$11,049	$12,929
Income (loss) (millions)	$313	$630	$560	$410	$138	$242	($293)
Steel products shipped (million net tons)	26.1	25.5	17.5	19.5	19.7	20.8	21.0
Employees	184,800	187,500	172,800	166,600	165,800	166,800	171,700

Sources: US Steel annual reports for 1973–1979.

severely, and make what survived afterward conform as nearly as possible to the ideal pattern that would have been adopted if one had been building the industry from scratch. In the case of US Steel, it took a few years and a change in the top command for this to be fully recognized.

Since its 1905 decision to build at Gary, Indiana, US Steel had constructed no fully integrated works on a greenfield site that was wholly free from government involvement. Local, state, or federal officials had been involved in Duluth, Minnesota, and Geneva, Utah, and had provided the impetus if not the origins for Fairless works in eastern Pennsylvania. In the mid-1970s, US Steel at last gave serious thought to another integrated works for a number of reasons. There was a general opinion that large expansion of national capacity would be needed over the next few years and that at least some of this would have to be in greenfield projects. Plans for expansion, and especially a willingness to undertake the long and costly business of building completely new mills, might be expected to encourage government to provide more effective restraint on imports. In the case of US Steel, executives recognized that the huge existing capacity in Pittsburgh or Youngstown was not only old but also inferior in location, and the costs of production in both areas were consequently high. Ever more exacting environmental standards, in particular, were pushing up operating costs in such heavily urbanized industrial districts. There was also an important gap in the deployment of US Steel capacity. It possessed three major concentrations of hot-strip-mill capacity in the western part of the manufacturing belt, at Gary, Pittsburgh, and Youngstown. By far the biggest market for thin flat-rolled steels was in southern Michigan. Major domestic rivals had strip mills either within that market area—as with National, Ford, and McLouth—or better placed either for land or lake delivery to it, as was the case with Republic and Jones and Laughlin in Cleveland or Bethlehem Steel at Lackawanna and over the last few years at Burns Harbor too. A final, imponderable, but significant factor was the allure, particularly to a "steelman," of the grand gesture of building a state-of-the-art plant. US Steel had constructed two works, which had been named after chairmen; was not this perhaps an appropriate time for a third?

In his November 1973 "Chairman's Newsletter," Speer managed to combine assertions about excellence and modernization with references to unfavorable circumstances and a necessity for expansion:

> The American steel industry has extensively modernized its facilities over the last several years and is technologically competitive with producers anywhere in the world. Clearly demand levels beyond those ex-

perienced during 1973 can only be met by further investment for added production capacity unless this nation is willing to depend even more heavily on the uncertainty of foreign sources which in time of short supply tend to be available only at prices well above the general level of the market. Such expansion to serve the nation's growing needs can occur only when profits of the steel companies are improved. Such profit improvement is currently arbitrarily inhibited by government restrictions. With several years required for the design, construction, testing, and break-in of major production facilities, such expansion programs should be initiated immediately.[8]

These points, together with complaints about environmental concerns, were to be the leitmotif of the next few years, during which there was little weeding out of plant. A year later he returned to the theme: "[The] only dependable source of supply for the steel needed in this country is to produce it in this country using American labor and the latest technology. Moreover we believe that steel produced here will cost less than steel produced in almost any country of the world for sale in U.S. markets."

At 17.5 million tons, US Steel's steel shipments in 1975 were lower than in any year since 1958. The following year was better, and it was at this time that a plan for a major new works was announced. It involved the huge site at Conneaut, Ohio, inherited from Carnegie Steel but undeveloped throughout the whole history of the Corporation. In February 1903 US Steel had announced it would build its new tube plant at Lorain, and two years later it was also committed to the Gary project, after which the only backward glance at Conneaut seems to have come in Andrew Carnegie's comment that he thought US Steel should have chosen his old Lake Erie site instead of Lake Michigan. During World War II, in a round of disposals of surplus properties at various plants, the executive committee considered a suggestion that the tract known as the Conneaut Farms and the improvements on them near the little community of the same name should be put up for sale. The committee drew back from this decisive step, instead recommending retention for "possible future use." Later it was said that Conneaut had been "in line" for a new works when the Trenton, New Jersey, area was chosen instead. In fact, throughout the next decades, a few cottages were built on leasehold along the lakeshore, but the rest of the area remained "a flat, vacant expanse of woods and brush . . . and nearly impenetrable hawthorn thickets."[9]

In late March 1976, Chairman Speer made a press announcement about expansion. He now reckoned that nationally 30 million tons of new capacity would be needed over the next seven years. Two-thirds of this would be in the form of extensions at existing works; the other 10 million tons would be in new plants. The Corporation wished to build a 4-million-ton integrated mill on a virgin site. Its projected cost was $3.25 billion, but to offset in part this extremely high capital outlay the new works would use 40 percent less energy and only half as many man-hours per ton of product as existing mills. Three locations were said to be under consideration—Baytown, Texas; a site on the Delaware River near Fairless works; and Conneaut.[10] In fact, it already had been decided that the new plant would be on Lake Erie. The Corporate Policy Committee considered a document entitled "Preliminary Evaluation and Request for Advance Engineering, Lake Front Steel Plant" on 29 April. A simple statement was made of the points in the plant's favor. The case began with the expected increase in demand, the need for new capacity to meet it, and for US Steel to build in order to maintain its share of the domestic industry. For a 30-million-ton increase in finished steel products, and assuming a 90 percent operating rate, about 50 million tons of additional raw-steel capacity would be needed. In order to retain its present 23–24 percent share, US Steel would require about 12 million tons more steel capacity. Of all consumption, 75 percent was in the eastern and central regions, which took 65 percent of the plate and 83 percent of the thin flat-rolled products. A new mill at Conneaut would produce about 18 percent of the national increase in finished steel by 1985, of which, at this stage, it was envisaged one-third would be shipped to other US Steel plants for finishing. With the report went a map of the proposed mill, indicating a three-step development. By the second half of the 1980s, it was anticipated Conneaut would enable US Steel to reverse its long-term declining share in the national market for hot-rolled sheets and plate, stabilize the situation in cold-reduced sheet, but not stop erosion of its position in tinplate. Within a year, the Arthur D. Little Company was undertaking a study of the project for US Steel, whose own Engineering Services Department had delivered a "General Description of Proposed Lake Front Plant." The scheme was now divided into two stages, the second of which would be completed by 1987. Each stage would involve 3.25 million tons of iron capacity (in the form of one 9,000-ton-per-day blast furnace) and 3.4 million tons of raw-steel capacity (two 300-ton-per-heat BOF converters). A 3.05 million ton hot-strip mill would be built in phase one and extended by 50 percent in phase two, which would also provide 1.38 million tons of plate capacity. But once completed, would even such a state-of-the-art plant be able to check the flow of imports? Speer claimed that at this time Japanese mills had a production-cost advantage over American producers of about $20 a ton of finished products. An American Iron and Steel Institute study had put the advantage at $70 per ton. In summer 1977 Merrill Lynch reckoned it might be as much as $83.65 a ton.

Speaking to the New York meeting of the AISI in spring 1977, Speer made

TABLE 21.3

US Steel share of U.S. markets for sheets, tin-mill products, and plate, 1955, 1965, 1974, and 1985–1988 as projected in 1976 (percentages)

	1955	1965	1974	1985–1988[a]
Hot-rolled sheet				
in Northeast	n.a.	n.a.	13	21
in all USA	24	17	15	20
Cold-rolled sheet				
in Northeast	n.a.	n.a.	14	13
in all USA	19	16	14	14
Tin-mill products				
in Northeast	n.a.	n.a.	25	23
in all USA	37	29	30	24
Plates				
in Northeast	n.a.	n.a.	32	42
in all USA	41	34	30	36

Source: US Steel Corporate Policy Committee, "Preliminary Evaluation and Request for Advance Engineering Lake Front Steel Plant," April 1976, USX.

[a] 1985 for tin-mill products and plate, 1988 for hot-rolled and cold-rolled sheets.

clear that a decision whether or not to build the Conneaut mill would depend on four points: restriction of imports, trade-offs in national energy policy, reform of the double taxation structure, and availability of financing.[11] A year later the 1977 annual report was guarded in its reference to the new plant. The need for the capacity was reaffirmed and conditions for proceeding briefly alluded to, but it also allowed ample scope for pigeonholing the whole thing: "Looking ahead, this nation will need new 'greenfield' plants if a substantial part of this country's steel requirements is to be provided by domestic producers. Such installations will benefit from economies of scale and from adoption of technology that can only be fully implemented in a new integrated operation. In anticipation of this, US Steel has engaged in extensive planning for a 'greenfield' plant. The first step is the development of an environmental impact study, which is nearing completion. Preliminary engineering is proceeding. Assuming favorable results from these steps, the Corporation will be well positioned to consider proceeding with this project when the cost-price relationship justifies the investment."[12] That May at the Minneapolis annual meeting, Speer was still diverting attention from current problems by reference to a new works. Circumstances and subsequent events seem to suggest that one of the reasons for reiterating the proposal was to exercise leverage on the federal government for more effective protection from imports. "We have already done considerable preliminary work on plans for a totally new steel plant along the shores of Lake Erie. Such a plant, using the most up-to-date technology, could increase productivity by perhaps 40 percent, requiring 30 percent less energy." If built, this works would have large blast furnaces and continuous casting. Its centerpiece would be a strip mill supplying the Michigan markets. Ultimately, it might have a steel capacity of 10 million tons, though the first phase might be only 3.8 million tons. After mentioning that environmental impact studies were almost complete, Speer added that tax legislation and protection from foreign steel would be necessary to make the project viable.[13] In 1978 the Army Corps of Engineers filed a draft environmental-impact study for Conneaut, and by the following spring the company had a permit from the Corps for the initial stages of construction—notwithstanding the forecast discharge of 11,000 cubic meters of effluent into Lake Erie every hour. Cost was now put at $4 billion for a 4-million-ton capacity, keeping the same $1,000 per ton relationship.

Most of Speer's senior colleagues were doubtful or opposed to the scheme. The finance committee time and again examined the likely costs and returns, concluding that the project would not be viable and would not even pay the interest on invested capital. But for the chairman, construction of the plant had become a personal crusade; as a senior colleague recalled—and emphasized—Speer desperately wanted the Conneaut mill. When defeated by cold commercial logic, he proved adept at suggesting one or another cost-cutting modifications—perhaps railing coke from the Mon Valley rather than building coke batteries, building a plate mill rather than a strip mill, or even installing electric furnaces and so avoiding the more costly blast furnace–BOF production route. Many of his colleagues recognized that modernization of existing works could be obtained for perhaps as little as one-quarter or even one-fifth of the proposed outlays. External circumstances, in the form of falling demand, and factors internal to US Steel were soon moving against the scheme as well, though eventually perhaps only an unforeseen accident was conclusive.

By late summer 1977 US Steel executive David Roderick was quoted in the commercial press as boldly declaring, "The cost-price relationship of our steel product won't justify such a $3 billion investment." In the course of the year, he summed up things even more concisely, saying he could see "no economic viability" for the project.[14] He pointed out that as compared with the projected outlay of $1,000 per ton of finished products, Bethlehem had spent roughly half as much at Burns Harbor. There was continuing dispute over the size of any compensating savings on the operating account. An outsider, P. E.

Schneider of the consulting firm International Ventures Management, put figures to the probable economies (at 1976 values) of $31 per ton on labor costs and $20 a ton for energy and materials. Substantial though these were, they would not cancel out the charges on the huge capital outlay. At this time the Canadian steel maker Dofasco, considering a major expansion of its own, reckoned that to make a greenfield plant such as Conneaut commercially acceptable an advance of 50–60 percent would be needed in steel prices.[15] It must have been obvious that no such increase was likely. In November 1978 Roderick stated that for Conneaut to be a practicable proposition there was a need to improve cost-price relationships, going on to make the interesting comment, "Foreign steel producers that are owned, supported, or controlled by their governments are, in effect, non-profit sources of supply." He continued to stress that serious and complex problems governing the economics of steel making would determine when construction got underway.[16] Roderick succeeded Speer in spring 1979, and in his introduction to the 1979 Annual Report gave what was effectively the *coup de grâce* not only to the Conneaut project but to all further expansion: "Until the economics of steel investments improve, the only responsible choice is to direct discretionary capital spending in steel toward maximizing return on investments already in place—not for major expansion of capacity to meet growing domestic steel demand."[17] The approval for Conneaut that followed the Environmental Impact Study contained a provision that work must begin by the end of 1983 and be finished in six years. Accordingly, when US Steel took no action, the project formally lapsed at the earlier of those dates. Instead of designing and building new plants, it was now necessary for US Steel to tackle the pressing issues of crisis management.

One or two final issues connected with the Conneaut project deserve brief attention. If built, such a plant, possessing both the finest technology and extremely heavy overheads, would have survived the recession and the root-and-branch rationalization program of the 1980s. However, it would probably have caused US Steel to join the list of steel companies that failed during the period. For the Corporation, it would have become the Lake Erie counterpart of Gary, and the resulting opportunity costs would have been dramatic. As its modern strip mill supplied southern Michigan markets, the more than forty-year-old Irvin works in Pittsburgh would have become marginal capacity during normal trading conditions and almost certainly redundant in the bad times and consequent retrenchment. With Irvin was bound up the longer-term hope for continuing steel production in Pittsburgh. In other words, if a works had been built at Conneaut, it seems likely that by 1990 Pittsburgh would have followed Youngstown as a once-great steel city in which it would have been found necessary to close all operations. In fact, as its first century ended, US Steel still operated Irvin and the 125-year-old Edgar Thomson works in which Carnegie Steel had begun its amazing career. Meantime, the lakeside tract on which the Speer works might have been built remained as it had always been, unindustrialized. Then in 1991, with striking irony, the largest part of it—over 3,000 acres—was dedicated as the David M. Roderick Wildlife Reserve.[18]

By the mid-1970s, levels of production, imports, and the cost implications of pollution controls were becoming ever more important and sometimes intertwined issues. To the Pittsburgh annual meeting in May 1977, Speer had noted the hoped-for increase in production had not occurred. He attributed this to weakened demand and imports, every million tons of which, he reckoned, cost 6,000 American steel jobs. He also targeted environmental costs for severe criticism, pointing out that over the last three years 11 percent of US Steel spending had been for pollution abatement. Emissions from the Clairton coke ovens and chemical plants had been a particular source of difficulty. In February 1972, the State of Pennsylvania and Allegheny County filed a suit to compel US Steel to comply with pollution laws there. Negotiations dragged on and the company embarked on large capital outlays to improve the situation. By the time Number 1 battery was relit in September, $6 million had been spent to enhance emissions control. Now, after years of court hearings, the 1977 annual meeting was told that the control measures had caused a dramatic cut in Clairton's effective annual coke capacity from 7.2 million to about 4 million tons. Speer traced out what he saw as the implications: "Ultimately that will reduce US Steel's position in the Mon Valley. It has to." Reduction of coke capacity by almost 50 percent might cause a comparable cut in steel.[19] By summer 1979 US Steel had negotiated an agreement on air and water quality in relation to its blast furnace operations in the area. Under its terms, emission controls would be fitted on six of the thirteen furnaces, the other seven would be shut down, and a new 5,400-ton-per-day furnace would be installed for Duquesne.[20] Worsening economic circumstances meant the last was never built.

In November 1977 David Roderick pointed out that steel imports were now almost 50 percent more than four years earlier. Falling demand in the second half of the year had led to steelworker layoffs. The situation was changing with bewildering rapidity, sometimes rendering marginal those operations previously profitable and regarded efficient. An example was the Wheel and Axle Division of the Homestead works located at McKees Rocks outside Pittsburgh, which made forged railroad axles on what US Steel characterized as "a high production, automated, precision forging machine" installed in 1966. This exceeded expected production rates and yields until 1974 when output began to fall due to physical deterioration. It made 41,000 tons in 1969 but only 31,000 tons 1974, and by late 1977 the division was operating at an annual rate as low as 13,000 tons. Its market share fell from 16 percent in 1974 to 9 percent in 1977. It was decided it needed rebuilding.[21] Soon, however, for most threatened plants, closure was a more common solution. In May 1978 in Minneapolis, Speer again complained about spending for environmental betterment, but above all he blamed imports for shutting American plants, causing the layoffs of thousands and the stranding of whole communities. He still insisted the long-term outlook was bright and that this justified their $6 billion capital spending program over the last ten years, of which one result had been that electrical and oxygen steel had increased from 20 percent of their capacity in 1968 to 70 percent—having replaced 123 higher-cost open-hearth furnaces. Continuous casting, requiring less investment, saving energy, and giving higher yields than conventional primary mill operations, was now used with 20 percent of their steel.[22]

On 24 April 1979, after struggling for well over a year against the advance of cancer, his increasing ill health caused Speer to resign as chairman, chief executive, and director of US Steel. The record of the Corporation during his chairmanship had been—for reasons in the main quite beyond his control— a dismal one. Raw-steel output and shipments of finished products had followed a variable course but had generally declined. Sales increased, but income after tax had drifted away and in 1979 turned into serious loss. As steel sales became more depressed, the share of US Steel earnings dwindled, so that in both 1977 and 1979 that aspect of the overall financial situation was rescued by the nonsteel sectors of corporate operations. As this occurred there was only a relatively slow decline in employment, and thus shipments and income per man fell. Dividends remained at high levels, but the income reinvested in the business underwent a relative decrease.

The Rationalizing of US Steel
after 1979

In April 1979, when the rapidly failing health of Edgar Speer forced his retirement, David Roderick replaced him as chairman of US Steel. The Annual Report for 1978 had referred to a continuing "weeding out process" that had brought about the permanent shutdown of "several marginal units"; there had, in fact, so far been relatively little of this. US Steel policy soon made a discernible shift, though, involving large reductions of capacity and disposal of assets, including reserves of raw materials. Although steel production and sales were now falling, the timing of the change in policy indicates that the transfer of chairmanship was an important factor in strengthening the company's resolve to make cutbacks. Twelve years later, two after his own retirement, Roderick looked back to 1979 and the changes he spearheaded after taking control. When he took the post, steel capacity was 35 million tons. One of the first things he had done was put in place a strategic-planning group bigger than before and with more "analytical strength." It was quickly realized that it would be impossible to modernize and therefore make viable more than half the existing capacity. Roderick described the situation: "In 1979–1980 [US Steel] was a non-sustainable steel division with marginal competitive aspects, non-competitive internationally, probably marginally competitive domestically. We ended that decade with a smaller group of operating clients, but one that was extremely competitive in steel domestically and internationally."[1] In retrospect, it had been an impressive success, but when the process began the outcome was uncertain. As time went on, those directing the process were forced to cut deeper than they had anticipated in its earlier stages, resulting in times of especially drastic surgery separated by quieter periods of stabilization.

TABLE 22.1.
US Steel assets, sales, and operating income, 1980,
by industry "segments" (millions of dollars)

	Assets	Sales	Operating income
Steel	6,022	8,738	58.3
Chemicals	788	1,487	70.5
Minerals & Ocean transport	1,100	664	76.3
Miscellaneous manufacturing/engineering	1,232	2,490	126.6
Domestic transport and utilities	787	644	144.5
Corporate assets, etc	1,819	(1,531)	1.5
Total	11,748	12,492	477.7

Source: US Steel, "Annual Report for 1980."

In the third-quarter report of 1979, which also carried notification of Speer's death, Roderick outlined the Corporation's business situation to stockholders: "We are continuing to study the economics of all our operations to determine their longer term viability. Many of these studies will be completed in the fourth quarter and appropriate action will be taken."[2] It marked a break with the previous practice of spreading new investment around the various plants, a process that all too often had vitiated the aim of making the finished product cheaper, better, and more saleable. From now on, investment for betterment would be highly concentrated on those plants judged to have a long-term future, the remainder marked for eventual closure through a process strikingly and chillingly characterized as "capital starving." A "Strategic Planning" group replaced the old so-called long-term planning process and gradually identified the sectors of business and individual plants with which it believed US Steel could do well. A first step was a listing of the plants by their costs of producing raw steel, but this proved of limited relevance since most of the man-hours and value added in production were downstream in the work of the rolling mills and finishing shops. Instead, the mark targeted was the world's best standards for blast furnace and BOP (basic oxygen process) yields and rolling mill outputs and qualities. Fairfield, a core of Mon Valley operations centered on Edgar Thomson and Irvin, Lorain, and above all Gary were the favored works; all the others were expendable if and when commercial circumstances suggested the axe must again be utilized. The process of concentration that resulted was drawn out over many years.

The first critical assessment of operations was carried out through the summer of 1979. Recommendations based on the analysis were taken to the board in the form of twenty-seven separate submissions for closure or disposal of units, and these were approved on 27 November. Next month the public learned that thirteen unprofitable or marginal operations would be closed, the write-off amounting to $293 million. Later, the full measure of the inefficiency of what was being abandoned was summed up in the statement that if these facilities had remained open in 1980, they would have lost US Steel about $100 million, equal to one-fifth of its income that year. During the early 1980s, further market changes and a fundamental shift in corporate structure made new rounds of retrenchment necessary. There was also an unprecedented diversification away from steel.

For many years US Steel had operated in a variety of activities, its interests in coal, chemicals, and construction, among others, overlapping into the spheres of other industries. From Roger Blough to Speer, policy had favored even more diversification, the general aim being to reach a balance of investment of roughly 50 percent in steel and 50 percent in other lines of business. In fact, by 1980 the "steel division" already made up only a little more than half of US Steel assets, accounted for more than two-thirds of sales, but provided only a very small share of operating income. Unfortunately, diversification had proceeded piecemeal, and as a result most of the operations acquired were closely related neither to steel nor to each other. For example, in 1969 US Steel formed US Steel Realty Development with the aim of building up a range of real-estate operations, including in 1970 two US Steel–owned hotels for Walt Disney World in Florida. In the early 1970s it added two financing subsidiaries. Interests in other fields of manufacturing took the Corporation in directions that, like its widespread investments in steel plant modernization, seemed likely to dissipate funds without focusing on good long-term prospects—with the added problem that for the first it had absolutely no expertise. In the latter half of the decade, there was an outstanding example of the temptations of this almost accidental spreading of interests. Senior members of the Rockwell Organization, whose headquarters was in the US Steel building in downtown Pittsburgh, suggested that the Corporation might consider interesting itself in the sorely troubled aircraft company Lockheed. Although Lockheed could then have been cheaply acquired, executives wisely decided that the proposed acquisition would not fit in with the existing structure.

For a time, an important line for expansion seemed to be in chemicals. Between 1963 and 1969, whereas all products and services sold by US Steel had increased by 33 percent, sales of chemicals had grown fourfold. Early in 1968 US Steel bought Armour Chemicals, a producer of chemical fertilizers, and formed US Steel Agri-Chemicals. In 1978, negotiations were said to be in an advanced stage with an unnamed partner for an investment of some $500 million for large-scale ethylene production. Unfortunately, world capacity for this basic "building block" chemical was already in excess of demand, and talks did not go further.[3]

During the early 1980s, by which time steel production was falling sharply, various companies were "monitored" as possible purchases. Oil and natural gas output was faltering at the same time, but there seemed every reason to expect domestic consumption of both fuels to remain fairly steady or even increase, making involvement in these fields a possibly effective counterweight to steel. A number of major petroleum and gas operations were considered such as Pennzoil of Houston and the Cities Service Company of Tulsa, which was also involved in industrial chemicals. Then attention was diverted suddenly in another direction. The Mobil Corporation made a bid to acquire its rival Marathon Oil. Its offer of $85 a share was rejected as inadequate by the Marathon board, which feared losing control of its own operations and of its best men. Almost on the rebound, it was captured by US Steel, whose very different lines of business meant that Marathon could retain control and personnel, within its own fields. At the end of 1981, US Steel bid $125 for each Marathon share as the first step to gaining overall control. By March 1982 the oil company's stockholders had approved the sale, and it became a wholly owned US Steel subsidiary. The link brought many benefits. In his report on 1982 operations, Roderick spoke of it as an "energy hedge for our energy dependent steel business" as well as a source of supply for chemicals operations. More important was the company's function as a counterbalance in earning power. In 1982 and 1983, while steel operations recorded an operating loss of $1.486 billion, Marathon's oil and gas sector brought in operating income of $2.348 billion. Moreover, losses made on steel could be used to cover some of the tax bill on the profitable Marathon operations. In 1986 the USX Corporation was formed to include the whole of the operations, which now extended so far beyond anything that could reasonably be included under the title of United States Steel Corporation. However in 1991, partly in reaction to the unwelcome takeover aspirations of Carl Icahn, steel operations and the oil and gas businesses were split into two separate groups with distinct stocks, though both controlled by USX; the steel sector was named the US Steel Group.

When the 1980s began, the "profile" of US Steel was 66 percent steel and 33 percent other business; in 1988, 59 percent of USX revenues came from energy, 34 percent from steel, and 7 percent from "diversified business."[4] After that there were further ups and downs in the fortunes of the two main divisions, with steel gradually regaining a more prominent position.

Some critics asked both at the time of the Marathon acquisition and subsequently what competitive benefits US Steel might have secured if the $6 billion purchase price had been invested instead in a widespread modernization of its steel sector. Given the capital outlay that had been envisaged for the proposed Conneaut, Ohio, plant, an investment of this amount could at least have given it on the order of six million tons of supermodern plant, though this would have represented no more than one-fifth of the Corporation's existing capacity.

Now an immediate effect of the Marathon acquisition was to further increase the urgency of cutting losses in some parts of the steel and associated divisions. Two further factors pressed the Corporation into a hastened reconstruction in steel. First, the U.S. government refused to provide greater protection for the industry, making cost reduction for these businesses more imperative than ever. At the end of 1982, announcing suspension of some of their plans, Roderick denounced the "predatory, discriminatory, and illegal" trading practices of Japan and the European Economic Community (EEC), which he suggested were "flagrantly victimizing" the steel industry of the United States by carving up the world's steel markets into "spheres of influence."[5] Second, there was a change in the top operational management of US Steel. William Roesch, president of the steel operation, had since 1978 pushed ahead vigorously with the program of closures and cost reduction. He died in December 1983 after a yearlong illness. His passing coincided with a national and corporate crisis in steel, but since May his successor as chief executive officer had already been in post. Tom Graham had been brought in from a

TABLE 22.2
Steel and energy sectors in US Steel and in USX, 1984, 1990, and 1997
(millions of dollars)

Year		Assets	Revenues	Operating Income	Op. income as a portion of revenue
1984					
	Steel	5,733	6,488	142	2.2%
	Oil and Gas	9,949	10,185	1,274	12.5%
1990					
	Steel	3,211	5,473	255	4.6%
	Energy	11,332	14,616	1,138	7.8%
1997					
	US Steel Corp.	6,694	6,941	704	10.1%
	Marathon Group	10,565	15,754	866	5.5%

Sources: USX Corporation annual reports for 1984, 1990, and 1997.

successful career at Jones and Laughlin. Over the next few years, his purposeful action as direct controller of USX steel operations helped concentrate investment in key areas and put a new emphasis on local initiatives. Coming in from outside the Corporation, Graham was appalled by the complicated hierarchical structure of US Steel and attempted to simplify it, aiming as far as possible to ensure there should be only one level of command between himself as chief executive and the plant superintendents. He also believed that few over age fifty could appreciate the changes needed in USX and, therefore, that new younger blood was needed among the top executives. As a result there was a large redundancy of older management, including those in charge of almost all their integrated works. His direct, no-nonsense approach, short-circuiting older, slower, and more bureaucratic systems of decision making, speeded the wholesale improvement in US Steel performance.

Rationalization of steel now proceeded over a wide front. In 1982 there was a dramatic fall in output. At 74.6 million net tons, national output was 61.7 percent of the 1981 level; US Steel made 51.7 percent as much as the previous year, and its steel sector recorded an operating loss of $82.20 per ton of steel shipped. The sudden deterioration of the market situation was especially severe in pipes and tubes. In 1981 sales of these products had amounted to 26 percent of all product groups, and the mills at McKeesport, Pennsylvania; Lorain, Ohio; Chicago; and Baytown, Texas, were highly profitable. It was then that the decision was made to go ahead with a new 600,000-ton seamless-pipe mill, Fairfield, Alabama, being chosen as its location in preference to older northern centers of production. Next spring, demand fell so sharply that only 20 percent of that year's US Steel sales were made up by pipes and tubes; in 1983 they were under 6 percent. It was resolved to go ahead with Fairfield development supported by supplier-financing arrangements, the oil compa-

TABLE 22.3

Share of the steel sector in USS–USX total operating income in years that steel recorded a profit

1981	52.8%	1994	36.3%
1984	9.1%	1995	79.6%
1987	13.4%	(1995[a]	81.5%)
1988	35.4%	1996	22.6%
1989	27.4%	1997	44.8%
1990	16.4%		

Sources: US Steel annual reports from 1981–1997.

[a] According to an alternative assessment.

nies contributing to the capital cost of the plant and later receiving its products in partial payment. Meanwhile, operations at McKeesport, threatened with closure or reductions time and again in previous years, now at last became a casualty. From 1979 to 1983, the US Steel share of national steel production had fallen from 21 percent to 16 percent, probably a steeper rate of decline than at any time in its history.

Production and shipments in 1983 were both only about half their 1979 levels. Late in December of that year, USX announced another set of plant and other facility write-offs. This was a far more radical program than that of four years before, involving closure of the whole or part of twenty-eight works and mines, cutting steel capacity from 31.3 to 26.2 million tons, and eliminating almost 15,500 jobs. Naturally, a favorable gloss was put on the process. Over the last two years raw-steel production from the various plants to be closed had been under 23 percent of their capacity. Eliminating them was expected to reduce 1984 shipments by less than 4 percent. This sort of paring at the margins still had some way to go. By late 1985, steel capacity was already almost a third below its 1979 level of 38 million tons. Sale of over $3 billion of assets, including withdrawing altogether from the once huge involvement in cement manufacture, had helped provide the finance to keep the whole operation afloat. In reaction to a sixth-month strike in 1986, there was a further series of closures including "various coke, sintering, blast furnace, and ingot processing facilities at Fairfield, Gary, and the Mon Valley," as well as indefinite idling for Geneva and the Texas works, except for its plate mill. Altogether in 1986–1987, 7.2 million tons of capacity were cut.[6] By 1990 raw-steel capacity had been reduced to 16.4 million tons. Seven years later it was 12.8 million, only a shade over one-third the 1999 figure.

At the same time as plant was being closed or sold, there were large-scale disposals of the raw materials, transport facilities, and other auxiliary operations, which had been built up and cherished over decades to provide long-term support for the immense size and width of the US Steel iron and steel operations. It was now recognized that these sectors unprofitably tied up huge amounts of capital and that many could no longer be claimed to be "strategic" resources. Their sale not only reduced costs but also produced more funds for upgrading surviving operations. Iron-ore reserves, accumulated from the earliest days and time and again extended after that, provided a notable example. In the case of Venezuelan holdings, withdrawal was forced on US Steel by unwanted action on the part of the host government. As late as the early 1970s, expansion had been underway at Cerro Bolivar to increase annual ore capacity to 24 million tons. Each year about 7 million tons was then being received by US Steel, where, especially if mixed with Quebec ore, it provided a furnace

TABLE 22.4
***US Steel annual raw-steel output and steel shipments, 1970,
by five-year averages in 1971–1995, and in 1997 (million net tons)***

Year	Raw steel	Steel shipments
1970	31.4	21.0
1971–75	30.6	21.9
1976–80	28.3	19.6
1981–85	16.4	12.5
1986–90	12.9	10.4
1991–95	11.2	9.9
1997	12.3	11.6

Based on: US Steel annual reports for 1970–1997.

burden of desirable chemical composition. As well as supplying Fairless, it had replaced Red Mountain ore in Birmingham furnaces and in considerable tonnages had penetrated as far into the interior as Pittsburgh and Youngstown. Then in 1974 the Venezuelan government nationalized the concessions, which in the normal course of affairs would not have expired until 1998 or 1999. There were transitional arrangements, but after losing control of the source, US Steel gradually reduced purchases of this ore until they were less than 3 million tons a year. As late as 1975 the Corporation was examining the costs of using ore from Carajas, in the far interior of Brazil, at Fairless, Baytown, and even at Fairfield and its Monongahela Valley works.[7] Soon afterward, in part because of a disagreement with the Brazilians over the form of development, US Steel decided to withdraw from further involvement. During the 1980s the Corporation also disposed of ore reserves in the western states and in Canada. In the mid-1970s great hopes had been entertained of the Quebec Cartier Mining Company (QCM), which having depleted its Lac Jeanine mine was opening a new complex of 20-million-tons annual capacity of concentrated ore at Mount Wright, Quebec. By 1977 Quebec supplies had become so important that their output was greater than that of the Corporation's Mesabi and western mines. Then in 1989 QCM was sold. Largely as a result of this, ore reserves fell in one year from 3.3 to 1.8 billion tons. By 1994 they were cut to 746.4 million tons, at which figure they were still equal to nearly forty-seven years' requirements at current production levels. From that point, though, the only important source of domestically produced ore, well able to support the lower levels of iron output, was the Minntac pellet operation in Mesabi, Minnesota; in 1980 executives decided that it was no longer necessary to go ahead with the final phase of development, Minntac Step IV.[8] Iron-ore production in 1970 was 54.9 million tons; by 1997 it was 16.3 million tons of pellets. Another means of securing short-term savings involved various sale or lease-back arrangements. Of these, a symbolic example in summer 1980 was the decision to sell the new ore carrier *Edgar B. Speer* on its delivery for $62 million and then charter it for a term of thirty-two years.[9]

US Steel controlled huge coal reserves, and those who were responsible for this section of its business found it hard to break from the timeworn conviction that sooner or later their output would be needed. At a rough estimate, the Corporation controlled the equivalent of four hundred years of supply; it had been paying taxes on this tonnage in the ground since its earliest days. Coal production fell from as high as 19.6 million tons in 1970 to below 7.5 million in the late 1990s. Into the mid-1970s new mine capacity was being developed. There was, for instance, expansion for the first time into steam coal for sale, investments being made at the Cumberland mine in Pennsylvania to supply Canadian power stations. As late as 1978, investment of $87 million in a 2.6-million-ton-per-year output to supply power-station needs was planned for Westville, Illinois. Eventually, it was shown that these mines could not compete in the markets for steam coal with western strip mines, and the argument that reserves were necessary to support this line of trade became untenable. Three coking-coal developments—in Alabama; Wyoming County, West Virginia; and the long-established Dilworth mine in Pennsylvania— were then expected to add 5 million tons of high-grade metallurgical coal to annual capacity and some 30 percent to all types of coal production.[10] But as trading conditions worsened and a new top direction took over, large-scale disposals began. Time and again proceeds from such sales made important contributions to US Steel's income. For example, in 1981 when third-quarter

TABLE 22.5
US Steel mineral holdings and production, 1980–1998 (million tons)

Year	Iron ore		Coal	
	Reserves	Production	Reserves	Production
1980	4,860	32.4	2,966	14.6
1985	4,331	27.8	1,924	11.9
1990	833	14.7	1,561	11.5
1995	731	15.5	863	7.5
1998	739	15.8	790	7.3

Sources: US Steel annual reports for 1980–1998.

profits were over $500 million, a large part of this income was attributable to
the sale of coal properties to Standard Oil Company, Ohio.[11] In 1984 coalfields
in the Lynch district of eastern Kentucky, bought in 1917 and accounting for
10.5 percent of all US Steel reserves, were sold; nine years later production of
steam coal ceased. Also in 1984 USX sold off the Robena and Dilworth mining complexes along the Monongahela, once regarded as vital to the success
of its regional coking operations. By the late 1990s, it had no coal properties
in Pennsylvania, Clairton by then being wholly dependent on mines farther
south. Limestone properties in Michigan were sold in 1987, the partnership
that bought them agreeing to supply US Steel's requirements of that material
for the next ten years.

During this period US Steel gradually disposed of a large number of interests in manufacturing, transportation, and services, many of which were
deeply embedded in its history and, though previously perceived as essential,
had been by no means always commercially beneficial. It sold the Universal
Atlas Cement Division to a German company for over $100 million. Time and
again attention focused on operations or purchases of goods or services that
represented intracorporate transactions and had hitherto not been properly
costed or had deterred outsiders from using the same sources. An interesting
instance from the service sector was the Corporation's major warehouse operation, US Steel Supply Division, which in a more confident age had been
called "department stores for steel." (Customers who bought direct from the
mills sometimes felt—wrongly—that those who bought from Steel Supply
might be getting their materials cheaper.) The division was sold off in 1986.
American Bridge had been an important part of US Steel from the beginning.
It had fabricating yards across the nation and usually a department within
each of the major US Steel works so that it would be the natural choice for any
building operations there. As a result of the former, many independent fabricators had been reluctant to purchase materials from US Steel and the latter
had often meant that plant managers failed to seek out the cheapest solution
to their construction problems. In 1987 American Bridge was sold to a company owned by the division's president at the time of sale. The Corporation
had bought its natural gas supplies from the Carnegie Natural Gas Company.
It became clear in the mid-1980s that this was a high-price source, and the
new rationalizing top management accordingly stopped purchases and
bought more cheaply from outsiders. Later, Carnegie Natural Gas was transferred to Texas Oil and Gas, a newly acquired USX subsidiary. In 1988 US Steel
sold majority interests in its Great Lakes fleet, railroads, and barge lines.

The chemicals division contributed an operating income of $141 million
to US Steel in the three years 1979–1981. Over the next four years it lost $2

million. In 1989 USX sold off US Steel Agri-Chemicals. Another casualty of the enforced rationalization program was much of the inflated research operations. The Monroeville Center, started in the mid-1950s, conducted a good deal of basic research, work not expected to have any immediate practical application. A bigger problem was that the center suffered from the delusions of grandeur not uncommon in various other parts of the Corporation, in this instance taking the form, "If we haven't found it, its not worth finding" (as one senior executive put it). In fact, some Monroeville work, for instance that on continuous casting, proved to have been misdirected. In 1983 employment there was about 3,500; in the course of the following eight years, that number fell to about 500.

As well as being involved in sweeping closures and disposals, US Steel during this period made important moves toward new alignments in steel. There was some consideration of a link with Republic Steel, but further examination showed that this would increase the size of the task of rationalization. The most radical possibility was frustrated by an outside decision. In the mid-1970s, the Conneaut project had seemed at last to promise improved access to prime Detroit-area markets. A few years later US Steel seemed about to follow a different route to the same end. Plans to buy National Steel for $575 million were made public in 1984. This would have brought in operations already slimmed by more than half their 1980 capacity, partly by the 1983 sale of Weirton works to its employees. National had a greater emphasis than US Steel on light flat-rolled products, lower costs of production—partly because of smaller fixed costs—and superior plant locations. At that time less than 30 percent of the Corporation's steel was concast, whereas at National the proportion was already between 60 and 70 percent. In the second quarter of 1983, US Steel lost $82 on every ton of products shipped, but National made a profit of $9; for the whole year National lost $154 million, while US Steel lost $1.160 billion. The Corporation also had something to contribute to a merger, particularly reductions in raw-material costs, above all for coke. Overall, it was estimated that the improved efficiency resulting from the "meshing" of the two operations could in a "normal" year increase their combined pretax income and cash flow by as much as $170 million. Unfortunately, the Department of Justice set unacceptable requirements. It wanted the new group to sell off two major units, Fairless works and the Granite City operations, which National had acquired in 1971. This was in order to ensure that the US Steel share of national capacity would not be increased, an argument that, as Roderick described it, "would have meant that we would be buying 5.6 million tons of capability, and disposing of 6 million." Under such circumstances, the deal fell through. Stepping into the breech in mid-1984, Nippon Kokan paid

$290 million for a half-share of National Steel. Thirteen years later US Steel and Inland Steel revealed that they had been discussing amalgamation for much of the last year but had called off further negotiations.[12]

Along a rather different line there was more success. An important step toward improved efficiency took the form of joint ventures, something new to the operations of US Steel, which in the past had always been large enough to carry through investments on its own account. In 1983 an attempt to link the future of the Fairless finishing mills with operations at the British Steel Corporation failed, but after the failure of the bid to merge with National Steel, US Steel succeeded in carrying through a number of arrangements with both foreign and domestic firms. A key instance was the sale of a 50-percent share in Lorain to Kobe Steel. This made it possible to finance the renovation of the Lorain blast furnaces, and while US Steel continued to market the plant's tubular products, Kobe developed higher quality bar production, which would supply Japanese automobile plants in the United States. By the end of 1995, the five-year, $500 million modernization program for Lorain was almost complete. A further joint venture with Kobe was for a new 600,000-ton plant to supply galvanized sheet to the same outlets as the Lorain bar mills as well as to American-owned automobile plants. This finishing operation was well located for these markets at Leipsic, Ohio, a small town some fifty miles southwest of Toledo, relatively depressed after the closure of a previous large manufacturing operation, and offering the additional attraction of a nonunionized labor force eager for new work. Leipsic further processed cold-rolled sheet from three strip mills: Gary, Irvin, and Fairfield. There were also new joint operations for sheet processing with two American firms, Rouge Steel and Worthington, both like Leipsic supplying the main southern Michigan plants of the motor industry.

The reconstruction programs involved major changes across the whole range of production units. Coke ovens were closed at Fairless, Fairfield, and Lorain, the coke for their furnaces now derived from Clairton, which even took over the task of supplying some of Gary's requirements. In the iron and steel departments, technical improvements spanned the whole range of the processes, including the more efficient production and use of energy and upgrading of blast furnaces, rolling mills, and finishing lines. Again, there was a new openness to cooperative action. For example, in 1985 US Steel joined Bethlehem Steel and the Department of Energy in research into the direct casting of one-inch-thick steel sections as an alternative to rolling them from slabs. Success in such a process was reckoned likely to eliminate 90 percent of the hot rolling involved with the product and save perhaps $45 a ton in energy, operating costs, and on capital equipment.[13]

Plans for a new US Steel have at last brought the abandonment of the Corporation's traditional comprehensive coverage of products so as to concentrate on those yielding higher returns. To this end, by 1983 its leadership had decided to withdraw from rail production, bringing mill closures at Gary and Fairfield as well as the end of the rail mill at South Chicago. Keen competition in bars and rods also forced closures. In 1983 10 percent of US Steel deliveries were of these products, but these figures dropped by 1986 to only 6.6 percent and two years later a mere 4.8 percent. In the expansive setting of the early 1970s, US Steel had put up four new bar mills to serve Midwestern markets; by 1987 all bar production was concentrated at Lorain, Ohio. During the early 1990s the Corporation withdrew from production of semifinished steels, and output thereafter was wholly confined to four product groups: sheet, strip, and tin mill products; plates; structural shapes and piling; and pipe and tubes. Even these four were not sacrosanct as the manufacture of structurals ceased in spring 1992. Sheet, strip, and tin mill products have grown greatly in relative importance, by the mid-1990s amounting to well over three-quarters of all shipments. The 1991 Annual Report recognized and claimed virtue in this radical change of direction: "We are no longer the 'super-market' for steel products of past years. Instead we are responding to selected markets."

By 1985 over 150 facilities had been closed and steel capacity reduced by over 30 percent since the start of Roderick's chairmanship. One after another once-famous works were abandoned or sold off.[14] In the 1970s and early 1980s, US Steel had operated with wide margins between its raw-steel capacity and raw-steel output and, partly because of the large use of primary rolling

TABLE 22.6
**US Steel finished-steel shipments by product, 1972, 1983, 1987, 1991,
and 1996 (portion of US Steel total)**

Product	1972	1983	1987	1991	1996
Sheet and tin mill products	40%[a]	64.8%	67.9%	73.6%	84%
Plate	–	16.6%[b]	12.1%[b]	19.2%[b]	9+%
Structurals	20%	–	–	–	n.a.
Tubular products	10%	5.1%	7.8%	7.0%	6+%
Bars and rods	18%	10.2%	4.5%	n.a.	n.a.
Semis and other	12%	3.2%	7.6%	0.3%	n.a.

Sources: US Steel annual reports for 1972, 1983, 1987, 1991, and 1996.

[a] Includes shipments of plate.

[b] Includes shipments of structurals.

mills, also between the latter and its output of finished steel products. Over the years it reduced steel capacity more than output, and the advance of continuous casting enabled it to narrow the gap between tonnages of raw steel and of finished products. In short, operations became tighter, making better use of the retained facilities.

Through all this protracted, painful process of rationalization there was a discernible geographical pattern. Many years ago, and in relation to another industry, S. Brubaker wrote, "Growth facilitates locational shifts, for it means that they can be accomplished without abandoning existing plants."[15] No doubt he was right, but industrial crises provide another, less happy context for locational change for, when they have to prune deeply, decision makers naturally take account of the most rational patterns for a future slimmer industry. The center of national steel production was still moving westward within the manufacturing belt, though in later years the huge growth and spread of mini-mills meant that this core area contained a smaller proportion of the total than had traditionally been the case. As the commercial situation pointed inexorably to the need for smaller capacity, the new technologies of large blast furnaces supplying highly productive BOP steel shops linked with high capacity continuous-casting units resulted in larger outputs from individual works and a focusing of production into fewer units. As late as the mid-1970s US Steel operated eleven fully integrated works. Fifteen years later three works survived, though another, Lorain, had passed into the hands of an associated company. Three main principles of action seem to have guided US Steel planners as they passed through this unprecedented commercial storm. Firstly, they chose to withdraw from those peripheral areas of the national economy in which, though consumption in total may be high, it is often scattered so that the "density" of demand is low. Such areas have attracted many of the mini-mills, and their position within the nation meant that they were particularly exposed to imports of foreign products. Secondly, the Corporation closed plants in poorer locations within the core area of the national industrial economy. Finally, within each of the main centers of continuing production, they concentrated operations in one major plant.

Those operations that survived the rationalization period as integrated operations were the ones that possessed oxygen steel capacity and a hot-strip mill. Homestead and South Chicago, without either and being primarily concerned with heavier products, did not survive. Youngstown, Fairless, and Geneva had strip mills but depended on open-hearth melting shops, Duquesne possessed a BOP plant but still delivered its steel to either its own bar mills or to the tube mills at McKeesport. For Homestead, Duquesne, and South Chicago, there was another factor against them: US Steel had other integrated

TABLE 22.7

**US Steel raw-steel capacity and production and shipments
of finished products, 1970, 1980, 1988, and 1997 (million tons)**

	1970	1980	1988	1997
Raw-steel capacity	n.a.	34.5	19.1	12.7
Raw-steel production	31.4	23.3	15.5	12.3
Shipments of products	21.0	17.1	12.2	11.6

Sources: US Steel annual reports for 1970, 1980, 1988, and 1997.

TABLE 22.8

Raw-steel production by selected states, 1956–1997 (portion of national total)

State	1956	1965	1977	1988	1997
Pennsylvania	26.4%	24.3%	20.5%	13.7%	7.5%
Ohio	19.4%	16.9%	17.1%	17.7%	14.6%
Michigan	6.1%	7.3%	8.0%	8.2%	6.7%
Indiana	12.5%	13.0%	17.1%	21.1%	23.2%
Illinois	8.4%	8.5%	8.7%	7.8%	7.3%
All other states	27.2%	30.0%	28.6%	31.4%	40.7%

Sources: AISI annual statistical reports.

works in the same areas, Edgar Thomson (ET) and Gary respectively. Gary was bigger than South Chicago and its layout was far superior. Although Homestead was bigger than ET, the latter had long been integrally linked with the Irvin strip mill. For a time the necessity of supplies of quality steel linked Irvin works with Gary, but the eventual decision to make a major outlay at ET in concasting was once again to save US Steel's longest established steel works in the Pittsburgh district.

The changes in the periphery were at plants that US Steel built or acquired relatively late in its career. Slow to expand into these regions, it withdrew from them early because they were marginal operations in terms of costs and size. Duluth was a special case. Elsewhere there were a wide variety of circumstances but a common theme of retrenchment, cut backs, or closures. In the West, where it had been growing since the late 1920s, US Steel suffered like other producers from the particularly high levels of penetration by foreign steel, especially from the Far East. By 1970 28 percent of the steel consumed in the seven western states was imported, twice the national average, while only 17 percent came from eastern mills. By 1977 the share of the latter in western supply had fallen to 11 percent and imports made up 38 percent, resulting in a narrowed scope for mills within the region. In flat-rolled products, the proportion of foreign steel was nearly 60 percent.[16] In the mid-1970s, the Western Steel Division of US Steel supplied about 20 percent of the steel consumed in the western states. Speaking in May 1975 at the annual meeting in San Francisco, Speer had made a virtue of what was in fact a serious cost problem: "The Pittsburg Works . . . produces the broadest range of steel products of any steel mill west of the Mississippi." Cold-reduced coil and sheet, tinplate, and galvanized-sheet production at Pittsburg, California, depended on hot-rolled coil from Geneva, Utah, 750 miles to the east. Mineral-oriented, the latter was using relatively poor ores and coking coal. In the mid-1960s, no more than about 17 percent of the Geneva output was delivered within the mountain states, and it was always disadvantaged by high costs of transportation to the coast. An increasing disadvantage was that its steel continued to come from open-hearth furnaces, which meant it could not supply hot-rolled coil of comparable quality at a price matching that of imported material. From the early 1960s, Geneva was sorely troubled by Japanese imports; a few years later it was "struggling to stay alive." Sheet and plate operations were modernized during the early 1970s in the hope of meeting the foreign invasion, and in 1976 a new blooming mill was put in to remove a production bottleneck.[17] These improvements failed to stem the tide running against Geneva and the whole western steel industry. Among the closures announced in February 1979 were the small steel works at Torrance and the Pittsburg rod mill, modernized only a few years before. In summer 1980, executives made the decision to sell Torrance for $19 million.[18] Pipe and wire operations at Pittsburg were closed four years later. From 1982 to 1986, Geneva recorded total operating losses of $40 million. Altogether between 1975 and 1984, efforts to modernize operations that it now recognized were "in the wrong place" cost US Steel two and a half times as much as Geneva earned during that period. A mid-1980s estimate of the expense of a wholesale Geneva modernization program was about $1 billion.[19]

Already during these difficult times, a review of the resource base of US Steel western operations had begun. By 1983 this analysis had resulted in the closure of the taconite mine and pellet plant at Atlantic City, Wyoming, and the direct-shipping ore mines near Cedar City, Utah. To replace the latter source, Geneva was supplied with pelleted ore from the Minntac mines in Minnesota.[20] Pittsburg operations were also in bad shape. Pressure from competitors with superior plants and better products meant that, in 1986, operations were only at 71 percent of capacity. In ten years its share of the West Coast finished-steel supply had fallen from 27 to 22 percent, and projections indicated that if Pittsburg's cold-reduction facilities were not updated within ten years, its share might sink as low as 5 percent. About this time, a possible solution was found to avoid huge outlays at Geneva and at the same time contain foreign competition thereby saving the future of the sheet and tin-plate sections of the Pittsburg works. However, it would require the eventual sacrifice of Geneva, two-thirds of whose steel was at this time delivered to its coastal associate. The Korean company POSCO paid $90 million for a half-share in Pittsburg, and it was agreed that POSCO and US Steel should each invest $150 million over four years to modernize operations there. Emphasis was placed on the fact that the Korean steel supplying Pittsburg would be concast and consequently of higher grade than that which had been utilized. To sidestep restrictions on imported steel, until fall 1989 POSCO would provide only 40,000 tons of coil from its Kwangyang works and Geneva would deliver the rest. After that, POSCO would become the prime source. The United Steelworkers of America denounced the link as "unfair competition," and 10,000 workers demonstrated when US Steel–POSCO used a nonunion construction company for Pittsburg's modernization. In fact, the new arrangements were applied much more quickly than expected, the decision coming in January 1987 to "idle indefinitely" Geneva's operations. Later the site was sold to an independent group that operated it under the name Geneva Steel. By 1990 some $400 million had been spent to improve Pittsburg operations, but some have suggested that differences in management style resulted in less than expected success for the joint POSCO–US Steel venture.[21]

More dramatic still, in so far as it involved withdrawal from an area until so recently regarded sufficiently attractive to merit large capital outlays on wholly new plant, was US Steel policy in the southwestern Gulf Coast region. By 1971 the works at Baytown, Texas, was becoming established, but it was already feeling the adverse impact of foreign competition in the form of imported Mexican plate, allegedly financed in part by subsidies from the Mexican government. Armco, whose plant at Houston had long been the region's largest, reckoned that imports that year took 40 percent of the markets served from their operation. Next year Gulf Coast markets were sluggish due to a combination of slower growth in demand, competition from imports, and a too rapid extension of local capacity. By midsummer the Baytown plate mill was operating at only one-third of its capacity.[22] Eventually, it was recognized that though the plant was good, it was not located in a satisfactory general market for plate given the competitive conditions of the times; the prime market for this product was the Chicago area. Despite these difficulties, US Steel embarked on expansion rather than retrenchment. By 1974 work was begun on two more electric furnaces and two concast units in order to double plate production. In 1976 the Corporation decided to spend $75–80 million to add a large-diameter pipe mill. By then raw-steel capacity at Baytown was about 2 million tons, but that year Gulf ports handled about 6 million tons of imported steel, almost two-thirds of it from Japan and delivered at prices 10–20 percent below most domestic quotations. Further competition in the region came from regional mini-mills, which had a combined capacity of about 1.6 million tons.[23] In the crisis conditions that marked the early 1980s, 34 percent of the steel bought in Texas in 1980 was imported. At this difficult time, Baytown experienced operating problems associated with computer control of its plate mill resulting in a high reject rate, poor yields, and lowered productivity. For a time it was hoped that a licensing agreement for Sumitomo technology would put the matter right.[24] By early autumn 1981, 2,000 employees were working there, but 650 of them were temporarily laid off.[25] Although the 1981 annual report referred to quality and customer service there as "second to none," difficulties continued and the pipe works was closed for two years beginning in 1983. There were also problems with unions, and though it eventually reopened under a new less restrictive labor agreement, after an analysis of its situation, the US Steel board decided in January 1987 to "idle indefinitely" the hot end and pipe mill. In fact, the Texas project, begun twenty years before with such high expectations, was being abandoned completely, the Corporation's remaining production of tubular products being focused once again in distant but fully integrated works. The following summer, a 117-acre business park opened near Baytown; it was designed for light industry, warehousing, and other such enterprises.[26]

Along the Atlantic Coast, what had promised to be a market area with great prospects proved disappointing. By the early 1980s, Fairless was gravely weakened by its dependence on open-hearth furnaces, and its hot-strip mill, installed in 1953, badly required overhaul. Indeed, as Roderick told the 1983 US Steel Annual Meeting, a particular problem with this plant was that, as it had been built all at one time, the whole of this works was becoming obsolete simultaneously, in contrast to the continuous if piecemeal updating that had gone into the various Chicago or Pittsburgh works—two weeks earlier to a Congressional meeting, he summed up the Fairless problem as a case of "terminal obsolescence." (Roderick also pointed out that, though only thirty years old, the engineering for it "came off the drawing boards more than 35 years ago.")[27] By the early 1980s, the outlay needed for modernization there was set at a much higher figure even than that at Geneva: $1.5 billion or possibly even $1.9 billion. New sinter plant, rebuilt coke ovens and blast furnaces, an oxygen steel unit, and continuous casting would be required as well as the upgrading of finishing mills. In 1982 deeply depressed trading conditions made the situation worse, the industry's capacity utilization rate in the last week of the year being the lowest since 1932; in December and January 1983, 165,000 workers throughout the domestic industry were laid off as a result. It was at this time that a possible radical alternative solution for the Fairless problem was explored. As US Steel faced up to the problems of sweeping closures, the steel industry in Europe was suffering equal distress. Between 1979 and 1983, the British Steel Corporation (BSC) had been forced to reduce its capacity by 7 million tons, and it was interested to explore ways of avoiding even more downsizings. Officials realized that there might be mutual gain from linking the fortunes of BSC and of Fairless works—with, as Roderick stressed, US Steel maintaining majority control. The plan under consideration from November 1982 was for the rather more modern iron and steel plant at Ravenscraig in central Scotland, which had oxygen steel plant, to supply slabs to the Fairless hot-strip mill; iron- and steel-making operations at the latter would be phased out. It was estimated such an arrangement might reduce mill costs at Fairless by as much as $20 a ton. Eighteen hundred jobs would be lost, but the remaining five thousand would be made more secure. An important part of the plan was that the first $300–400 million of profits from the arrangement would be used to improve the Fairless finishing mills.[28] Unfortunately, though predictably, other steel companies and unions opposed the link. Then in September 1983, the Commerce Department declared the arrangement would wreck the import restraint agreement made with the European Economic Community eleven months earlier. Other problems included the facts that BSC could not unequivocally guarantee the price of the coil and that Ian MacGregor, a friend of Roderick, was moved from the chairmanship of BSC to that of the British coal industry. The plan foundered and a slow rundown of Fairless operations followed. By June 1984 the coke plant had been idled, supplies after that being railed from Clairton. In 1987, following a long strike, US Steel arranged a new four-year labor contract starting 1 February 1987. One of its conditions was that the hot end at Fairless should not be shut down during this period. In July 1991, six months after this contract expired, all iron and steel making at Fairless ended. After this, steel for the continuing output of thin flat rolled products was supplied by other works. By the end of the 1990s, after another onslaught from imported steel, yet more departments there were closed, and Fairless produced only tinless and hot-dip galvanized products. In short, much less survived than had been planned at the time of the discussions of the link with Ravenscraig.

In Alabama, blast furnace improvements and the late 1971 decision to install two oxygen converters at Fairfield increased US Steel capacity by about

600,00 tons, or 20 percent. Under Speer, a good deal of money was invested at Fairfield, although not all of his senior colleagues agreed with the policy. In late winter 1975 the last Fairfield open-hearth furnaces were closed, and a little over a year later abandonment of its melting shop meant the effective end of operations on the nearby Ensley site. With all this occurring, the vice president of the Southern Steel Division of US Steel was optimistically anticipating that southern and southwestern markets might need another 7 million tons of steel products by 1980. During 1981 executives decided to build a 600,000 ton seamless-pipe mill at Fairfield designed to produce high-strength pipe able to withstand the particularly hostile conditions of deep wells.[29] However, the slowdown in market growth and following recession stopped expansion, and the Fairfield blast furnace, oxygen steel, slabbing, and hot-strip operations were idle for almost two years, restarting in February 1984 under a new local agreement with the workforce that provided for less-restrictive practices. If not for the commitment made only a year or so earlier for the new pipe mill, it seems likely that Fairfield would have been swept away in the rationalization pushed through under Tom Graham's presidency. However, as US Steel had made clear that it regarded Fairfield as a long-term viable plant, in 1985 plans were announced for a new slab caster, bringing total recent spending there to $1 billion; as a result of which, as the Corporation's quarterly review put it, "the company will now begin to see a return on the modern facilities that it has put into the plant in recent years to make it process-competitive."

In the manufacturing belt, there was a high mortality of smaller specialized works. As in the reconstruction during the 1930s, some of these were in communities whose economies had been overwhelmingly dependent on a single mill. The once-major sheet operation at Vandergrift, Pennsylvania, was one example. At about the time US Steel was incorporated, that plant provided work for some 4,000 men; by 1950, having been converted into a specialty sheet operation, its payroll had dropped to 2,100. In 1980 Vandergrift had only 400 workers, and by the time it was idled in January 1988 this had fallen to only about 100. Later that year Vandergrift was sold to Allegheny Ludlum.[30] Another casualty was the small Johnstown plant, inherited along with Lorain from the late-1880s enterprise of T. L. Johnson, and which had later concentrated on castings and forgings for steel-mill equipment, mine cars, coke-oven doors, and other industrial items. Here again the threat of layoffs was used to gain concessions from the workers. Early in 1984 US Steel announced plans to close the works, which then employed about 1,400 people. After this, the Corporation tried to negotiate lower wages—a cut from about twenty dollars an hour to around fifteen dollars—and a reduction of benefits. When the workers rejected this, US Steel sold the plant to a new profit-sharing company set up by three former employees and a local businessman.[31]

Demolition of the blast furnaces at the Ohio works, Youngstown. *Reprinted from* The Vindicator, *Youngstown, Ohio.* © *1982, The Vindicator Printing Company*

US Steel operated seven integrated works in the western part of the man-ufacturing belt in the early 1970s, the Pennsylvania mills at Donora and most of Clairton having closed ten years before, and McKeesport and Duquesne being already effectively linked into one operation across the Monongahela. Of the remaining seven, that at Youngstown—the two linked operations of the iron and steel plant and primary mills of the Ohio works inside the city of Youngstown and the finishing mills at McDonald works seven miles to the north—was widely recognized as least likely to survive. It was landlocked, and by the early 1970s some commentators were already recognizing that with-out an extremely unlikely investment in a federally built canal providing the area's works cheap transport for raw materials, Youngstown and its satellite towns were destined sooner or later to become light-industry communities. A few years later it was calculated to cost about seven dollars a ton to rail coking coal to this region; at the same time the charge for carrying coal to Pittsburgh by barge was roughly one dollar per ton.[32] Already, during the generally good trading conditions of the 1950s and 1960s, the Mahoning Valley had suffered heavy losses of steel jobs; by 1972 it was estimated that 15,000–20,000 jobs had disappeared over the previous twenty years or so.[33] For a time prospects at the US Steel plants seemed to brighten. During the winter of 1971–1972, when these were shut down for seven months, extensive repairs and improvements were made to the blooming and hot-strip mills. But production remained wholly dependent on open-hearth furnaces, which were both a reason for continuing higher costs and a telling indication that decision makers at the highest level did not believe a very large outlay was justified. In fact, the trying times of the decade were yet to come and would throw doubt on the claims that US Steel had put its works there "in first rate condition."[34]

Although by late fall 1973 the high cost of installing pollution controls was said to mean that US Steel was on the point of closing both its Youngstown plants, which then employed 6,000 workers, operations in fact continued.[35] Four years later, when there was much talk of the new Conneaut mill, Roderick made clear that together the expense of pollution control and

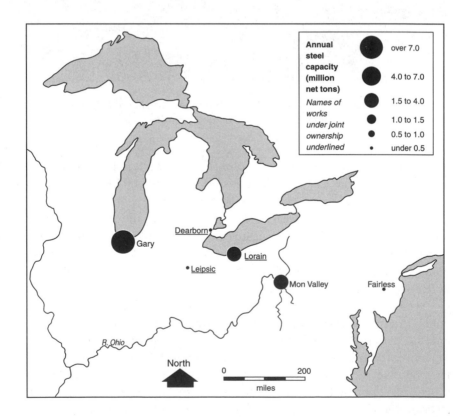

US Steel in the Northeast, 1999

increasing imports meant Youngstown's prospects were "quite dismal." At the beginning of 1978 the Corporation confirmed that operations would end with a loss of the now 5,000 jobs at the works. The district seemed to be threatened with a midwinter catastrophe, for only four months earlier Youngstown Sheet and Tube had closed its operations there, again with 5,000 jobs lost. Interestingly, the US Steel closure announcement came a day after the Carter administration had announced details of its trigger-price system for controlling the flow of imports, the timing making obvious that it was not merely foreign competition but also outdated facilities that were causing the industry's troubles.[36] Even now the end was delayed. When asked about this at the May 1979 annual meeting, Roderick, newly appointed chairman, was more circumspect in his response than he had been two years earlier but made clear that the Corporation was prepared to use the threat of closure to help fight troublesome restrictions on company freedom to conduct operations with a minimum of constraints: "We have phased down certain aspects of that plant. We have said that we will continue to operate that plant provided it can contribute to US Steel's bottom line or until we are forced to make substantial investments for environmental or other purposes that would not be in the interests of stockholders."[37] On 7 January 1980, the decision to close was at last made. An internal assessment made at that time revealed how uneconomic operations at Youngstown had become. Ohio works was an open-hearth plant of 1.5 million tons annual capacity. Its primary mills had been installed between 1895 and 1909. The bar and narrow-strip mills at McDonald dated from 1918 to 1931 and its strip mill from 1935. More modern bar- and strip-mill installations by US Steel and by other companies had rendered such plants "job shop" suppliers, and the Youngstown operations had been conducted at a loss after 1974. The requirements for environmental protection were the last straw: "considering necessary expenditures mandated for air and water quality facilities, continued operation is not viable." In turn, the costs of closing down were high. The 1979 shipments had been 607,000 tons. Of this it was projected that no more than about 175,000 tons could be shifted to other US Steel plants; the rest of the business would be lost. The one-time cost of closure and dismantling was computed as high as $252 million, but there would be annual cost savings on the profit-and-loss account of $7 million. Additionally, otherwise essential future spending on environmental control of as much as $60 million would be avoided. The workforce had been diminished during the years; now the remaining 3,913 employees lost their jobs.[38] According to union figures, there had been 26,250 primary-metals jobs in the Youngstown area in 1954–1955 and 15,957 twenty years later. By 1984 only 1,753 remained.[39]

During this period Chicago increased its share of US Steel capacity and production so as to become its unquestioned principal center. Development there was by no means painless, but the pain was very unequally distributed. After a period of unprofitability, all Joliet operations except for a rod mill, modernized in 1965 at a cost of $12 million, were discontinued in the late 1970s. Though it had shipped 116,000 tons in 1979, this unit had not been profitable since 1974. Early in 1980 its long-term commercial viability was assessed and found wanting. The decision to close was made even though only 24,000 tons of its business could be transferred to another works within US Steel—Cuyahoga in Cleveland—and the other 92,000 tons would be lost to competitors; 714 employees would be affected. At the same time, it was resolved to abandon the high-quality wire making plant at Waukegan, over sixty miles away on the other side of Chicago, which had drawn its rod supply from both Joliet and South works. In this case a little over 40 percent of the business was expected to relocate to Cuyahoga while the rest would be lost. Combined costs of closure at Joliet and Waukegan were $74.6 million; likely annual savings on the profit-and-loss account were anticipated to be $13.7 million.[40] Even at Gary there were occasional temporary closures of various departments. In one instance a permanent change demonstrated what a high price had been paid for the continuing operation of obsolescent plant. The plant's 80-inch hot-strip mill was installed in 1936, modernized in 1948, and partially upgraded in 1976 at the cost of another $7.6 million. But a new 84-inch-strip mill was commissioned in 1967 and from that time the 80-inch mill operated only to meet peak demand. After 1974 it was unprofitable and operations were discontinued in mid summer 1979. When it was decided five months later to dismantle the mill, the one-off cost was put at $14.5 million; annual savings on the profit-and-loss account would be $15.5 million.[41] Usually, solutions for problems at Gary called for more expenditure rather than for closure, but technical advance could quickly render even relatively new plant at least second rate. Over its first eighteen years, the 84-inch-strip mill produced almost 55 million tons of hot-rolled band. Early in 1985 a comprehensive study was made to determine what was needed to make it a "world-class" operation in terms of productivity, quality of product, and reliability.[42]

South works was a particularly vulnerable operation. Decline was clearly underway by 1978, and conditions worsened during the collapse of national steel production from 1981 into 1983. By the end of 1983 most of it was idle. At this time South Chicago's future prospects were again largely linked with the trade in which its earliest successes had been achieved, rail manufacture. Eventually, US Steel decided not to build a new rail mill on site. After this, decline in structural-steel consumption became so serious that executives concluded it best to withdraw from the market completely. As a result, in April 1992 operations at South Chicago ended, 110 years after they began. In the mid-1970s, up to 10,000 people had worked there; when it closed nearly twenty years later, there were only 690 employees. Despite the loss of this plant, the district increased in importance. Chicago had 39.6 percent of the 1973 steel capacity at US Steel integrated works—by 1993 Gary alone made up 62.6 percent. This degree of concentration was achieved at the cost of a particularly high rate of mortality in Pittsburgh-district mills.

Over the quarter century from 1973, US Steel raw-steel capacity fell by 71 percent. In Chicago the shrinkage was 48 percent, but for the Pittsburgh district it was 74 percent. Two out of three steel-making plants and three of the four major mill complexes that had earlier sprawled alongside the Monongahela were gone. Not only was the overall change dramatic, but the course of events that brought it about also was complicated. Two important themes were first the inheritance and subsequent abandonment of a great deal of old equipment and later the changing patterns of interplant linkage. A surprising amount of the plant around Pittsburgh was old, though often well maintained and occasionally "modernized." For instance, of the thirteen US Steel blast furnaces in the area in 1979, seven were of less than twenty-six feet in diameter and were not competitive in either output or fuel consumption with modern furnaces. At Homestead, the 48-inch plate mill dated from 1896, though it had been overhauled and modernized as late as 1971. At National works there was a blooming mill originally installed in 1908, and though it had been improved greatly since, not until 1980 had the decision been made to replace its steam-drive power plant with an electrical one.[43]

From the time of the Irvin project in the late 1930s and the major

TABLE 22.9

US Steel steel capacity and production in total and in the Chicago, Birmingham, and Pittsburgh areas (thousand net tons)

	Capacity			Production	
	1960	*1973*	*1989*	*1993*	*1997*
Total	41,616	43,360	20,100	11,850	12,350
of which:					
Chicago plants	13,588	15,350	8,200	8,700	7,428
Pittsburgh plants	12,339	10,500	2,700	2,960	2,561
Birmingham plants	3,997	4,000	2,900	2,240	2,361

Sources: Iron Age; R. Cordero, Iron and Steel Works of the World, 6th ed. (London: Metal Bulletin, 1974); and US Steel records, USX.

TABLE 22.10
**Annual raw-steel capacity of US Steel's Pittsburgh works
in the early 1970s (thousand tons)**

Homestead (open hearth)	3,430
McKeesport (open-hearth reserve, effectively abandoned)	c.1,400
Duquesne (BOF)	3,000
(Electric furnace)	260
Edgar Thomson (BOF)	2,900

Source: US Steel Corporate Policy Committee, 6 December 1977, USX.

wartime extensions at Homestead, it has been common to graft new finishing operations onto existing iron- and steel-making facilities. In the mid-1960s, as the oxygen converter revolutionized technology, a process of linked developments was applied in cutting out open-hearth capacity. After the melting shop at National works closed, the rolling mills there obtained steel, delivered as red-hot ingots to its McKeesport blooming mill, from the more modern plant across the river at Duquesne. As company publicity cheerfully proclaimed, "From an efficiency standpoint this is a logical merger." It was not then recognized that the same logic might eventually lead to the concentration of the whole of the district's capacity into one complex.[44] Demolition of Edgar Thomson's rail mill in the mid-1960s made more space available at a plant that until then had received little recent expenditure. Indeed, by this time ET was classed as a "semi marginal" plant. Only nine of its sixteen 205-ton open -hearth furnaces (originally built in 1913) were at work in summer 1966, a year of record national output and the fifth best in US Steel history.[45] In summer 1971, a bad year, US Steel announced a temporary closure of ET, blaming this on a low level of orders. Then, after the mill and 2,000 workers had been idle for six months, it was announced that the open-hearth melting shop would be replaced by a 3-million-ton BOF unit. The plant's future now depended on its continuing as the source of slabs for the Irvin hot- and cold-strip mills, where by 1972 capacity for cold-reduced sheets was doubled to 2.5 million tons.[46] Even then the way ahead was by no means smooth. In 1980 the US Steel Board approved the expenditure of $42 million for a slab-casting plant at ET, which would also reduce costs at the strip mill. Two years later the plan was suspended because of "soft" market conditions. A splendid illustration of the economies—and complicated flows of materials—made possible by massed local capacity came in the 1982 consideration of the procurement pattern for steel to support Irvin if it was fully loaded, now that concasting at ET had been deferred. At capacity Irvin required 2.7 million tons of slabs. The

44-inch primary mill at ET could provide 1.9 million tons, and the rest would come from Duquesne and Homestead. But the existing BOF shop at ET had a 2.9-million-ton steel capacity, and therefore spare tonnage to be sent on as ingots to other local mills. Sheet and tin products were expected to be of ever-higher quality, and this alone would eventually require the Corporation to go ahead with the ET slab caster. When this happened it would permit the reloading of the primary mill at Duquesne and closure of that at McKeesport.[47]

By this time the general development context had become extremely unfavorable. National raw-steel production fell from 120.8 million tons in 1981 to 74.6 million the next year. US Steel output in 1982 was 12.1 million tons, only 36.1 percent of capacity, little more than half the level of the previous two years, and less than at any time since 1938. It was at this awful time that long-term prospects for the Pittsburgh area suddenly seemed to worsen when the Irvin strip mill was unlinked from its long-term connection with ET and made into a department of Gary works. Even Irvin itself was temporarily stopped because of "plant realignments and consolidations and reappraisal of hot strip mill capabilities," during which the cold-reduction mills there worked in part on hot-rolled coil from the Gary 84-inch strip mill, where costs were lower than those locally.[48] In such circumstances it was only logical that US Steel now merged continuing Pittsburgh operations into a composite "Mon Valley Works."[49] The outlook for ET and the whole valley looked even bleaker when in 1983 it was announced that US Steel would spend $300 million building new slab-casting plants at both Gary and Fairfield but made no such commitment for the local area. That year the open-hearth plant at Homestead and the Carrie blast furnaces closed permanently.

Pittsburgh-area closures announced in December 1983 meant US Steel employment there would fall to about 9,000 people, less than one-third the 1979 total. Within four months another round of cutbacks, expected to cost 4,744 jobs widely scattered through the area's various operations, was tabled. In spring 1984 the blast furnaces and steel plant at Duquesne closed, and the Homestead structural and plate mills now drew their steel from Edgar Thomson. At Clairton the light structural mills, the last finishing facilities, were abandoned. During 1986 the Homestead and Carrie sites were sold, and the following year USX turned over those at Duquesne and McKeesport to Allegheny County. Not until 1989 was there at last some assurance of long-term security for the area's industry in the form of a decision to go ahead with a $250 million, 2.6-million-ton slab caster at Edgar Thomson. Once more, acceptance of a new labor agreement was a precondition for the investment.

During the early postwar years, US Steel had operated twenty-five blast furnaces in the Monongahela Valley; as late as 1979 there were still thirteen, but after 1983 there were only two. In the six-county Pittsburgh area, there had been over 100,000 steel jobs in various companies during the mid-1960s, but by the middle of 1994 only 28,000 remained. Year after year, as conditions and prospects degenerated, the mill towns along the Monongahela became more obviously stricken. By 1986–1987, the whole industrialized section of the valley presented a succession of depressing vistas of decay. At Clairton the steel works still stood, though abandoned long before. National and Duquesne plants were idle and empty. Downriver, the once great works at Homestead was silent, and on the opposite bank the furnaces that had supplied it with molten metal were inactive. The blast furnaces and BOF plant at Edgar Thomson were operating, but even then a visitor found the region miserable: "Braddock, in particular is a sad and dislocated community. . . . [A] settlement that looks to be physically decomposing." As late as 1980, the population in the Monongahela Valley mill towns was 653,000; in 1990 the same towns contained 579,000.[50]

Apart from a shared interest in Lorain, US Steel now operated a complete cycle of iron and steel production at only three works: in ascending size, Fairfield, Mon Valley, and Gary. Their respective shares of the company's rated ingot capacity for 1993–1994 were 16.1 percent, 21.3 percent, and 62.6 percent.[51] In 1994 they accounted for 19.2 percent, 22.8 percent, and 58 percent of corporate output. Belatedly, by narrowing its product range, by drastic reduction in the number of mills and units of plant, and as a result of improvements in technology, US Steel made spectacular progress in efficiency and especially labor productivity. During the 1970s, progress in these areas had been very uncertain, but in the middle 1980s it went on at a startling pace. Ten man-hours were required for every ton of steel in 1979; by 1988 the figure was under six, and in the early 1990s each ton of steel produced and shipped required only four man-hours. By this time and judged by labor productivity, US Steel had become one of the world's most efficient producers of steel. In 1997 it operated at 96.5 percent of its savagely pruned raw-steel capacity; by 1998 all its steel was concast.

The achievement of such a dramatic improvement was the product of a relentless and often ruthless reduction through the Roderick years. As he advanced toward retirement, the success and exceptional character of his regime was well summed up in one of the commercial journals: "For the last decade, Roderick has fought to dismantle USX's legacy, what Morgan and Gary called a 'rounded proposition,' a fully integrated steel giant with seemingly inexhaustible supplies of iron ore, limestone, and coal, which were mined and transported to the mills by the company's large shipping fleet and its own rail lines" This had been a time-honored structure, but in commercial terms the giant had become increasingly heavy-footed. In the course of the Roderick chairmanship, US Steel had sold $4.7 billion of assets from coal reserves to timber and had written off a further $3 billion.[52]

Two years after Roderick's retirement in 1989 at the usual age of sixty-five, his key collaborator in the rationalization program, Tom Graham, resigned from US Steel at age sixty-four. In eight years as chief operating officer and then as president, the closure program and new manning practices that he had pushed through had been largely responsible for the remarkable reduction of the Corporation's breakeven point from 75 percent to 60 percent of its capacity. Graham's career at US Steel had also been exceptional in that he had come in from another firm.[53] Interestingly, in neither the USX annual report on 1990 operations nor the United States Steel Group 1991 summary report did his achievements receive any special mention.

From June 1989 Roderick was succeeded as chairman by Charles A. Corry. Corry had begun his career with US Steel thirty years before, but most of the companies or divisions in which he had worked had already disappeared. AS&W, his first employer, had gone with all ten of the plants that it then operated, and the eight plants of American Bridge, to which he moved afterward, had either been closed or had passed out of US Steel control when that division was sold off in 1987. By the time his chairmanship began, most of the US Steel Group's major closures had already occurred. The exceptions were the ending of all iron and steel operations at Fairless in summer 1991 and the final stage of run-down of South works the following spring. The important achievement of the Corry years was to maintain the momentum toward higher efficiency. Productivity continued to increase. This continuing success was gained by confining capital expenditure to cost reduction, improvement of the product, or unavoidable outlays for replacement of plant and equipment while rigidly excluding any spending for expansion of capacity. During the six years of the Corry chairmanship, the industry followed an uneven course in which US Steel experienced widely varying operating conditions. From a high of 11.5 million tons of steel shipments in 1989, the Corporation reached a low of 8.8 million in 1991 but by 1994 had recovered to 11.4 million tons. In the mid-1990s US Steel usually accounted for just below one-third of USX revenues. In 1995 Thomas Usher became USX chairman and Paul Wilhelm president of the US Steel Group—only the second time in US Steel history (the first was in the early 1970s when Edwin Gott was chairman and Edgar Speer president) in which both of the two chief posts have been held by men with direct technical and operational experience in steel making. Though it had varied a good deal from year to year, in the US Steel Group the employee hours required from the whole workforce per ton of products shipped had fallen from 4.94 in 1989 to 4.01 in 1998.

US Steel 1997 production was 12.3 million tons of crude steel; shipments of products were 11.6 million tons. Nevertheless, both the industry and its leading company have lost their once preeminent place in America's industrial structure. At its formation US Steel was the giant of the entire manufacturing economy; by the mid-1980s its stock market valuation was a mere $3.3 billion as compared with $41 billion for Exxon and $83 billion for IBM.[54] By 1997 the Fortune 500 list ranked General Motors in its top position with a total revenue of $178.17 billion; USX's revenues of $21.06 billion ranked it fifty-fifth. Despite the decline in national status, during this last period of its first century, US Steel, that part of USX concerned with what until the early 1980s was the core activity to which all the components of the Corporation were related, has undergone far more changes than at any time in its history. It has managed at last to free itself from the almost institutional and inhibiting burdens of being "Big Steel." The course of its advance to this position has

TABLE 22.11

***Annual US Steel production of crude steel and shipments of steel
products per employee, 1902, 1920, 1937, and 1950–1999 (tons)***

Year	Ingots and castings	Steel products
1902	64.95	53.01
1920	80.56	57.96
1937	79.43	53.95
1950	109.11	78.51
1960	121.28	83.07
1970	156.45	104.63
1975	152.78	103.01
1980	156.16	115.28
1985[a]	262.05	196.53
1990	552.18	447.57
1994	547.96	495.92
1999	624.52	551.70

Sources: US Steel and USX annual reports.

Note: Figures for the 1970s and 1980s underrepresent productivity at that time, for more workers were then in nonsteel activities than was the case in the United States Steel Group figures for the 1990s.

[a] 1985 figures are based on estimates of the US Steel share of USX employment.

been a far from easy one. It was as the leader of an industry only just beginning to be aware of the range and severity of its problems that US Steel set out
on the business of the 1970s. After an uncertain start, it then showed unexpected resolution in rationalizing its great collection of plants. Its course
through these twenty-five years may be summarized as involving major reductions in capacity and production and the even more extreme pruning of its
workforce. In net income the business has not performed well, but in this respect its experience has been typical of the whole industry. However, judged by
net income in relation to sales or employment costs per ton of product
shipped, it has achieved a great deal. The pioneers of US Steel would have had
difficulty in recognizing the company operating under more or less the same
title in the late 1980s and through the 1990s. In 1901 the United States Steel
Corporation made 50.1 percent of the nation's finished rolled iron and steel;
ninety-five years later its share was 10.9 percent. A 1997 Dun and Bradstreet
directory chose to list the activities of USX as "Crude petroleum production;
Natural gas production; Petroleum refining; Petroleum bulk stations and ter-
minals; Gasoline service stations; Blast furnaces and steel mills; Natural gas
transmission."[55] But, though confused by many aspects of the circumstances
and structure of the descendant of the corporation they had created, J. P.
Morgan and Elbert Gary would surely have praised the achievements of their
latest successors.

Labor on the Defensive
during the Rationalization of the
1980s and 1990s

During the last two decades of the twentieth century, there have been quite remarkably changes in the labor situation at US Steel. The shrinking capacity; the installation of new, high-capacity, low-labor-input equipment in both furnace operations and above all oxygen steel making and continuous casting; and the use of more highly automated rolling mills have reduced the work force and the significance of labor costs in the overall expenses of production. Employee numbers, after varying relatively little through the late 1960s and the 1970s, fell very sharply as a result of these changes.

In the early 1980s, as both US Steel and the rest of the industry closed plants and laid off men on an unprecedented scale, the employers' negotiating body, the Steel Companies Coordinating Committee (SCCC), chaired by

TABLE 23.1
Production, financial returns, and labor in US Steel, 1980, 1990, and 1998

	1980	*1990*	*1998*
Steel shipments (th. net tons)	17,100	11,039	10,686
1. Sales (millions)	$12,492	$6,073	$6,283
2. Operating income (millions)	$478	$437	$330
3. Overall operating costs (millions)	$12,014	$5,636	$5,704
4. Wages, salaries, and other labor costs (millions)	$3,647	$1,118	$1,305
5. 4. as portion of 3.	30.38%	19.8%	22.88%
Avr. no. of employees in year	149,172	24,664	20,267

J. Bruce Johnston of US Steel, frequently deplored the refusal of the United steelworkers of America (USWA) to provide the operators some relief from the pressure of rising wages and benefits. As SCCC put it in a release of July 1982, "with the steel industry's very existence threatened, it is regrettable that the Steelworkers chose to increase wages for a few at the expense of the many." In relation to imports, Johnston stated that union leadership "well knows...that steelworker employment costs are the single leading factor in our competitive vulnerability."[1] Three years after this, the united front of the industry was broken and the SCCC was disbanded.

In the middle of 1986, USWA rejected a "final proposal" from US Steel for a new basic labor agreement. Johnston then wrote to each of their unionized employees to point out that the Corporation was "facing an economic shutdown with non-union competitors, bankrupt competitors [some of which he noted had claimed that they had been able to lower their costs as much as $85 a ton as a result of filing for bankruptcy], and foreign competitors."[2] The strike began on August 1. Two months later the US Steel president, Tom Graham, wrote to 3,000 chief executives of US Steel customer firms in the hope of making clear to them the Corporation's position in the continuing dispute. He explained that before the strike, average employee cost per steelworker was $25.35 an hour; US Steel was asking for a $2-an-hour reduction. He put the matter into a wider context: "A major issue at stake is whether or not US Steel would retain the right to manage its own business, and to take reasonable measures necessary to operate in an efficient and competitive manner. These proposed steps are in no way unprecedented. They are simply a recognition on our part of the new economic realities in today's market place, these measures also help US Steel maintain its position as the low-cost, high-quality supplier to you—our customer."[3] Early in December Johnston again contacted each employee stressing that since the strike began, the Corporation had offered improved terms including, if USWA had been willing to accept a four-year contract, a legal commitment to build a concast plant at Edgar Thomson and to modernize Irvin works. The union had refused the improved offer and as a result was favoring the failed and reconstructed steel companies.[4] In mid-January 1987, after 184 days of striking, an agreement between US Steel and USWA was at last reached. Steel output in the United States in 1986 had been 93.1 percent the 1985 level; due to the strike, US Steel managed only 57.4 percent of the previous year's total.

During this period, another survey of all US Steel properties was undertaken. It revealed a continuing surplus of coke, iron, steel, and primary mill capacity and led to proposals to close facilities involving a further 6,600 jobs. As the annual report for 1987 recognized, the result was increased productivity. Five years before, 10.8 man-hours were required for each ton of steel shipped; for the remainder of the calendar year from April 1987 it was only 3.8 man-hours. The human impact of this otherwise highly desirable result was blurred by the unemotional wording in which it was addressed in the report: "The concessionary contract with the steelworkers, reached only after a 184-day strike, produced a favorable impact on steel segment operating income. Continued emphasis on cost control resulted in improved margins." In other parts of the report, the matter was taken a little further. The contract had provided for a profit-sharing arrangement resulting in "additional compensation to eligible steelworkers," but "reduced wages and benefits provides for a one-time remanning agreement effectively eliminating 1,346 jobs."[5]

As the 1987 experience indicated, in addition to closure of marginal plants and the loss of jobs, a distinctive feature of the late 1980s was the way in which management used the promise to install new capacity, or more commonly the threat of probable closure of existing plant, in order to secure their workers' acceptance of employment conditions more favorable for the Corporation than in the past. It must be stressed that such devices have been by no means confined to US Steel, or even to the American steel industry. Indeed, in contemporary market-oriented economies such methods of bargaining are the common stuff of "wise" labor management. Nowhere in the United States was the process more clearly illustrated than at South Chicago.

In January 1979 US Steel authorized an engineering study of a proposal to modify the South works from a 34-inch to a 46-inch structural mill so as to also make rails of an adequate quality to meet the more exacting standard now being demanded by railroads. Roderick stressed the attraction of the plant's favorable location in relation to markets. In March 1981 the central policy committee considered the plan. It was noted that two domestic competitors were installing facilities to produce up to 900,000–950,000 tons of better quality rails, resulting in the closure of the Ensley rail mill as well as lower sales and increased costs at Gary. In two years, if no action was taken, rail output by US Steel might fall from 479,000 to 250,000 tons. Conversely, if they invested in improved plant, their market share could increase to 750,000 tons, the new plant becoming far and away the nation's largest producer.

It was believed that a better product and improved practice might not only secure US Steel involvement in a "strong, profitable rail market" but also result in a premium of $15 a ton over the present proceeds. Construction might begin later that summer with production starting in spring 1983.[6] When the plan was made public in April 1981, the likely employment in the new mill was put at 1,000. However, it soon became clear that the workers and the

TABLE 23.2
**Actual supply of rails in the United States, 1980, and projected figures
for 1983–1990 (thousand tons)**

	1980	*1983–90 with improved plants*
US Steel		
Ensley	185	n.a.
Gary	294	23
South Chicago	n.a.	727
Bethlehem	190	200
Colorado Fuel and Iron	438	400
Wheeling-Pittsburgh	n.a.	200
Total domestic supply	1,107	1,550
Imports	229	50
Total market for rails	1,336	1,600

Source: US Steel Corporate Policy Committee, 25 March 1981, USX.

local community alike would have to make important concessions to ensure that plans went ahead. When no start had been made by summer 1983, Roderick informed Sen. Charles H. Percy of Illinois that the recent depression had made it necessary to reevaluate the project.[7] At a press conference that fall, T. C. Graham explained that three preconditions had been set for the new mill: union acceptance of "competitive" manning, exemption of rails from the Illinois state sales tax, and relief from environmental agreements. The first two had been agreed upon, but the third was not yet accepted, though there were positive signs it would be eventually. A consequence of the delay was that thinking moved on from building an integrated operation to establishing a mini-mill. To ensure that the whole plant would be a low-cost producer, corporate officials had started discussions with the union about general manning in addition to the arrangements already accepted for the rolling mill itself. If all went ahead, the "New South Works," as it was now referred to, would employ 2,000 men.[8]

On 27 December 1983—presumably after keeping the matter in limbo over the Christmas holidays—Roderick announced that the unions had not accepted these new arrangements and as a result US Steel would permanently cancel the plan for the new mill and close most of South works. The union responded angrily to news of the closure, releasing an outspoken statement to the press: "Enough is enough. US Steel has ripped off the American taxpayer, the state of Illinois, the city of Chicago, and the United Steelworkers of America."[9] A few weeks later, USWA decided to take US Steel to court over what it described as "assaults" on steelworkers' rights, and in August 1984 a Cook County judge issued an order restraining the company from demolishing one of South works' blast furnaces, a slabbing mill, and plate mill. Eventually, though, a federal court ruled that US Steel had not broken an agreement with USWA to build the rail mill. It was now revealed that the Corporation had succeeded in getting the rescinding of a pollution-control order that would have cost it $33 million. The implication was that it was the refusal of the union to agree to new terms of work for the whole of the plant that had caused the scheme to founder. The failure of this project marked the beginning of the end for South works. In the early 1970s, its rated annual steel capacity had been over 7 million tons; in 1990 it made only 714,000 tons, the following year 533,000 tons, and before closing in April 1992, only 44,000 tons.

At Duquesne, the USWA went so far as to commission consultants in order to arm itself to fight the arguments of US Steel for closure. In the early 1960s it had seemed that Duquesne was being equipped to be a leading player in the Corporation's long-term operations. The "Dorothy" blast furnace, capable of 850,000 tons of iron annually, was commissioned in 1963. Late that year the works was chosen for US Steel's first two oxygen converters. Optimists pointed out that room had been left for a third vessel that might increase the plant's BOF capacity to 3 million tons. Twenty years later, in the midst of deep depression and a thoroughgoing plant rationalization program, the situation at Duquesne looked very different. In May 1984 US Steel closed the blast furnace and the BOF shop. The men resolved to argue their case. Locker-Abrecht Associates of New York delivered a preliminary report on the plant to the USWA in January 1985. Its broad conclusions were not very favorable to the workers' cause: "In general, Duquesne has been a middle range cost producer compared to other steel facilities in the Mon Valley region and a middle to high cost producer relative to integrated facilities throughout the United States."[10] But another report commissioned from Eichley Engineers came out with a much more aggressive assessment: "The shutdown of Duquesne is one of US Steel's continuing plans to abandon the Mon Valley. This has been prompted by material transportation costs, labor costs, and significant local environmental constraints." Eichley analysts estimated that rehabilitating and restarting the whole works would cost between $12.5 million and $17.5 million, the higher figure applying in order to ensure that blast-furnace operations would be fully successful. Roderick immediately agreed to a full review with the intention of rebutting the USWA's consultants point-by-point on the engineering, marketing, and financial aspects of the case. The outcome was a US Steel document entitled "Response to the Locker/Abrecht Duquesne

Works Feasibility Study." The dispute about different interpretations of the situation came to a head on 15 April 1985 in a meeting with Allegheny County commissioners and a subsequent press conference at which Roderick summarized the company's analysis. The consultant's figures of costs and savings were comprehensively refuted. None of the firms suggested by the USWA reports as potential customers for slabs produced at Duquesne could realistically be expected to become so; indeed, some were contemplating installing their own slab casters. One, California Steel—the remnant of the old Kaiser plant—was so far from Pittsburgh that freight costs alone would amount to at least $65 a ton, and in any case that company was under contract to foreign sources of slab supply. With such dramatically different projections of costs and profits, it was hopeless to expect any meeting of minds. Duquesne was no longer viable, remained closed, and was eventually dismantled.

TABLE 23.3

Estimates of costs and profits involved in the rehabilitation of Duquesne works, 1985

	Locker-Abrecht projection	US Steel figure
Costs of restarting and operating the blast furnace and BOF plants for three years	$90 million	$243 million
Cost of slabs in best of three years' operation	$209 per ton	$258 per ton
Profit (loss) over three years	$35 million	($110 million)

Source: US Steel press conference, 5 April 1985, USX

Meanwhile, the steel industry base, and therefore the manufacturing economy, of the main areas, was collapsing on a far greater scale even than in the 1930s. The city of Pittsburgh adjusted extremely well to these changing circumstances, but some of the mill towns were devastated by the collapse in steel. By the early 1980s, the main streets of Homestead were already eloquent of acute deprivation. Edgar Thomson remained in operation, but even so Braddock suffered terribly. Between 1950 and 1970, that town's population had already fallen from 16,000 to 8,000, and by the mid-1990s its residents numbered only 4,500. Median family income in Allegheny County was then about $35,000; in Braddock it was no more than $20,000.[11]

TABLE 23.4

Manufacturing jobs in the United States, Pittsburgh metropolitan area,
and Allegheny County as percentage of 1919 total

Region	1975	1980	1985	1990	1995
United States	91	101	96	96	94
Pittsburgh MSA	88	85	53	47	44
Allegheny Co.	86	82	51	43	38

Source: Regional Economic Information Services, University of Pittsburgh, 1998.

TABLE 23.5

Employment in steel operations in the Pittsburgh area,
1974 and summer 1996

Operation	1974	1996
US Steel		
Homestead	9,250	0
Edgar Thomson/Irvin	3,800	2,600
National/Duquesne	12,250	0
Clairton	4,520	1,500
Christy Park	1,413	0
J&L (LTV)		
Pittsburgh	5,140	800
Wheeling-Pittsburgh		
Monessen	2,300	230
Allenport	1,700	430
Total	40,373	5,560

Source: New Steel, September 1996, 37.

Conclusion

United States Steel in the Long View

The United States Steel Corporation came into being as a wonder of the man-ufacturing world, its largest industrial firm and dominating what was then the key sector of the economy. Yet, from the start it was not a rationally con-ceived, rounded creation but rather an agglomeration of businesses. It was commercially successful because of its huge size and predominating power in the marketplace. Even in the early years, the Corporation was outmaneuvered by some of its smaller competitors and as a result underwent a sustained if slow decline in market share. For the well over one-quarter of its history dur-ing which Elbert H. Gary was in control, there seems to have been only a rela-tively half-hearted program for streamlining production. Myron Taylor's chairmanship coincided with major changes of direction in the national economy, the industry, and at US Steel itself—in structure of demand, tech-nology, and even more dramatically in commercial circumstances due to the Great Depression. His attempts to rationalize US Steel were followed by re-newed large-scale growth in demand as a result of wartime activity initially and later from an economy reconstructing and expanding rapidly as the first and greatest "affluent society." In global context the United States was the in-dustrial leader for another generation. For US Steel the dangers of its preemi-nent position in world steel derived from the fact that for many years it had seemed that a straightforward extension in capacity was more essential to continuing success than any rationalization and reconstruction of existing plant. As a consequence, when slower growth and foreign competition even-tually required new ways of thinking, both the industry and its leading company, conditioned by decades of easy supremacy, for a time failed to recognize the necessities of their situation. After a decade or so in which there was gradual realization of the size of the challenge, US Steel was at last thoroughly rationalized and reorganized. The interesting outcome was that, though losing its proud title as the world's greatest steel company, US Steel gained the perhaps greater distinction of being one of the most efficient as it approaches its centennial.

Some more general conclusions may be drawn. In the first place, as Ford, Bacon, and Davis recognized more than sixty years ago, company size in itself is an ambivalent feature. It gives power, but in a twentieth-century setting successful use of this power also required a cleverly crafted corporate structure and exceptional quality in top direction. Inevitably, it was difficult to find both of an appropriate standard for the task. The exercise of responsibility—to employees and society as well as to stockholders—is another prerequisite and by definition almost certainly rules out a narrowly "business-like" response to situations. It may also mean that there is a slower reaction both to opportunities and challenges than in smaller firms. Being more flexible, the latter have grown more and faster, though this has meant that they in turn became less adaptable and newcomers are always available to take their place as pacemakers. In these respects at least, as a preeminent corporate giant US Steel can be reckoned to have been almost fatally handicapped from the start.

Second, there remain unresolved problems about top management. In the structure that US Steel adopted from its early days, following the power struggle between Charles Schwab and Elbert H. Gary, the chairmanship became the all-powerful office. Throughout its history the character, personality, and business convictions of twelve chairmen have had a remarkable influence in steering and giving tone to the whole nature of the huge organization. Nevertheless, one may wonder if it is better to have technical men at the helm or instead to be led by financiers, accountants, and others unfamiliar with the details of production. Schwab and William Corey, men from the production side, were superseded by a series of leaders from the wider business world or other professions. Gary and Taylor were the essential driving forces at US Steel from 1903 to 1938. In the mid-1950s for a short period a steelman, Benjamin Fairless (after a long period as president), again obtained the top direction, but he was succeeded by the lawyer Roger Blough. A generation later the financial expertise of David Roderick replaced other steelmen, Edwin Gott and Edgar Speer. It would be difficult to decide from the record which of the two avenues to the top proved most beneficial to the Corporation. Perhaps the best solution was a good working partnership between a chairman and a president drawn from the two traditions, as with Taylor and William Irvin, Irving Olds and

Fairless, or Roderick and Tom Graham. Another aspect of top management is perhaps easier to clarify. The influence of the New York headquarters at 71 Broadway as compared with on-the-spot control in the steel centers was probably disadvantageous, but in a multiplant company like US Steel it is difficult to see how in some form or other the problems it epitomized could have been avoided. Pittsburgh is currently the control center, but Chicago and Birmingham together have by far the preponderant share of capacity. Certainly the experience of Bethlehem Steel, with a long tradition of steel—or shipbuilding— men in supreme control and with operations firmly centered in Bethlehem itself does not give strong support to any contention that its different form of organization is a superior model of command.

Finally, there is the corporate relationship with government. As a great trust, US Steel was at a disadvantage from the start, attracting in turn the often hostile or at best critical attentions of the Roosevelt administration, Congress, the Federal Trade Commission, and the Temporary National Economic Committee. Later still, in price or labor negotiations, US Steel was usually the representative and "spokesman" for the industry, attracting public and government attention, usually ill favored. This has been important in relation to the problem of foreign competition and government response. In the nineteenth century, the iron and steel industry was given tariff protection on the basis of the infant industry argument, and this was maintained long after steel could no longer make any pretensions to infancy. The leading grounds of defense for protection in the latter part of the twentieth century were that other countries operated with low-cost labor, benefited from government subsidies, and were dumping their steel in the American market. The experience of the last twenty-five years or so suggests that mature steel industries may be just as much in need of some sort of protection and perhaps even of financial assistance from government as young ones in order to keep up technologically with newer plant and reequip to meet the challenge. When such reconstruction has occurred, oversight and protection may again be relaxed, but it seems undeniable that any government standing aloof from the problems of basic industries such as steel helps cripple its domestic producers in the world's markets. Indeed, to do so is to espouse doctrines of state-business relationships that are outdated in what has become the global competitive arena at the end of the twentieth century—once again in terms of political economy rather than merely economics alone. Not only the experiences of Japan and Korea but also those of the more comparable "mature" industrial economies of Western Europe prove that an old-style standoff between steel and government is a dangerous anachronism.

For government, the well-being of the domestic steel industry is no longer of such importance as in the past when steel was unquestionably both the supplier of the material foundations of the whole national manufacturing economy and in time of international tension or conflict its prime source of the means of military power. This fall from industrial preeminence is marked in the less than vital way in which various administrations have dealt with the industry's complaints about the unfairness of importers. Steel no longer has a high priority in the thinking of modern administrations, whatever their political color. The practice of foreign companies "dumping" their steel on the domestic market is a serious nuisance to the American industry, but as the world's political leader—and chief activator of the World Trade Organization —the government of the United States must maintain a large degree of openness to trade with the rest of the world. Moreover, whereas the secretary of the treasury is a key government official—so that banking or financial crises would receive urgent attention—the Department of Commerce does not rank so high.

On some occasions steel industry actions have run foul of government in other ways. The Fairless rod mill was successfully sold by US Steel to the Anshan works in China, but negotiations to sell the Texas works plate mill to Iraq failed when government recognized that it might be used to roll armor plate. From a completely different perspective, US Steel requests for government assistance in retaliation against alleged unfair competition from Spanish steel makers were once frustrated because the State Department was at the same time concerned to renegotiate the lease of military bases in Spain.

By the late 1990s, as it neared its centenary, US Steel had a raw-steel capacity of about 12.8 million net tons. In 1997 it operated over 96 percent of this capacity. It had three fully integrated works strategically located in relation to each other and for distribution to the major markets in the eastern half of the nation as well as two other major operations in joint ownership with leading Japanese and Korean companies. It shipped 11.6 million tons of products, just under 11 percent of the national total. Net income was $452 million, or $38.82 per ton of shipments. The increase in labor productivity since the early 1980s had been almost revolutionary. There seemed every reason to conclude that the long and extremely difficult rationalization program of the last twenty years had been carried to a logical conclusion and that the prospects were bright. But, as the 1990s passed, the Corporation experienced continuing and powerful pressure from three major directions: overseas producers, other reconstructed integrated companies, and mini-mills. In relation to each of them there is some evidence that US Steel was still losing out.

The problem of the inflow of foreign steel is one shared with all home producers. Consumption of steel in the United States has been growing less rapidly since the early 1970s; in some years levels are no higher than they were at that time. An exceptional year was 1998 with a demand of some 140 million tons, 17–18 million tons more than in 1973. But whereas 1973 imports were 15.2 million net tons, they amounted to 41million tons in 1998, almost 30 percent of the domestic consumption. By early fall 1998, penetration of foreign steel was running at 35 percent of total consumption in the United States. A few months later, the 1998 US Steel annual report recognized that shipments and prices would continue to be affected by high levels of low-priced imports. Even if the new restraints introduced in spring 1999 work successfully, there seems little likelihood that there will be any great reduction in pressures from the wider steel-producing world; estimated world capacity that year was expected to be about 100 million tons in excess of world requirements.[1]

As US Steel pared and modernized its operations, other integrated producers busily did the same. In relation to this section of the national industry, some of its leading member firms, and the whole of the steel produced in the United States, US Steel suffered a decline. It is perhaps well to take a longer view. For instance, in 1960 Bethlehem Steel produced 58.1 percent as much steel as the Corporation, in 1976 66.5 percent, and by 1997 79.7 percent. The share of all integrated companies in total American output fell from almost 90 percent in 1960 to only a little over 50 percent by the late 1990s. This is a striking indication of the advance of the mini-mill, or as it is now sometimes called the EAF (electric arc furnace) sector. As late as 1991, under 23 percent of United States steel came from mini-mills; within another five years they controlled almost half the capacity. The 1997 output of Nucor, the largest of these EAF operations, made it the nation's third largest producer with a tonnage 91.6 percent as large as that of Bethlehem and almost three-quarters that of US Steel. Its advance has been dramatic, and most key indi-

TABLE 24.1
US Steel's steel production, 1960, 1976, and 1997

Year	Share of total U.S. steel output	Share of output by all U.S. integrated companies	Share of output by a group of major companies[a]
1960	27.5%	30.6%	35.6%
1976	30.6%	n.a.	n.a.
1997	12.1%	23.3%	25.2%

[a] US Steel, Bethlehem, the constituent companies of LTV, National-Weirton, Inland, and Armco (later AK Steel).

cators are highly favorable, though they also indicate how much US Steel has improved its own performance. Most threatening of all to the Corporation, as the 1998 report recognized, mini-mills are now invading fields the major companies long regarded as their own preserve and have therefore in recent decades concentrated more of their finishing capacity. By 1996 sheet- and tin-mill products made up 83.9 percent of all tonnage shipped by US Steel mills—a far higher proportion even than for rails a century before. With the introduction and rapid adoption of thin-slab casting, mini-mills have made headway in these lines as well. The smaller operators experienced difficulties in producing the finest deep-drawing steels that, when finished, are free from surface imperfections, but through the 1990s have been improving both technology and product. After the mid-1990s progress in this direction seems to have become a flood. As late as 1995 EAF-based sheet mills shipped only 3 million tons of hot-rolled bands and related thin flat-rolled products. By fall 1998 forecasts indicated that within another two years their shipments would reach 14 million tons.[2] With mini-mills upgrading their product and most integrated producers busily investing to improve existing hot-strip mills, competition for the available business threatens to become so keen that it will lead to either general underuse of facilities or some costly failures and withdrawals from business. At the same time, prices in the world markets have been tumbling—in summer 1997 the world spot export price for hot-rolled band was $350 a metric ton; by the end of 1998 it was only $185 per metric ton.[3] In April 1999 yet another indicator was provided of the increasing competition from the mini-mills when General Motors announced four-year contracts to buy an annual total of almost 5 million tons of steel from the industry. It was negotiating with four mini-mills to supply 2–2.5 million tons of sheet a year to its stampings suppliers.[4]

In technology, the integrated industry seems to be poised on the edge of further major changes. Over the last quarter-century, a new orthodoxy has superseded an older one: large, highly productive blast-furnace operations linked with BOP or electrical furnaces and continuous casting installations feeding finishing mills have replaced the blast furnace–open hearth–primary mill–finishing mill route. But the indispensability of the blast furnace is now under question. Directly reduced iron (DRI), serving electric furnaces, or even direct steel processes as foreshadowed in the US Steel–Nucor plans for processing iron carbide in the mid-1990s, may make it possible to dispense with much of the capital expensive and labor intensive upstream end of the production process—coke ovens, blast furnaces, and oxygen converters. Already there is a national swing away from the BOP process to the electric furnace, though this is largely due to the advance of the mini-mills. In 1992, for the first time, no open-hearth steel was produced. That year 62 percent of steel came from oxygen steel plants and 38 percent from electric furnaces; four years later the proportions were 57.4 and 42.6 percent. Now it is seen that the large-scale casting of slabs to be rolled down and finished in the increasingly important hot-strip mills, though so recently brought to full success, may be replaced by the continuous casting of thin shapes, which require much less further rolling. If as seems likely these advances fully prove themselves, over the next ten or fifteen years as the natural life of major conventional plant units ends, it may well be that some of the alternative processes will be inserted in the production sequence instead—for instance, a relatively cheap DRI plant instead of a costly new blast furnace or even a major refit. If this process proceeds piecemeal, much of the old distribution pattern of the industry will be preserved. However, these new conditions may not only reshape the industry—and thereby provide a much more effective rebuttal of the challenge from the mini-mills—but also upset much of the conventional wisdom about scale and location. Economies of scale and concentration may be much less with a mill made up of a combination of all these new processes. Instead there may be many smaller units. For this reason, and because of increased freedom from dependence on large tonnages of various minerals, the future steel industry will be much more market oriented. Where does this leave a major operation from the past like US Steel?

One interesting pointer to another possible important trend has been joint ventures in the domestic market. Given the highly competitive situation in the industry, more such ventures or even mergers between the majors may be expected. Many questioned the rationale of talks about the possible association

TABLE 24.2
Raw-steel production by major companies 1960, 1976, 1997, and 1999
(million net tons)

Company	1960	1976	1997	1999
US Steel	27.3	26.3	12.3	12.4
Bethlehem	15.9	17.5	9.8	9.4
LTV companies	17.7	n.a.	8.3	8.4
National-Weirton	5.7	10.0	8.6	n.a.
Inland	5.1	7.3	5.4	n.a.
Armco-AK Steel	5.0	6.9	4.4	6.2
All integrated operations	89.1	n.a.	52.7	n.a.
Total US raw-steel output	99.3	128.0	101.5	106.0

of US Steel and Inland on grounds of duplication of facilities in the same area, but there might have been important economies in energy, raw-materials supply, administration, and sales costs, in addition to which various units of plant in the Gary and Indiana Harbor works could have become complementary. At a lower level of linkage, Bethlehem has now become an important participant in the Lorain bar operations, already jointly operated with Kobe, thus reducing to a minority share the Corporation's involvement in a works that was until the late 1980s one of its prime operations. Investments in cold reduction and galvanizing lines, located in small communities much nearer the automobile industry's centers at Leipsic or even in southern Michigan, are instances of the steel industry tailoring its products even more carefully to suit the needs of major consumers. In the next few years, producers might have to go even further into the processing of steel, though not necessarily all the way to the operation of stamping works. Regardless, there is a new fluidity in company structure, in integration of processes, and in location of the industry. There seems no reason to expect that in the near future this will all crystallize into a new fixed pattern.

Another change, perhaps in the short term, is a new internationalism in steel, not only in trade but also in company organization. In the mid-1990s, US Steel took an interest in the finishing operations of the Kosice works in Slovakia. By 1999 it was moving to fuller involvement in the whole of that plant as a means of getting a foothold in both a good site and a strategic location to serve the expanding markets of central and eastern Europe. In other words, having earlier disposed of interests in Europe such as Altos Hornos of Spain and after decades of competition from European Union countries, US Steel is now taking the offensive not far from the heartland of its commercial adversaries. More widely, with domestic pollution control standards becoming stricter and with many developing countries still relatively lax about such matters, the Corporation may relocate more of the relatively "dirty" end of the industry out of the United States, where it might then concentrate on preparing imported semifinished material for its final markets.

In light of continuing competition from its integrated peers, from mini-mills, and from imports; of technical change; and the fact that consumption seems most unlikely to increase on a scale sufficient to accommodate them all, the much improved position of US Steel should be seen not as a tidy, finished success but only a stage on a difficult journey. As with all the other participants in the struggle for business, its position is vulnerable and therefore must always be actively defended. A primary consideration will be whether the market can support its present three integrated operations. If not, which of these will close or be drastically curtailed? Gary is the biggest, best-equipped works and occupies the prime national location, being the principal US Steel supplier to the automobile markets. Fairfield is not only the Corporation's foothold in the South but also serves the Gulf Coast. The Mon Valley operations play a vital role in US Steel strategy, supplying sheet and coil to further processors in western Pennsylvania and Ohio and railing material for finishing at Fairless. The large outlay on slab casting at Edgar Thomson in the early 1990s was a good augury for continuing operations there. However, there is a good deal of overlap between the three plants. If future competition requires further cutbacks, ET's size, character, and location, and the fact that US Steel has more rivals in this area than in the South, suggests that this operation is the most vulnerable.

Laying aside speculation about the future, some firm statements can at least be made about the present. A Corporation that began the century with two-thirds of the nation's raw steel-making capacity is ending it with scarcely more than 10 percent. Today the size, structure, and operating context of US Steel means that it is scarcely recognizable as a descendant of J. P. Morgan, Charlie Schwab, and Judge Gary's creation of 1901. Looking ahead, all that is fairly certain about the early part of US Steel's second century is that its future will be no less tumultuous, colorful, difficult to predict, or interesting than its past.

Appendix A

Statistical Tables

TABLE A1

**US Steel's and independent companies' shares of production,
1902 and 1910 (percentages)**

Product	1902		1910	
	US Steel	Independents	US Steel	Independents
Lake Superior ore (shipments)	60.4	39.6	51.1	48.9
Coke	37.5	62.5	32.7	67.3
Pig iron	44.8	55.2	43.3	56.7
Steel	65.2	34.8	54.3	45.7
Steel rails	67.7	32.3	58.8	41.2
Structural shapes	57.9	42.1	51.3	48.7
Plates and sheets	59.4	40.6	48.0	52.0
Wire rod	71.6	28.4	67.3	32.7
Total, all finished rolled steel	51.3	48.7	48.1	51.9
Wire nails	64.9	35.1	55.4	44.6
Tinplate and terneplate	73.4	26.6	61.0	39.0

Source: AISI quoted in E. Jones, *The Trust Problem in the United States* (New York: Macmillan, 1922).

Appendix A

TABLE A2

Steel ingot capacity of US Steel's Pittsburgh and Chicago integrated works in selected years, 1901–1999 (net tons in thousands)

Year	Pittsburgh[a]	Chicago[b]
1901	4,700	1,876
1904	5,695	2,038
1920	7,792	6,856
1930	8,537	9,303
1938	9,709	9,987
1945	12,023	10,244
1954	10,941	11,609
1960	12,339	13,588
1973	10,500	15,350
1976	9,521	13,299
1989	2,700	8,200
1999	c.2,800	c.7,600

[a] Includes the Edgar Thomson, Homestead, Duquesne, and McKeesport works to 1904, and additionally Clairton and Donora thereafter.

[b] Includes the Joliet and South Chicago works to 1904; after 1904 also includes the Gary works.

TABLE A3

US Steel's share of national ingot steel and finished hot-rolled products capacity and production, 1901–1936

Year	Ingot steel		Finished hot-rolled products	
	% capacity	% production	% capacity	% production
1901	47.2	65.7	46.9	60.1
1902	n.a.	65.2	n.a.	58.8
1903	49.8	63.1	49.8	56.5
1904	51.1	60.6	51.1	56.5
1905	51.7	59.9	51.7	54.8
1906	51.8	57.7	51.8	54.0
1907	51.4	57.1	51.4	53.2
1908	50.5	55.9	50.5	52.5
1909	50.7	55.7	50.7	50.2
1910	50.2	54.3	50.2	49.6
1911	49.5	53.9	49.5	49.8
1912	47.4	54.1	47.4	50.7
1913	47.9	53.2	47.7	49.9
1914	46.6	50.3	46.6	49.1
1915	45.5	50.9	45.5	48.2
1916	44.4	48.9	44.4	47.7
1917	42.3	45.0	42.3	45.2
1918	41.0	44.0	41.0	44.4
1919	40.2	49.6	40.2	47.8
1920	39.5	45.8	39.9	44.0
1921	38.8	55.4	39.4	53.2
1922	38.9	45.2	39.5	44.5
1923	38.4	45.2	39.3	44.2
1924	37.8	43.4	38.5	41.7
1925	39.3	41.6	39.1	39.7
1926	38.6	42.0	38.1	40.4
1927	38.7	41.1	37.8	39.5
1928	37.9	39.0	37.4	37.1
1929	38.6	38.7	37.7	37.3
1930	37.8	41.2	37.9	39.3
1931	n.a.	n.a.	38.8	37.5
1932	n.a.	n.a.	38.2	34.4
1933	n.a.	n.a.	37.6	33.1
1934	n.a.	n.a.	38.1	31.7
1935	n.a.	n.a.	37.7	31.2
1936	n.a.	n.a.	36.4	32.6

Source: Ford, Bacon, and Davis, "Reports," vol. 200, pt. 2.

TABLE A4

**US Steel's share of industry production of various finished products,
1920–1936 (percentage)**

Year	Rails	Plates	Heavy structs.	Pipe & tubes	Wire rod	Black sheet	Tin Plate	HR Strip
1920	58.1	46.4	46.4	43.4	17.0	26.7	43.6	n.a.
1925	55.0	51.1	42.1	39.2	15.2	17.1	47.4	9.5
1930	51.1	54.5	46.5	33.5	17.5	19.4	38.1	12.2
1936	54.8	36.8	46.4	28.3	18.8	18.0	32.5	20.7

Source: Ford, Bacon, and Davis, "Reports," vol. 200, pt. 2.

TABLE A5

**Prices for No. 1 heavy melting scrap in Pittsburgh, Chicago, and Eastern Pennsylvania,
1905–1937 (annual averages, dollars per gross ton)**

Year	Pittsburgh	Chicago	E. Pennsylvania
1905	15.77	13.96	16.54
1910	15.34	13.43	14.76
1915	13.23	10.91	12.51
1920	25.74	22.82	22.99
1925	18.60	16.27	16.44
1930	15.20	12.08	13.12
1935	12.72	11.45	11.30
1937	18.85	17.11	17.78

Source: Iron Age, 6 January 1938, 112–14.

TABLE A6

Consumption of steel by metal-fabricating industries in economic areas consuming over 0.4 million tons, 1954 (thousands of tons)

Area	Steel used
Detroit	4,976
Chicago	3,788
Pittsburgh	2,000
Cleveland	1,649
Philadelphia	1,580
Los Angeles	1,330
NE New Jersey	1,162
Milwaukee	1,151
Buffalo	977
Flint, Jackson, Owosso	962
New York	905
San Francisco–Oakland	749
Youngstown	719
St. Louis	663
Baltimore	641
Hamilton-Middletown	618
Lansing	596
Gary, Hammond, East Chicago	573
Muncie, Anderson, Richmond, Kokomo	570
Houston	556
Grand Rapids	542
Birmingham	473
Canton	454

Source: *Business Review of the Federal Reserve Bank of Cleveland*, April 1958.

Note: The assessment was based on the 1954 U.S. Census of Manufactures and includes all consumption except that by primary manufacturers themselves. It was estimated to be accurate within 2 percent.

TABLE A7
**US Steel plants and the plants of major overseas steel companies,
1960, 1967, 1974 (millions of short tons)**

Plant	1960	1967	1974
USS, Gary	8.0	8.5	9.5
USS, South Chicago	5.6	4.2	5.7
USS, Homestead	4.6	4.6	4.5
USS, Fairfield	2.2	3.9	3.1
USS, Fairless	2.7	3.0	3.0
USS, Lorain	2.7	2.7	2.7
USS, Geneva	2.3	2.5	2.6
USS, Edgar Thomson	2.5	2.5	2.5
USS, Youngstown	2.7	2.2	2.2
USS, Duquesne	1.7	1.8	1.9
Nippon Kokan, Fukuyama	n.a.	4.2	15.1
Nippon Steel, Kimitsu	n.a.	2.2	13.8
Kawasaki, Mizushima	n.a.	1.8	11.0
Sumitomo, Wakayama	n.a.	7.2	8.3
Thyssen, Beeckerwerth/Bruckhausen/Ruhrort	n.a.	9.0	15.7
Italsider, Taranto	n.a.	3.1	11.5
Usinor, Dunkerque	n.a.	3.1	8.2
Hoesch, Dortmund	n.a.	3.6	6.3
Hoogovens, Ijmuiden	n.a.	3.2	6.1

Source: Hans Mueller correspondence with author, c. 1976.

TABLE A8
US Steel and USX Steel Division key indicators, 1970, 1980, 1990, 1994, 1996

Year	Raw steel produced (million tons)	Steel shipped (million tons)	Employees	Net income (millions)	Net income as % sales	Employment cost per ton shipped
1970	31.4	21.0	200,734	$147	3.0%	$107.17
1980	23.3	17.1	149,172	$504	4.0%	$280.22
1990[a]	13.6	11.0	24,664	$310	5.1%	$101.27
1994	11.7	10.6	21,310	$201	3.4%	$132.66
1996	11.4	11.4	20,831	$273	4.2%	$120.65
1997	12.3	11.6	20,683	$452	6.6%	$121.70

Sources: US Steel annual reports.
[a] From 1990, figures exclude Marathon Oil and other USX oil and gas operations.

TABLE A9
**US Steel production, shipments, employment, and financial performance,
1902–2000**

Year	Steel Output (million tons)	Steel Shipped (million tons)	Sales (million $)	Income (loss) (million $)	Employees	Income (loss) as % of sales
1902	10.9	8.9	423.1	90.3	168,127	21.3
1903	10.3	8.1	398.2	55.4	167,709	13.9
1904	9.4	7.3	324.9	30.2	147,343	9.3
1905	13.4	10.1	409.2	68.6	180,158	16.8
1906	15.2	11.3	484.0	98.1	202,457	20.3
1907	14.9	11.5	504.4	104.6	210,180	20.7
1908	8.8	6.8	331.6	45.7	165,211	13.8
1909	15.0	10.6	441.1	79.0	195,500	17.9
1910	15.9	11.8	491.8	87.4	218,435	17.8
1911	14.3	10.3	431.7	55.3	196,888	12.8
1912	18.9	13.8	533.9	54.2	221,025	10.2
1913	18.7	13.4	560.8	81.2	228,906	14.5
1914	13.2	9.9	412.2	23.4	179,353	5.7
1915	18.3	12.8	523.7	75.9	191,126	14.5
1916	23.4	17.1	902.3	271.5	252,668	30.1
1917	22.7	16.9	1,284.6	224.2	268,058	17.5
1918	21.9	15.6	1,344.6	125.3	268,710	9.3
1919	19.3	13.5	1,122.6	76.8	252,106	6.8
1920	21.6	15.5	1,290.6	109.7	268,004	8.5
1921	12.3	8.8	726.0	36.6	191,700	5.0
1922	18.0	13.1	809.0	39.6	214,931	4.9
1923	22.8	15.9	1,096.5	108.7	260,786	9.9
1924	18.5	12.7	921.4	85.1	246,753	9.2
1925	21.2	14.8	1,022.0	90.6	249,833	8.9
1926	22.7	15.8	1,082.3	116.7	253,199	10.8
1927	20.7	14.3	960.5	87.9	231,549	9.2
1928	22.5	15.4	1,005.3	114.1	221,702	11.4
1929	24.5	16.8	1,097.4	197.5	254,495	18.0
1930	18.8	12.8	828.4	104.4	252,902	12.6
1931	11.3	8.4	548.7	13.0	215,750	2.4
1932	5.5	4.3	287.7	(71.2)	164,348	(24.7)
1933	9.0	6.3	375.0	(36.5)	172,577	(9.7)
1934	9.7	6.5	420.9	(21.7)	189,881	(5.1)
1935	12.5	8.1	539.4	1.1	194,820	0.2
1936	18.9	11.9	790.5	50.5	222,372	6.4
1937	20.7	14.1	1,028.4	94.9	261,293	9.2
1938	10.5	7.3	611.1	(7.7)	202,108	(1.3)
1939	17.6	11.7	846.0	41.1	223,844	4.9
1940	22.9	15.0	1,079.1	102.2	254,393	9.5
1941	29.0	20.4	1,622.3	116.2	304,248	7.2

TABLE A9 (CONTINUED)

Year	Steel Output (million tons)	Steel Shipped (million tons)	Sales (million $)	Income (loss) (million $)	Employees	Income (loss) as % of sales
1942	30.0	20.6	1,863.0	71.2	335,866	3.8
1943	30.5	20.1	1,972.3	62.6	340,498	3.2
1944	30.8	21.0	2,082.2	60.8	314,888	2.9
1945	26.5	18.4	1,747.3	58.0	279,274	3.3
1946	21.3	15.2	1,496.1	88.6	266,835	5.9
1947	28.6	20.2	2,122.8	127.1	286,316	6.0
1948	29.3	20.6	2,481.5	129.6	296,785	5.2
1949	25.8	18.2	2,301.7	165.9	291,163	7.2
1950	31.5	22.6	2,956.4	215.5	288,265	7.3
1951	34.3	24.6	3,524.1	184.3	301,328	5.2
1952	29.4	21.1	3,137.4	143.6	294,263	4.6
1953	35.8	25.1	3,861.0	222.1	301,560	5.8
1954	28.4	20.2	3,250.4	195.4	268,142	6.0
1955	35.3	25.5	4,097.7	370.1	272,646	9.0
1956	33.4	23.9	4,228.9	348.1	260,646	8.2
1957	33.7	23.4	4,413.8	419.4	271,037	9.5
1958	23.8	17.0	3,472.1	301.5	223,490	8.7
1959	24.4	18.1	3,643.0	254.5	200,329	7.0
1960	27.3	18.7	3,698.5	304.2	225,081	8.2
1961	25.2	16.8	3,336.5	190.2	199,243	5.7
1962	25.4	17.8	3,501.0	163.7	194,044	4.7
1963	27.6	18.9	3,637.2	203.5	187,721	5.6
1964	32.4	21.2	4,129.4	236.8	199,979	5.7
1965	32.6	22.5	4,465.0	275.5	208,838	6.2
1966	32.8	21.6	4,434.7	249.2	205,544	5.6
1967	30.9	19.8	4,067.2	172.5	197,643	4.2
1968	32.4	22.5	4,609.2	253.7	201,017	5.5
1969	34.7	22.4	4,825.1	217.2	204,723	4.5
1970	31.4	21.0	4,883.2	147.5	200,734	3.0
1971	27.2	19.3	4,966.7	154.5	183,940	3.1
1972	30.7	20.8	5,443.4	157.0	176,486	2.9
1973	35.0	26.1	7,031.1	313.0	184,794	4.6
1974	33.9	25.5	9,339.2	630.3	187,503	6.8
1975	26.4	17.5	8,380.3	559.6	172,796	6.7
1976	28.3	19.5	8,607.8	410.3	166,645	4.7
1977	28.8	19.7	9,609.9	137.9	165,845	1.4
1978	31.3	20.8	11,049.5	242.0	166,848	2.2
1979	29.7	21.0	12,929.1	(293.0)	171,654	(2.3)
1980	23.3	17.1	12,492.1	504.5	149,172	4.0
1981	23.4	16.5	13,940.5	1,077.2	141,623	7.7
1982[a]	12.1	10.3	18,919.0	(332.0)	119,987	(1.9)
1983[a]	14.8	11.3	17.539.0	(1,161.0)	98,722	(6.6)

TABLE A9 (CONTINUED)

Year	Steel Output (million tons)	Steel Shipped (million tons)	Sales (million $)	Income (loss) (million $)	Employees	Income (loss) as % of sales
1984[a]	15.1	12.1	19,104.0	414.0	88,753	2.6
1985[a]	16.7	12.9	19,283.0	313.0	79,649	2.1
1986[a]	9.6	8.5	15,025.0	(1,833.0)	63,915	(12.1)
1987[b]	11.5	8.6	4,899.0	199.0	32,865	4.1
1988[b]	15.5	12.2	6,509.0	703.0	34,327	10.0
1989[b]	14.2	11.5	6,509.0	540.0	27,173	8.3
1990[b]	13.6	11.0	6,073.0	310.0	24,664	5.1
1991[b]	10.5	8.8	4,864.0	(507.0)	22,234	(10.4)
1992[b]	10.4	8.8	4,919.0	(1,606.0)	21,479	(32.6)
1993[b]	11.3	10.0	5,612.0	(238.0)	21,527	(4.2)
1994[b]	11.7	10.6	6,066.0	201.0	21,310	3.3
1995[b]	12.2	11.4	6,475.0	301.0	20,845	4.6
1996[b]	11.4	11.4	6,547.0	273.0	20,831	4.2
1997[b]	12.3	11.6	6,941.0	452.0	20,683	6.5
1998[b]	11.2	10.7	6,477.0	364.0	20,267	5.6
1999[b]	12.0	10.6	5,470.0	51.0	19,266	0.9
2000[b]	11.7	10.8	6,132.0	(21.0)	19,353	(0.3)

[a] The Marathon Group and the whole of USX are included with the steel operations.

[b] Refers only to the US Steel Group, excluding not only Marathon but also some sectors previously included in US Steel.

TABLE A10
Ingot steel capacity of US Steel works (net tons in thousands)

Works	1901	1920	1930	1938	1945	1957	1960	1973	1989	2000
Alabama										
Bessemer	n.a.	5								
Ensley	336	1,254	1,142	1,075	1,568	1,770	1,770	rundown		
Fairfield	n.a.	n.a.	840	943	1,092	2,227	2,227	4,000	2,900	2,400
California										
Pittsburg	n.a.	37	215	290	393	380	380	closed		
Torrance	n.a.	42	191	204	211	222	237	250	closed	
Illinois										
Joliet	672	820	739	closed						
S. Chicago	1,204	2,626	3,588	4,341	4,525	5,441	5,589	7,350	closed	
Union	364 (idle)									
Indiana										
Gary	n.a.	3,410	4,976	5,646	5,719	7,204	7,999	8,000	8,200	7,600
Massachusetts										
Worcester	n.a.	202	213	213	280	287	closed			
Minnesota										
Duluth	605	605	336	610	973	973	closed			
Ohio										
Bellaire	n.a.	470	closed							
Cleveland	672	1,086	896	closed						
Columbus	n.a.	224	closed							
Lorain	616	1,450	1,680	1,747	1,944	2,565	2,678	3,000	3,300	(USS/Kobe)
Mingo Junction	n.a.	672	672	672	n.a.	sold				
Youngstown	728	1,568	1,904	2,229	2,344	2,943	2,712	2,500	closed	
Oregon										
Portland	n.a.	3	2							
Pennsylvania										
Ambridge	n.a.	10								
Braddock	1,120	1,546	1,618	1,820	2,297	2,179	2,529	2,500	2,700	2,800
Clairton	n.a.	728	806	805	805	1,064	1,064	closed		
Donora	n.a.	841	703	702	842	1,015	1,015	closed		
Duquesne	1,120	1,519	1,692	2,038	2,147	1,521	1,741	3,000	closed	
Farrell	n.a.	672	874	812	1,050	sold				
Homestead	2,128	2,386	2,930	3,267	4,732	4,043	4,598	5,000	closed	
Johnstown	n.a.	17	21	21	24	25	25	60	sold	
McKeesport	336	772	788	1,077	1,200	1,446	1,392	closed		
New Castle	560	862	560							
Pencoyd	n.a.	272	272	242						
Pittsburgh	n.a.	464								
Sharon	123	168								
Trenton	n.a.	n.a.	n.a.	n.a.	n.a.	2,200	2,687	4,400	3,000	
Vandergrift	n.a.	309	376	403	500					
Texas										
Baytown	n.a.	n.a.	n.a.	n.a.	n.a.	n.a.	n.a.	800	closed	
Utah										
Geneva	n.a.	n.a.	n.a.	n.a.	1,283	2,077	2,300	2,500	sold	
W. Virginia										
Benwood	n.a.	255								

Appendix B

Chief Officers of US Steel, 1901–2000

Chairmen

1901–1927	Elbert H. Gary
1927–1932	J. P. Morgan Jr.
1932–1938	Myron C. Taylor
1938–1940	Edward R. Stettinius
1940–1952	Irving S. Olds
1952–1955	Benjamin F. Fairless
1955–1969	Roger M. Blough
1969–1973	Edwin H. Gott
1973–1979	Edgar B. Speer
1979–1989	David M. Roderick[a]
1989–1995	Charles A. Corry[a]
1995–	Thomas J. Usher[a]

[a] After 1986, chairmen of USX Corporation

Presidents

1901–1903	Charles M. Schwab
1903–1911	William E. Corey
1911–1932	James A. Farrell
1932–1938	William A. Irvin
1938–1953	Benjamin F. Fairless
1953–1959	Clifford F. Hood
1959–1967	Leslie B. Worthington
1967–1969	Edwin H. Gott
1969–1973	Edgar B. Speer
1973–1975	Wilbert A. Walker
1975–1979	David M. Roderick
1979–1983	William R. Roesch
1983–1991	Thomas C. Graham[b, c]
1991–1994	Thomas J. Usher[c]
1995–	Paul J. Wilhelm[c]

[b] From 1983 to 1986, officially vice chairman and chief operating officer.
[c] Presidents of the US Steel Group of USX Corporation.

Chairmen of Finance Committee

1901–1903	Robert Bacon
1903–1907	George W. Perkins
1907–1927	Elbert H. Gary
1927–1934	Myron C. Taylor
1934–1936	William J. Filbert
1936–1938	Edward R. Stettinius
1938–1956	Enders M. Voorhees
1956–1970	Robert C. Tyson
1970–1973	Wilbert A. Walker
1973–1975	David M. Roderick
1975–1991	W. Bruce Thomas[d]
1991–	Robert M. Hernandez[d]

[d] Officially chief financial officer.

Notes

1. Origins

1. A. Berglund, *The United States Steel Corporation* (New York: Columbia University Press, 1907), 12–16.

2. F. Popplewell, *Some Modern Conditions and Recent Developments in Iron and Steel Production in America* (Manchester: Manchester University Press, 1906), 24; J. S. Jeans, *American Industrial Conditions and Competition* (London: British Iron Trade Association, 1902), passim.

3. H. H. Campbell, *The Manufacture and Properties of Structural Steel* (New York: Scientific Publishing, 1896), v.

4. See "The Tendency to Concentrate," *IA*, 13 December 1900, 29.

5. P. Temin, *Iron and Steel in Nineteenth Century America: An Economic Enquiry* (Cambridge, Mass.: MIT Press, 1964), table C9.

6. Based on U.S. Department of Commerce, *Historical Statistics of the United States* (Washington, D.C.: GPO, 1957), tables P233–49, P187–232, M195–210.

7. E. H. Gary, "Evidence to Stanley Committee," quoted in *IA*, 5 June 1913, 1394–95; 27 June 1913, 1494.

8. A. Carnegie letter, 28 September 1889 [*sic*, 1899], quoted in *IA*, 5 June 1913; AISA annual report, 1897.

9. E. S. Meade, "The Genesis of the United States Steel Corporation," *Quarterly Journal of Economics* 15 (August 1901): 517; E. Hubbard, "American Trusts and English Combinations," *Economic Journal* (June 1902): 168; N. R. Lamoreaux, *The Great Merger Movement in American Business, 1895–1904* (Cambridge: Cambridge University Press, 1985), passim.

10. Quoted in Jeans, *American Industrial Conditions and Competition*, 196.

11. Meade, "Genesis of the United States Steel Corporation," 539–40.

12. *IA*, 3 August 1899, 7; 23 May 1912, 1320; 13 June 1912, 1492.

13. *IA*, 30 May 1912, 1349.

14. *IA*, 21 December 1899, 21.

15. H. O. Evans, *Iron Pioneer: Henry W. Oliver, 1840–1904* (New York: E. P. Dutton, 1942), 266–67.

16. *IA*, 22 February 1900, 18; Gary, "Evidence," quoted in *IA*, 5 June 1913, 1395.

17. J. F. Wall, *Andrew Carnegie* (New York: Oxford University Press, 1970), 782; quoted in W. T. Hogan, *Economic History of the Iron and Steel Industry in the United States* (Lexington, Mass.: D. C. Heath, 1971), 465–67.

18. A. Carnegie to A. Cassatt, 31 December 1900, ACLC.

19. C. M. Schwab to A. Carnegie, 24 January 1901, ACLC (present writer's emphasis); A. Carnegie to C. M. Schwab, [late January 1901], ACLC.

20. *IA*, 26 December 1901, 26–27.

21. C. M. Schwab to Stanley Committee, 4 August 1911, in U.S. Congress, House, *Hearings before the Committee on Investigation of U.S. Steel [Stanley Committee]* (Washington, D.C.: GPO, 1912), 1276–79 [cited hereafter as *Stanley Committee*]; *BAISA*, 1 September 1911.

22. Schwab to Stanley Committee, 4 August 1911, 1278.

23. A. Carnegie to G. Lauder, "Saturday" [December 1900], Collection 81, ACLC.

24. Wall, *Andrew Carnegie*, 786–89; A. Carnegie to C. M. Schwab, 7 January 1901, ACLC.

25. I. M. Tarbell, *The Life of Elbert H. Gary: The Story of Steel* (New York: Appleton, 1925), 111–18.

26. D. Reid in *USA v. USS*, 1:482–83, 498.

27. Temin, *Iron and Steel in Nineteenth Century America*, 190–93.

28. G. Harvey, *Henry Clay Frick: The Man* (New York: Charles Scribners, 1928), 260–61.

29. A. Carnegie to C. M. Schwab, 7 March 1901, ACLC.

2. Early Years of Industry Leadership, 1901–1904.

1. J. W. Gates quoted in *ITR*, 10 August 1911, 233–34.

2. Berglund, *United States Steel Corporation*, 113; R. Vangermeers, "The Capitalization of Fixed Assets in the Birth, Life, and Death of US Steel, 1901–1986" (N.p., n.d.), 6; E. S. Meade, "The Steel Corporation Bond Conversion," *Quarterly Journal of Economics* 18 (1904): 42–43.

3. Jeans, *American Industrial Conditions and Competition*, 300; *BAISA*, 23 January 1901.

4. *ITR*, 31 May 1900; 20 June 1901, 26b; 11 December 1902, 34c; 27 November 1902, 38a, 38b; and 18 December 1902, 34c; *Pittsburgh Dispatch*, 17 December 1902.

5. US Steel Executive Committee minutes, 4 March 1902, USX.

6. US Steel Executive Committee minutes, 21 April 1903, USX; S. Whipple, "Notes of Conversations with C. M. Schwab," 1935–36, BP; Hogan, *Economic History*, 487.

7. U.S. Commissioner of Corporations, *Report on the Steel Industry*, 2 vols. (Washington, D.C.: GPO, 1911, 1913), 269.

8. AISA, *Statistical Report*, (New York and Washington, D.C.: AISI, 1909).

9. *BF&SP*, January 1938, 85–87; *Iron and Steel Engineer*, December 1953, 87.

10. J. Gayley quoted in *IA*, 22 June 1911.

11. US Steel Executive Committee minutes, 1, 2 July 1902, USX.

12. J. Garraty, *Right Hand Man: The Life of George W. Perkins* (New York: Harper, 1957), 117.

13. Berglund, *United States Steel Corporation*, 78.

14. *IA*, 23 August 1901, 27.

15. R. S. Craig, "A History of the United States Steel Corporation and Predecessor Companies: A Case Study of American Business," report for US Steel, 1966, USX, passim; American Steel and Wire, Thirtieth Anniversary (N.p.: American Steel and Wire of New Jersey, c.1930).

16. W. Dinkey to Board of Carnegie Steel Company, 17 April, 8 May 1905, USX; Carnegie-Illinois list of abandoned properties, 29 May 1946, USX.

17. US Steel, "Preliminary Report to Stockholders," 17 February 1902, 6, 18, 19.

18. *BAISA*, 15 November 1906.

19. *IA*, 22 May 1913, 1242.

20. *BAISA*, 25 June 1904.

21. United States Industrial Commission, *Reports*, 13 vols. (Washington, D.C.: GPO, 1899–1901), 13:xcix.

22. J. P. Morgan, circular, 2 March 1901, quoted in Hogan, *Economic History*, 477.

3. Judge Gary's "Umbrella"

1. W. B. Dickson to S. C. Munoz, 19 March 1939, WD.

2. Quoted in A. R. Burns, *The Decline of Competition* (New York: McGraw Hill, 1936), 77 n. 1.

3. Tarbell, *Elbert H. Gary*, 257.

4. S. Whipple, "Notes of Conversations with C. M. Schwab," 1935–36, BP, 94.

5. "US Steel Preliminary Report [to Stockholders]," February 1902.

6. *New York Tribune* quoted in *BAISA*, 25 November 1902.

7. Berglund, *United States Steel Corporation*, 140.

8. US Steel Executive Committee, 9 April 1901, USX.

9. U.S. Commissioner of Corporations, *Report*, 2:3.

10. *Stanley Committee*, 1:92, 97.

11. F. Taussig, *Some Aspects of the Tariff Question* (Cambridge, Mass.: Harvard University Press, 1915), 173.

12. Harvey, *Henry Clay Frick*, 271–73.

13. W. B. Dickson to W. E. Corey, 10 August 1904, WD.

14. W. B. Dickson to W. E. Corey, 16 February 1909, WD; W. B. Dickson, note on E. H. Gary, WD.

15. H. N. Casson, *The Romance of Steel* (New York: A. S. Barnes, 1907), 335.

16. J. G. Butler, *Recollections of Men and Events* (New York: Putnams, 1927), 150–58, 261.

17. Whipple, "Notes of Conversations," 113–20.

18. US Steel Finance Committee minutes, 17 November 1907, USX.

19. Butler, *Recollections of Men and Events,* 150–58.

20. US Steel Finance Committee minutes, 21 May 1908, USX (emphasis in original).

21. Carnegie Steel Company minutes, 8 February 1909, USX.

22. E. H. Gary to H. C. Frick, 18 February 1909, USX.

23. E. H. Gary to W. B. Dickson, 11 May 1909, WD; Burns, *Decline of Competition,* 79.

24. Butler, *Recollections of Men and Events,* 150–58; *BAISA,* 1 November 1909.

25. quoted *BAISA,* 1 March 1912.

26. W. B. Dickson, note, 28 January 1910, WD; W. B. Dickson, memo, 1 December 1910, WD.

27. *IA,* 21 December 1911, 1333; 4 April 1912, 890; 2 May 1912, 1084; and 10 October 1912, 936–37.

28. *IA,* 10 October 1912, 872–73.

29. Berglund, *United States Steel Corporation,* 167.

30. Burns, *Decline of Competition,* 305; *IA,* 6 January 1949, 170.

31. E. H. Gary, "Evidence to the Stanley Committee," quoted in Hogan, *Economic History,* 1101.

32. H. R. Seager and C. A. Gulick, *Trust and Corporation Problems* (New York: Harper and Brothers, 1929), 258; Taussig, *Some Aspects of the Tariff Question,* 181.

33. V. S. Clark, *History of Manufactures in the United States,* 3 vols. (New York: McGraw-Hill, 1929), 3:64.

34. Burns, *Decline of Competition,* 300.

35. *Fortune,* May 1931, 91.

36. *IA,* 22 May 1913.

37. *IA,* 2 February 1911, 304–5.

38. *IA,* 17 April 1919, 1032; U.S. Congress Transport Investigation and Research Board, *Economics of Iron and Steel Transportation* (Washington, D.C.: GPO, 1945), 133.

39. *IA,* 15 May 1919, 1287.

40. Proceedings before the U.S. Federal Trade Commission, Washington D.C., 9 July 1919.

41. J. W. Jenks and W. E. Clark, *The Trust Problem* (Garden City, N.J.: Doubleday, 1920), 393.

42. *IA,* 2 February 1922, 360; and 23 February 1922.

43. *IA,* 20 April 1922, 1070; and 15 March 1923, 741.

44. C. D. Edwards, "The Effect of Recent Basing Point Decisions upon Business Practices," *American Economic Review* 38 (1948): 833.

45. Inland Steel Company, "Reply to FTC," quoted in *IA,* 4 September 1919, 643.

46. *IA,* 1 January 1925, 56.

47. *IA,* 1 January 1925, 54.

48. E. H. Gary, "Address to AISI," New York City, 24 October 1924, appendix A (quote).

4. The Changing Balance of Locational Advantage and Expansion

1. A. Carnegie quoted in *Iron and Coal Trades Review,* 27 December 1895, 812.

2. J. H. Bridge, *The Inside Story of the Carnegie Steel Company* (New York: Aldine, 1903), 358; *IA,* 9 March 1911.

3. *BAISA,* 23 January 1901; *IA,* 4 April 1912, 892.

4. Jeans, *American Industrial Conditions and Competition,* 154.

5. *IA,* 5 March 1903; National Tube, *A History of National Tube* (New York: US Steel, 1953), 4.

6. Department of Commerce, *Historical Statistics* (1957); D. A. Fisher, *Steel Serves the Nation: The Fifty Year Story of United States Steel* (New York: US Steel, 1951).

7. US Steel Finance Committee minutes, 7 February 1905, USX.

8. *Engineering,* 7 June 1901, 741; and 6 June 1902. In addition, see the regular "letters" by the Philadelphia correspondent published in this British trade journal.

9. US Steel Finance Committee minutes, 30 December 1902, USX.

10. *IA,* 22 March 1906, 1022.

11. *BAISA,* 15 November 1906; and 31 December 1906.

12. Carnegie Steel minutes, 3 January 1911, USX.

13. *IA,* 14 April 1904, 18; U.S. Commissioner of Corporations, *Reports,* pt. 3, 461–65; *ICTR,* 13 May 1910, 786.

14. Carnegie Steel minutes, 6 March 1905, USX.

15. Carnegie Steel minutes, 25 September, 13 November 1905, USX.

16. Carnegie Steel minutes, 19 April, 24 May, 13 September, and 18 October 1909, USX; memo on Edgar Thomson works, [c. World War II], USX; Carnegie Steel minutes, 8, 15 January 1912, USX.

17. *IA,* 19 February 1925, 540–44; AISI, *Statistical Report.*

18. US Steel Finance Committee minutes, 5 September, 4 October 1904, USX.

19. E. J. Buffington, "Testimony during the Pittsburgh Plus hearings," quoted in *IA,* 15 February 1923, 480.

20. US Steel Finance Committee minutes, 13 June 1905, USX (author's emphasis).

21. *New York Tribune,* 23 November 1905, quoted in *BAISA,* 15 December 1905.

22. US Steel Finance Committee minutes, 29 December 1905, 15 January 1906, USX.

23. *IA,* 2 December 1909, 1715; and 26 May 1910, 1212; Temin, *Iron and Steel in Nineteenth Century America,* 285.

24. US Steel, "Annual Report for 1907," 30–31; "Gary Works of Illinois Steel Company" (Illinois Steel, 1922); *BF&SP,* August 1937, 871–83.

5. Government, Business, and Industrial Development

1. US Steel Finance Committee minutes, 16, 29 January, and 29 May 1923, USX.

2. K. Warren, *The American Steel Industry, 1850–1970* (Pittsburgh: University of Pittsburgh Press, 1988), 74–76.

3. Quoted in *Iron,* 2 December 1892, 500.

4. F. Hobart, "Iron and Steel in 1913," *EMJ,* 10 January 1914, 79–80; E. M. Hoover, *The Location of Economic Activity* (New York: McGraw Hill, 1948), 22.

5. Casson, *Romance of Steel,* 354–55.

6. US Steel Executive Committee minutes, 6, 20 May 1902, USX.

7. Seager and Gulick, *Trust and Corporation Problems,* 228–30.

8. *Proceedings of a Conference of Governors at the White House, May 1908* (Washington, D.C.: GPO, 1909), 41–42.

9. *ICTR*, 1 July 1910, 12.

10. US Steel Finance Committee minutes, 22 January 1907, USX (author's emphasis).

11. US Steel Finance Committee minutes, 1 April 1907, USX.

12. *EMJ*, 14 September 1907, 508.

13. US Steel, "Annual Report for 1907," 32.

14. US Steel Finance Committee minutes, 13 November 1906, 24 August, and 7 September 1909, USX.

15. US Steel Finance Committee minutes, 28 September 1909, USX.

16. *IA*, 27 September 1911, 1196–98; *ITR* quoted in *ICTR*, 14 April 1911; *IA*, 30 November 1911, 1198; US Steel, "September 1911 Tour of Upper Lakes," USX.

17. *IA*, 27 September 1911, 626.

18. US Steel Finance Committee minutes, 16 July, 8 October 1918, USX.

19. *IA*, 13 February 1919, 434.

20. US Steel Finance Committee minutes, 17 December 1920, USX; *IA*, 19 February 1925, 545, 547.

21. *IA*, 1 January 1925, 16.

22. S. B. Sheldon to C. L. Close, 31 August 1925, USX.

23. *IA*, 23 November 1922, 1367; 12 June 1924, 1726; and 9 October 1924, 918. For an interesting discussion of the "ore tax issue" by a top executive of another steel company, see C. B. Randall, "Mining Taxation in the Lake Superior District," in *AISI Yearbook* (1932).

24. G. Zeller of Duluth works to A. R. Alanen of the University of Wisconsin, 2 November 1977, USX.

25. US Steel Control Committee on Ford, Bacon, and Davis, 5 November 1937, USX, 21.

26. Ford, Bacon, and Davis, Report 37 of "Reports on the Operations of the United States Steel Corporation," 200 vols., special reports for the United States Steel Corporation, 1936–1938, Report 37, "American Steel and Wire" (April 1937), 292, 296, 309.

27. U.S. Temporary National Economic Committee, *Investigation of Concentration of Economic Power* (Washington, D.C.: GPO, 1939–40), pt. 19, 10542, 10554.

28. *IA*, 17 November 1898, 10; Campbell, *Manufacture and Properties of Structural Steel*, 194, 196.

29. E. Armes, *The Story of Coal and Iron in Alabama* (Birmingham, Ala.: Birmingham Chamber of Commerce, 1910), 435; J. Bowron, "Steel Making in Alabama," *TAIMME* 81 (1925): 400.

30. *IA*, 18 April 1912, 978–79; and 30 May 1912, 1349; Casson, *Romance of Steel*, 302; R. Gregg, "Origin and Development of the Tennessee Coal, Iron, and Railroad Company," report for US Steel, n.d., 15.

31. *IA*, 7 November 1907, 1327.

32. Casson, *Romance of Steel*, 304.

33. *Manufacturers' Record* quoted in *BAISA*, 15 May 1907; J. Fuller, "History of the Tennessee Coal Iron and Railroad Company, 1852–1907" (Master's thesis, Emory University, 1958), passim.

34. Casson, *Romance of Steel*, 297; *Stanley Committee*, 1: 402.

35. D. H. Bacon to Stanley Committee, quoted in *IA*, 18 April 1912, 978–79.

36. E. H. Gary, W. E. Corey, D. H. Bacon, and J. Farell to Stanley Committee, 2 June 1911, quoted in *IA*, 18 April 1912, 978; 30 January 1913, 313–14; and 22 May 1913, 1240. For slightly different figures, see *ITR*, 1 February 1912, 311.

37. Gregg, "Origin and Development," 17.

38. J. Gayley in *Stanley Committee*, 1:402; L. Ledyard in *Stanley Committee*, 1: 936.

39. E. H. Gary quoted in C. W. Barron, *They Told Barron* (New York: Harper, 1930), 79–80.

40. T. R. Roosevelt to Attorney General Bonaparte, 4 November 1907, quoted in *BAISA*, 1 February 1909.

41. Seager and Gulick, *Trust and Corporation Problems*, 234; Bowron, "Steel Making in Alabama," 402.

42. J. W. Gates to Stanley Committee quoted in *IA*, 1 June 1911, 1311.

43. J. Kennedy to Stanley Committee quoted in *IA*, 4 April 1912, 888.

44. W. E. Corey to Stanley Committee quoted in *IA*, 30 January 1913, 313; G. G. Crawford to Stanley Committee quoted in *IA*, 3 July 1913, 23; *IA*, 21 September 1911, 642; and 19 December 1912, 1436–38; Gregg, "Origin and Development," 19.

45. Barron, *They Told Barron*, 87.

46. P. Tiffany, *The Decline of American Steel: How Management, Labor and Government Went Wrong* (New York: Oxford University Press, 1988), 7, 9.

47. US Steel Finance Committee minutes, 18 November 1901, USX.

48. US Steel Finance Committee minutes, 7 February 1905, 11 and 26 April 1905, 5 June 1905, 19 June 1906, 16 November 1909, USX.

49. US Steel Finance Committee minutes, 19, 26 February 1918, USX; *JISI* 2 (1922): 299; *BF&SP*, May 1936, 433.

50. US Steel Finance Committee minutes, 23 October 1928, 25 July 1929, 15 October 1929, USX.

51. US Steel Finance Committee minutes, 22 November 1932, 16 February 1933, USX.

52. US Steel Finance Committee minutes, 21 May 1935, 22 June 1937, 7 September 1937, USX; M. Taylor, "Address to Stockholders' Meeting," Hoboken, New Jersey, 4 April 1938.

6. Entrepreneurial Failure?

1. W. E. Corey, 10 January 1911, WD.

2. *BAISA*, 10 November 1902; Jeans, *American Industrial Conditions and Competition*, 315.

3. US Steel Finance Committee minutes, 7 March, 4 April 1905, USX.

4. US Steel Executive Committee minutes, 23 May 1901, USX.

5. Carnegie Steel Company minutes, 6 June 1904, USX; W. B. Dickson to W. E. Corey, 1 January 1909, WD.

6. US Steel Executive Committee minutes, 24 June 1902, USX.

7. Carnegie Steel Company minutes, 27 February 1905, USX.

8. Carnegie Steel Company minutes, 3 July, 24 July 1905, USX.

9. US Steel Executive Committee minutes, 11 February 1902, USX.

10. "H" Column Steel Company, untitled pamphlet, March 1905, BP.

11. S. Whipple, "Notes of Conversations with C. M. Schwab," quoted in R. Hessen, *Steel Titan: The Life of Charles M. Schwab* (New York: Oxford, 1975; reprint, Pittsburgh: University of Pittsburgh Press, 1991), 173.

12. Carnegie Steel Company minutes, 28 May 1906, USX.

13. C. M. Schwab to A. Carnegie, 6 July 1898, ACLC.

14. W. E. Corey to H. C. Frick, [September 1908], FP; W. E. Corey to H. C. Frick, 2 October 1908, FP.

15. H. Grey to C. M. Schwab, 27 October 1908, BP; C. M. Schwab to H. Grey, 21 November 1908, BP.

16. Carnegie Steel Company minutes, 19 December 1910, 27 December 1910, 10 April 1911, USX.

17. US Steel Finance Committee minutes, 21 November 1911, USX.

18. A. P. Moore, ed., *The Book of Prominent Pennsylvanians* (Philadelphia: Leader Publishing, 1913), 110.

19. *IA*, 24 May 1923, 1481.

20. US Steel Finance Committee minutes, 3 October, 24 December 1923, USX.

21. *IA*, 28 April 1927, 1236; *ICTR*, 6 July 1928, 13; *ITR*, 3 January 1929, 83; Hogan, *Economic History*, 890.

22. S. L. Goodale, *Chronology of Iron and Steel* (Cleveland: Penton Publishing, 1931), 283–87; *IA*, 22 December 1927, 1740.

23. Quoted in Hessen, *Steel Titan*, 267; *IA*, 28 April 1927, 1231–32.

24. US Steel Finance Committee minutes, 30 November 1926, 1 December 1926, 11 January 1927, USX.

25. US Steel Finance Committee minutes, 14 June 1927, USX; Hessen, *Steel Titan*, 267–69.

26. *IA*, 18 December 1924, 1603; and 15 October 1925, 1032; *Fortune*, September 1930, 124, 126; and December 1930, 49, 136.

27. Ford, Bacon, and Davis, "Reports," 200:240; W. H. North and W. H. King, "Report of a Visit to the United States," prepared for Dorman, Long of Middlesbrough, Eng., 1946, 22.

7. The Interlude of World War I

1. Gary quoted in *ICTR*, 12 June 1914, 91.

2. *JISI* 1 (1914): 677.

3. US Steel, "Steel and the War," USX (typescript).

4. Goodale, *Chronology of Iron and Steel*, 254–73.

5. US Steel, "Annual Report for 1915."

6. Goodale, *Chronology of Iron and Steel*, 266; *Encyclopedia Britannica*, 12th ed., s.v. "Shipping."

7. Harvey, *Henry Clay Frick*, 319–22.

8. *IA*, 15 May 1919, 1287.

9. US Steel Finance Committee minutes, 21 June 1921, USX.

10. *IA*, 1 January 1925, 10–17; and 19 February 1925, 540–44.

11. US Steel Finance Committee minutes, 27 June 1922, USX.

8. Labor Conditions and Relations during the Gary Years

1. A manager to I. Tarbell, 31 October 1914, USX; H. D. Williams to C. L. Close, 14 September 1925, USX.

2. R. S. Baker, "What the Steel Corporation Really Is and How It Works," *McClure's Magazine* 18, no. 1 (November 1901): 9; Fisher, *Steel Serves the Nation*, 224–25.

3. A. Shadwell, *Industrial Efficiency: A Comparative Study of Industrial Life in England, Germany, and America* (London: Longmans, Green, 1906), 1:327; Hogan, *Economic History*, 451.

4. C. A. Gulick, *Labor Policy of the United States Steel Corporation* (New York: Columbia University Press, 1924), quoted in E. D. McCallum, *The Iron and Steel Industry in the United States* (London: P. S. King, 1931), 240–41.

5. W. B. Dickson, memo, n.d., WD.

6. J. A. Fitch quoted in W. Z. Foster, *The Great Steel Strike and Its Lessons* (New York: B. W. Huebsch, 1920), vii; S. Yellen, *American Labor Troubles* (New York: Harcourt Brace, 1936), 99; J. A. Fitch, "Old Age at Forty," *American Magazine* 71 (March 1911).

7. E. J. McGee, remarks at the US Steel stockholders' meeting, Hoboken, New Jersey, 7 May 1945.

8. Quoted in D. Brody, *Steelworkers in America: The Nonunion Era* (Cambridge, Mass.: Harvard University Press, 1960), 34.

9. W. B. Dickson to W. E. Corey, 10 May 1909, WD.

10. W. B. Dickson, notes of the AISI meeting, 27 May 1910, WD.

11. Fisher, *Steel Serves the Nation*, 224.

12. E. H. Gary, statement, 16 January 1912, quoted in *BAISA*, 1 March 1912.

13. E. H. Gary, statement, 20 April 1914, USX.

14. A. Cotter, *The Gary I Knew*, v, 28; *IA*, 26 April 1923, 1204.

15. C. M. Cabot, report and assorted letters from stockholders, March 1912, USX.

16. US Steel Finance Committee minutes, 28 May 1912, USX.

17. *IA*, 19 April 1923, 1125.

18. Hogan, *Economic History*, 443.

19. Quoted in Yellen, *American Labor Troubles*, 251.

20. US Steel Executive Committee minutes, 1, 2 July 1901, USX.

21. R. R. Brooks, *As Steel Goes* (New Haven, Conn.: Yale University Press, 1940), 28–29.

22. *BAISA*, 25 August 1901.

23. Hubbard, "American Trusts and English Combinations," 168; Hogan, *Economic History*, 443; *Engineering*, 6 September 1901, 323.

24. W. E. Corey, remarks at US Steel President's Dinner, 10 January 1911.

25. *Britannica Yearbook* (1913), 900.

26. Hogan, *Economic History*, 457.

27. E. H. Gary to J. Fitzpatrick, W. Z. Foster, et al., [1919], USX; E. H. Gary, quoted in Tarbell, *Elbert H. Gary*, 283.

28. Gulick, *Labor Policy*, 103 n. 3.

29. S. Lorant, "Chronology" in *Pittsburgh: The Story of an American City* (Lenox, Mass.: S. Lorant, 1988).

30. Brooks, *As Steel Goes*, 40. Excellent accounts of the 1919 strike can be found in Yellen, *American Labor Troubles*, chapter 8; and Brody, *Steelworkers in America*, chapter 12.

31. Quoted in S. E. Morison, H. S. Commager, and W. E. Leuchtenburg, *The Growth of the American Republic* (New York: Oxford University Press, 1969), 2:420–21.

32. Hogan, *Economic History*, 460 (quote); Fisher, *Steel Serves the Nation*, 74; Yellen, *American Labor Troubles*, 252.

33. *Encyclopedia Britannica*, 14th ed., s.v. "Gary, Elbert H."

9. The Changing Face of Competition and the End of the Gary Years, 1919–1927.

1. *IA*, 21 April 1927.

2. US Steel Finance Committee minutes, 7 June, 20 October 1925, USX.

3. Quoted in *ICTR*, 14 June 1929, 901.

4. Tarbell, *Elbert H. Gary*, 355.

5. McCraw and Reinhardt, quoted in B. E. Seeley, ed., *Iron and Steel in the Twentieth Century: Encyclopedia of American Business History and Biography* (New York: Facts on File, 1994), 442.

6. Quoted in supplement to *ITR*, 18 August 1927, viii.

7. US Steel Board of Directors minutes, 30 August 1927, USX.

8. *[UK] Stock Exchange Yearbook*, 1931.

10. ERW Pipe and the Wide Continuous Strip Mill.

1. S. Whipple, "Notes of Conversations with C. M. Schwab," 1935–36, WP.

2. *A History of National Tube;* Ford, Bacon, and Davis, "Reports," vol. 159, "National Tube"; *ICTR*, 25 January 1929, 156; and 31 May 1929, 831; G. A. V. Russell, "Notes on Visit to the United States and Canada, February–April 1941," report to United Steel Companies, Sheffield, Eng., 1941, 56.

3. Ford, Bacon, and Davis, "Reports," 200:11.

4. US Steel Finance Committee minutes, 10 May 1932, USX.

5. Pittsburgh correspondent of *ICTR*, 4 April 1930, 565.

6. J. W. Hall, *Mechanical Treatment*, vol. 2 of *The Metallurgy of Steel* (London: Charles Griffin, 1918), 599.

7. I am deeply indebted for much of this account to the late Joe Malborn, who worked at Vandergrift starting in 1912 and later for the United Engineering and Foundry Company for many years. His advice on the evolution of the sheet trade was fully and generously given and was invaluable.

8. D. Eppelsheimer, "The Development of the Continuous Strip Mill," *JISI* (1938).

9. US Steel Finance Committee minutes, 9 April 1929, USX.

10. *ICTR*, 3 June 1927.

11. W. A. Irvin, "Changing Trends in the Steel Industry," in *AISI Yearbook* (1938), 42–43.

12. US Steel Finance Committee minutes, 9 April 1929, USX.

13. *BF&SP*, February 1937; June 1937, 628; December 1938, 1168; January 1939, 73; and March 1939, 255.

11. Crisis and Response

1. C. R. Daugherty, M. G. de Chazeau, and S. S. Stratton, *Economics of the Iron and Steel Industry* (New York: McGraw Hill, 1937), vii; W. W. Rostow, *The Stages of Economic Growth: A Non-Communist Manifesto* (Cambridge: Cambridge University Press, 1971), passim.

2. F. L. Allen, *Only Yesterday* (1931; reprint, London: Penguin, 1938), 454.

3. *New York Times*, 16 August 1927.

4. M. Taylor, "Ten Years of Steel," statement at the annual meeting of US Steel stockholders, Hoboken, New Jersey, 4 April 1938, 1.

5. M. Taylor, "Address to the Conference of Major Industries," New York City, October 1928, 27; *IA*, 1 November 1928, 11–19.

6. Fisher, *Steel Serves the Nation*, 225; *IA*, 28 October 1937, 87. There is an excellent summary of the bonded debt issue in "The US Steel Story," *33 Magazine*, February 1968, 74.

7. "Myron Taylor: An Appreciation" (USS, 1956), USX; Seeley, *Iron and Steel in the Twentieth Century*.

8. *National Cyclopedia of American Biography* (1934).

9. US Steel Finance Committee minutes, 19 March 1929, USX.

10. Referred to in US Steel Finance Committee minutes, 16 June 1936, USX.

11. US Steel Finance Committee minutes, 12 December 1933, 26 June 1934, 24 September 1935, 29 October 1935, USX.

12. Taylor, "Ten Years of Steel"; US Steel, "Annual Report for 1935."

13. Ford, Bacon, and Davis, "Reports," 200:xxi, xxix.

14. Ford, Bacon, and Davis, "Reports," 100:8–9, 200:131.

15. Ford, Bacon, and Davis, digest of "Reports," vol. 32.

16. E. Jones, *The Trust Problem in the United States* (New York: Macmillan, 1922), 214.

17. Quoted in E. G. Nourse, et al., *America's Capacity to Produce* (Washington, D.C.: Brookings Institution, 1934), 269.

18. Ford, Bacon, and Davis, "Reports," vol. 32, passim.

19. Ford, Bacon, and Davis, "Reports," 150:36–39, 194, 196, 204, 263.

20. Ford, Bacon, and Davis, "Reports," vol. 100.

21. *Fortune*, June 1932, 39; and March 1936, 169–70.

22. Ford, Bacon, and Davis, "Reports," 100:4, 6.

23. Ford, Bacon, and Davis, "Reports," 200:323, and overall digest.

24. E. B. Aldefer and H. E. Michl, *The Economics of American Industry* (New York: McGraw Hill, 1942), 76.

25. Carnegie-Illinois Steel Company minutes, 7 April 1936, USX; Ford, Bacon, and Davis, "Reports," vol. 150.

26. Taylor, "Ten Years of Steel," 5, 12, 17; *IA*, 1 January 1925, 10–17.

27. See, for instance, *Harvard Business Review*, January 1933; C. R. Daugherty, de Chazeau and Stratton, *Economics of the Iron and Steel Industry*, 378–79; and M. Worthing, "Comparative Assembly Costs in the Manufacture of Pig Iron," *Pittsburgh Business Review*, January 1938.

28. *IA*, 6 January 1938, 112–14.

29. G. A. V. Russell, "Report on Observations Made during a Tour of Representative Steel Works in the United States, March–April 1938," report to United Steel Companies, Sheffield, Eng., 1938, 65, 220.

30. Quoted in G. W. Stocking, *Basing Point Pricing and Regional Development: A Case Study of the Iron and Steel Industry* (Chapel Hill: University of North Carolina Press, 1954), 26 n. 23.

31. Minutes of a special meeting of the Control Committee, 15 December 1937, USX; Taylor, "Ten Years of Steel," 3; *IA*, 7 April 1938, 706.

32. Ford, Bacon, and Davis, "Reports," vols. 27, 35, and 109.

33. TNEC 1939–40, Part 19, 1055, (author's emphasis).

34. Ford, Bacon, and Davis, "Reports," 100:79–80.

12. Labor Relations under Myron Taylor and Philip Murray

1. Brooks, *As Steel Goes*, 78; Obituary of A. H. Young, *New York Times*, 7 March 1946.

2. F. Perkins, *The Roosevelt I Knew* (New York: Viking Press, 1946).

3. W. B. Dickson to A. H. Young, 23 March 1936, WD.

4. *Worcester Telegram*, 17 July 1936, USX.

5. "It Happened in Steel," *Fortune*, May 1937, 93; Seeley, *Iron and Steel in the Twentieth Century*, 422.

6. Steel Workers' Organizing Committee, "Agreement with United States Steel Corporation," 2 March 1937; Seeley, *Iron and Steel in the Twentieth Century*, 328.

7. United Steelworkers of America, *A Brief History of the United Steelworkers of America* (USWA, 1999); *Fortune*, May 1945.

8. J. N. Blackman, remarks at US Steel stockholders' meeting, Hoboken, New Jersey, 3 May 1943.

9. US Steel of New Jersey, statistics 6 November 1948, USX; US Steel stockholders' meeting, Hoboken, New Jersey, 1 May 1950, USX.

10. There is a vivid account of Homestead in the late 1930s in W. A. H. Birnie, "I Make Steel," *New York World Telegram Weekend Magazine*, 20 March 1937.

11. US Steel, records of Homestead works, USX; *Fortune*, May 1945.

12. Brody, *Steelworkers in America*, 51.

13. *Fortune*, April 1936, 136; U.S. Temporary National Economic Committee, *Investigation of Concentration of Economic Power*, 17339.

14. Letter to H. Ruttenberg, 1 March 1939.

15. *Pittsburgh Post-Gazette*, 29 August 1945.

13. New Regions

1. *IA*, 5 May 1927, 1305; AISI, *Yearbook* (1929), 405.

2. *IA*, 8 August 1929, 351.

3. *IA*, 8 August 1929, 351; TNEC 1939–40, Part 19, 10619.

4. Daugherty, de Chazeau, and Stratton, *Economics of the Iron and Steel Industry*, 688.

5. *Fortune*, March 1938, 182, 194, 202, 204.

6. *Fortune*, March 1936, 202.

7. R. Gregg to TNEC 1939–40, Part 19, 10543, 45,51.

8. Daugherty, de Chazeau, and Stratton, *Economics of the Iron and Steel Industry*, 63; Stocking, *Basing Point Pricing*, 266.

9. "Evidence of B. Fairless," in U.S. House, *Subcommittee on the Study of Monopoly Power* (Washington, D.C.: GPO, 1950,) 640, 642, 971–72.

10. Campbell, *Manufacture and Properties of Structural Steel*, 159.

11. *IA*, 18 April 1912, 981–82.

12. *BAISA*, 1, 30 December 1905.

13. W. E. Corey, retirement speech, 10 January 1911, WD.

14. H. C. Estep in *ICTR*, 5 August 1921, 167; "Evidence of Asst. Traffic Manager Illinois Steel Company to Pittsburgh Plus Hearings," quoted in *IA*, 12 July 1923, 83.

15. *IA*, 1 January 1925, 17; *ITR*, 27 December 1928, 1637.

16. *IA*, 1 January 1925, 10–17.

17. *Transactions of the American Society for Steel Treating* 12 (July–December 1927): 148–49.

18. Columbia Steel Company, *A History of Columbia Steel Corporation* (US Steel, n.d.).

19. US Steel Finance Committee minutes, 30 July 1929, USX.

20. US Steel Finance Committee minutes, 25 August 1931, USX.

21. Ford, Bacon, and Davis, "Reports," vol. 131, "Columbia Steel" (2 parts).

22. *IA*, 30 November 1939, 54; J. L. Perry of US Steel, 12 February 1948, USX; Warren, *American Steel Industry*, 238.

23. *IA*, 18 January 1948, 149; Russell, "Representative Steel Works in the United States, March–April 1938," 39.

24. Evidence of B. Fairless in U.S. Temporary National Economic Committee, *The Basing Point System*, monograph 42 (Washington, D.C.: GPO, 1939–40), pt. 19, 10555–56.

25. *New York Evening Post*, 12 January 1901, reprinted in *The Empire of Business* (1912): 236–37.

26. US Steel Finance Committee minutes, 27 September 1904, 10 January 1905, 8 October 1907, 30 March 1909, 17 October 1911, USX.

27. US Steel Finance Committee minutes, 6 February 1923, 30 January 1924, 7 September 1926, 21 April 1931, 24 September 1931, 29 March 1932, USX.

28. Ford, Bacon, and Davis, "Reports," 200:118–19.

29. Ford, Bacon, and Davis, "Reports," vol. 165.

30. Ford, Bacon, and Davis, "Reports," 200:288.

31. Russell, "United States and Canada, February–April 1941," 7–9.

14. Government-guided Growth

1. W. S. Tower, presidential address to AISI, May 1941.

2. E. W. Hawley, *The New Deal and the Problem of Monopoly* (Princeton, N.J.: Princeton University Press, 1966), 394.

3. AISI, *Yearbook* (1941), 47–48, 50; *IA*, 3 June 1943, 93.

4. "Steel in the War Years," *Fortune*, May 1945; W. A. Hauck, "Steel Expansion for War," reprinted in *Steel*, June 1945, passim.

5. Fisher, *Steel Serves the Nation*, 50.

6. US Steel stockholders' meetings, Hoboken, New Jersey, May 1943, May 1944; US Steel, "Annual Report for 1945," 13.

7. See also R. R. Bowie, *Government Regulation of Business*, Cases from the National Reporter System (N.p., 1949), 1732.

8. US Steel Board of Directors, minutes, 28 October, 25 November 1941, USX.

15. Filling Out the Production Map

1. *IA*, 10 August 1944, 80.

2. US Steel Board of Directors, minutes, 31 July 1945, USX; US Steel, "The New Industrial West" (New York: US Steel, 1947), 6.

3. US Steel Board of Directors, minutes, 30 April 1946, USX.

4. Bowie, *Government Regulation of Business*, 1729–30; *Iron and Steel Engineer*, November 1946, 108.

5. A. G. Roach quoted in *Western Metals*, July 1950.

6. Bowie, *Government Regulation of Business*, 1727; S. N. Whitney, *Anti-Trust Policies* (New York: Twentieth Century Fund, 1958), 273–76.

7. *US Steel Quarterly*, May 1948, 6.

8. A. G. Roach, speech delivered in Salt Lake City, quoted in *Provo Herald*, 11 October 1956.

9. *Economic Review of Federal Reserve Bank of Cleveland*, October 1969, 10.

10. M. J. Barloon, "Steel, the Great Retreat," *Harpers Monthly Magazine*, August 1947, 145–55.

11. C. E. Williams, "Beneficiation of Iron Ore," *TAIMME* (1931): 186–88; *Fortune*, December 1945; AISI, *Yearbook* (1947), 337.

12. "Iron Ore Dilemma," *Fortune*, December 1945.

13. B. Fairless, *It Could Only Happen in the United States* (N.p., 1957), 37–38.

14. B. Fairless, statement to the Joint Committee on the Economic Report, Washington, D.C., January 1950.

15. Minutes of the forty-ninth annual US Steel stockholders' meeting, 1 May 1950, Hoboken, New Jersey; A. J. Berdis, "Development of the Fairless Works of United States Steel," (28 September 1951), USX.

16. *US Steel Quarterly*, August 1950.

17. I. Olds, remarks to US Steel stockholders' meeting, 7 May 1951, Hoboken, New Jersey.

18. Quoted in *Economist*, 6 December 1952, 693.

19. BISRA, *The Fairless Works* (London: British Iron and Steel Research Association, 1953).

20. Minutes of US Steel stockholders' meeting, Hoboken, New Jersey, May 1954, USX, 27.

21. Quoted in *Pittsburgh Press*, 2 May 1954.

22. For examples of expansive early thinking and an econometric analysis, see I. Olds, "Address to the 3rd Annual Conference of the Greater Philadelphia–South Jersey Council," 5 October 1951; R. L. Leffler, "The Future of the Delaware Valley," address to Philadelphia Mortgage Bankers Association, 14 November 1955; and W. Isard and R. E. Kuenne, "The Impact of Steel upon the Greater New York–Philadelphia Industrial Region," *Review of Economics and Statistics* 35 (1953): 289–301.

23. *BF&SP*, June 1947, 691–92.

24. *IA*, 21 January 1965, 28.

25. *Houston Chronicle*, 23 May 1966.

26. *Pittsburgh Press*, 28 November 1971.

27. US Steel Corporate Policy Committee, 13 August 1975, USX; *IA*, 29 April 1971, 49.

16. Triumph and Marking Time, 1945–1960

1. G. W. Wolf, "General World Steel Situation," special statement to the Joint Committee on the Economic Report, Washington, D.C., January 1950, 57–60.

2. T. Diamond, "This New Round of Steel Expansion," *Harvard Business Review* 34, no. 3 (May–June 1956): 89–90.

3. Note by Oliver Iron Mining Company, 1950, USX.

4. US Steel, "Annual Report for 1943," 7; W. O. Hotchkiss, "Iron Ore Supply of the Future," *Economic Geology* (May 1947): 207–9; H. S. Harrison of Cleveland Cliffs Iron Company, quoted in *BF&SP*, 1966.

5. Ford, Bacon, and Davis, "Reports," 200:113.

6. G. S. Armstrong, et al., *The Iron and Steel Industry*, vol. 10 of *An Engineering Interpretation of the Economic and Financial Aspects of American Industry* (N.p., 1952), 67; *Steel*, May 1955, 161; State of Minnesota, "Iron Ore Beneficiation," *Federal Reserve Bank of Minnesota Review*, October 1957, 8–11; *IA*, 9 November 1967, 63.

7. US Steel, "Annual Report for 1950," 8; US Steel, "Annual Report for 1952," 9.

8. M. W. Reed to U.S. Congress, Joint Committee on the Economic Report, 24 January 1950.

9. G. W. Stocking to U.S. House, *Subcommittee on the Study of Monopoly Power* (Washington, D.C.: GPO, 1950), 4a:971; *IA*, 5 October 1950, 108.

10. *Fortune*, January 1956.

11. B. Fairless to stockholders, 4 May 1953, Hoboken, New Jersey; L. Worthington and E. Gott quoted in *Forbes*, 1 August 1964.

12. B. Fairless to stockholders, 2 May 1955, Hoboken, New Jersey.

13. T. Campbell quoted in *IA*, 11 January 1962, 5.

14. R. C. Cooper, "Productivity in US Steel," included in U.S. Senate Judiciary Committee, *Hearings before the Subcommittee on Antitrust and Monopoly* (Washington, D.C.: GPO, 1957–58), August 1957.

15. Biennial raw-steel utilization rates for US Steel between 1951 and 1960 were 93.1 percent, 85.8 percent, 88 percent, 72.2 percent, and 61.7 percent.

16. Quoted in Knowlton Hill, "Account of the 1959 USWA strike for the S.C.C.C.," 1959–1960, USX.

17. R. Blough, "A Talk of Two Towns," April 1958.

18. United Steel Workers of America, "Foreign Competition and Steel Wages," April 1959, USX.

19. *Wall Street Journal*, 6 January 1960.

17. A Time of Transition, the 1960s

1. *Steel Facts*, October–November 1968, 2; *Mining Annual Review* (1971), 53.

2. *IA*, 16 August 1962, 67–68; and 7 May 1964, 46.

3. M. Borrus, "The Politics of Competitive Economies in the United States Steel Indus-

try," in *American Industry in International Competition: Government Policies and Corporate Strategies*, ed. J. Zysman and L. Tyson (Ithaca, N.Y.: Cornell University Press, 1983), 76.

4. *Fortune*, August 1962, 199; and September 1962, 99.

5. *Business Week*, 8 July 1967.

6. Brody, *Steelworkers in America*, 112–13.

7. *Chicago Sun Tribune*, 23 April 1963, passim; *US Steel Quarterly*, February 1928, 1.

18. Response to a Technological Revolution

1. *Iron and Steel Engineer*, December 1967, 95.

2. R. Cordero, *Iron and Steel Works of the World*, 6th ed. (London: Metal Bulletin, 1974).

3. T. C. Graham, letter to the author, October 1999.

4. O. Cuscoleca in *Transactions of the Metals Branch of AIME* 200 (1954): 826–27.

5. Diamond, "This New Round of Steel Expansion," 85–90.

6. Borrus, "Politics of Competitive Economies," 69.

7. Ibid., 71; W. Adams, ed., *The Structure of American Industry* (1950); ibid., 6th ed. New York: Macmillan, 1992), 111–12.

8. *IA*, 21 January 1964, 194; *Fortune*, October 1966; *Wall Street Journal*, 7 March 1963; "Oxygen in Steelmaking," *Scientific American* 218 (April 1968): 24–32.

9. U.S. Congress, *Hearings before the Joint Economic Committee on Steel Prices, Unit Costs, Profits, and Foreign Competition* (Washington, D.C.: GPO, 1963), 224.

10. *US Steel Quarterly*, February 1968, 6.

11. U.S. Federal Trade Commission, *The United States Steel Industry and Its International Rivals*, (Washington, D.C: GPO), 1977.

12. I am indebted to T. C. Graham for guidance in this matter.

13. R. Tyson quoted in *Business Week*, 8 July 1967.

14. *Fortune*, January 1965, 235; *Time*, 23 December 1957; *IA*, 3 July 1969, 9; *Iron and Steel Engineer*, January 1966, D3.

15. *US Steel Quarterly*, February 1968, 6.

16. *US Steel Quarterly*, May 1959, 7, 9.

17. *US Steel Quarterly*, May 1962, 11.

18. US Steel Executive Committee minutes, 13 August 1963, USX.

19. *US Steel Quarterly*, November 1963, 7; *Steel*, 30 September 1963.

20. *Fortune*, January 1965, 235–38.

21. *Business Week*, 8 July 1967.

22. U.S. Federal Trade Commission, *The United States Steel Industry and Its International Rivals*, 498–500.

23. Adams, *Structure of American Industry*, 113; *33 Magazine*, February 1968, 108; *Steel*, 28 April 1969.

24. R. Cordero, *Iron and Steel Works of the World*, 6th ed.

25. *Chicago Daily News*, 15 December 1971; *IA*, 11 November 1974, 23.

19. Long-term Changes in Corporate Organization and Location

1. W. E. Corey, 10 January 1911, WD.

2. E. H. Gary quoted in Seager and Gulick, *Trust and Corporation Problems*, 250.

3. United States Bureau of Census, *1954 Census of Manufactures: General Summary* (Washington, D.C.: GPO, 1957).

4. Ford, Bacon, and Davis, "Reports," 200:70.

5. J. B. Appleton, *The Iron and Steel Industry of the Calumet District*, University of Illinois Studies in the Social Sciences, no. 13 (Chicago: University of Illinois, 1925).

6. *IA*, 1 January 1925, 10–17; and 19 February 1925, 540–44.

7. *IA*, 14 June 1928, 1864; McCallum, *Iron and Steel Industry in the United States*, 138; *BF&SP*, March 1931, 438; *IA*, 7 January 1932, 26.

8. Taylor, "Ten Years of Steel," 11; *IA*, 7 April 1938, 70c; US Steel Finance Committee minutes, 16 June 1936, USX.

9. *BF&SP*, January 1923, 70.

10. *Fortune*, March 1936, 59, 194.

11. *Fortune*, September 1931; and October 1938, 46, 48.

12. U.S. Temporary National Economic Committee, *Basing Point System*; US Steel, "Some Factors in the Pricing of Steel," USX, 19; *Fortune*, July 1931, 62.

13. Ford, Bacon, and Davis, "Reports," 200:119.

14. Ford, Bacon, and Davis, "Reports," vols. 8, 11, 34, and 36.

15. *IA*, 9 May 1946, 129.

16. C. R. Cox, variously quoted in *Pittsburgh Sun-Telegraph*, 1 August 1946; *Chicago Times*, 15 August 1946; and *Calumet Record*, 17 September 1946.

17. *Fortune*, September 1948, 78.

18. *IA*, 10 March 1949, 153.

19. C. F. Hood, press conference, 19 January 1950.

20. M. J. Barloon, "Institutional Foundations of Pricing Policy in the Steel Industry," *Business History Review* 28 (1954): 214, 232.

21. *IA*, 8 August 1946, 104B; AISI, *Iron and Steel Works Directory of the United States and Canada* (New York: AISI, annually); *IA*, passim.

22. A. L. Rodgers, "Industrial Inertia: A Major Factor in the Location of the Steel Industry in the United States," *Geographical Review* 42 (1952): 60.

23. *BF&SP*, August 1955, 942; and August 1959, 888–89.

24. *BF&SP*, January 1957, 52, February 1963, 164.

25. *Economist*, 26 May 1956, 796; *IA*, 17 May 1956, 51.

26. *The Vindicator*, Youngstown, Ohio 29 June 1962, 1 March 1963.

27. *33 Magazine*, November 1966, 74.

28. Corplan Associates, "Technological Change: Primary Metals," 1964, USX, 27.

29. *New York Times*, 12 August 1961.

30. See, for instance, L. Worthington, "Beginning to See the Light," an address to the Service Club Assembly, Homestead, 24 September 1962.

31. *US Steel Quarterly*, May 1964, 9.

32. *BF&SP*, October 1965; and November 1965, 48.

33. Quoted in *Gary Post Tribune*, 12 June 1966.

20. The National Steel Industry since 1970

1. *Forbes*, 15 February 1970, 24; *Steel Times*, December 1969, 775.

2. *Financial Times*, 27 November 1972.

3. Quoted in *London Times*, 1 February 1980.

4. *New York Times,* 7 August 1989, 21–22.

5. G. Cloos, "Steel Fights Back," *Review of Federal Reserve Bank of Chicago,* (March–April 1985): 4–5.

6. *Financial Times,* 29 July 1986; *Iron and Steel Engineer,* December 1985, 54.

7. AISI, 1968; *Financial Times,* 4 June 1971, 27 November 1972.

8. *Fortune,* January 1976, 108–9.

9. *Mining Annual Review,* 1988: c.65.

10. D. M. Roderick to Congressional Steel Caucus, 19 April 1983, USX, 2; *US Steel News,* July 1982.

11. *Financial Times,* 3 September 1982, 7 June 1993.

12. *Iron and Steel Engineer,* December 1985, 54.

13. *Iron and Steel Engineer,* August 1988, 71–72.

14. *Fortune,* February 1978, 123–24.

15. For an early discussion of mini-mill economics, see "Iron and Steel in America," in *IA,* June 1955 (special anniversary issue), D-12.

16. Quoted in Adams, *Structure of American Industry,* 1992, 85.

17. *New Steel,* January 1994, 29.

18. N. B. Schwartz, "Electrics Turn to Carbon Steel," *IA,* 13 March 1969, 65–72; J. R. Mills, "Steel Mini Mills," *Scientific American* 250, no. 5 (May 1984): 29–35.

19. Mills, "Steel Mini Mills," 36; H. W. Paxton, "Future Ferrous Technologies," (unpublished paper, Carnegie Mellon University, 1991), 5; T. Allen, formerly of J&L, in conversation with the author, October 1999.

20. *New Steel,* January 1994, 56.

21. *New Steel,* October 1993, 8.

22. *Iron and Steel Engineer,* February 1996, D1; Steel Manufacturers Association website, 20 October 1998.

23. *Iron and Steel Engineer,* February 1996, D24.

24. *New Steel,* January 1997, 11.

25. Ibid.

26. *Mining Annual Review,* 1985, 59.

27. *Financial Times,* 22 July 1975.

28. *Financial Times,* 4 June 1971, 1 April 1980, 25 October 1983.

29. *Mining Annual Review,* 1986, 63; *Iron and Steel Engineer,* August 1988, 5.

21. The Chairmanships of Edwin H. Gott and Edgar B. Speer

1. *Newsweek,* 30 December 1968, 42.

2. *Business Week,* 8 July 1967; "Steel: Signs of Springtime?" *Forbes,* 15 February 1970, 24.

3. US Steel, "Annual Report for 1970," 4.

4. *Pittsburgh Post Gazette,* 27 February 1973.

5. *Duluth Herald,* 12 February 1964; *American Metal Market,* 13 September 1971, 1; *33 Magazine,* April 1972; *US Steel Quarterly,* August 1973; *Pittsburgh Press,* 8 August 1975; Zeller n.d.

6. Quoted in *Johnstown Tribune Democrat,* 11 August 1975.

7. *US Steel Quarterly,* November 1974, 1, 5.

8. *US Steel Quarterly,* November 1973, 1.

9. US Steel Finance Committee minutes, 8 December 1942, USX; E. Speer quoted in *Conneaut News-Herald,* 22 April 1977; USX, account of Conneaut project, 17 February 1998, USX.

10. *Pittsburgh Press,* 25 March 1976.

11. Quoted in *Conneaut News Herald,* 26 May 1977.

12. US Steel, "Annual Report for 1977," 5.

13. US Steel stockholders' meeting, 1 May 1978, USX; *Fortune,* 13 February 1978, 128.

14. Quoted in USX, account of Conneaut project, 17 February 1998.

15. *IA,* 12 September 1977.

16. US Steel, "3d Quarter Report 1978," USX; *Financial Times,* 21 June 1979.

17. D. Roderick in US Steel, "Annual Report for 1979," 30.

18. USX, account of Conneaut project, 17 February 1998.

19. *American Metal Market,* 18 August 1972, 5; *Metal Bulletin,* 26 September 1972, 29; US Steel, "1st Quarter Report for 1977," USX.

20. US Steel Corporate Policy Committee, 20 August 1979, USX.

21. US Steel Corporate Policy Committee, 6 December 1977, USX.

22. US Steel stockholders' meeting, 1 May 1978.

22. The Rationalizing of US Steel after 1979

1. *Pittsburgh Business Times,* 26 August 1991.

2. US Steel., "3d Quarter Report for 1979," 2.

3. US Steel, "Annual Report for 1969," 14; *Financial Times,* 7 October 1978.

4. US Steel, "1st Quarter Report for 1989," 2.

5. *Financial Times,* 16 December 1982.

6. US Steel Corporate Policy Committee, 22 September 1987, USX.

7. US Steel Corporate Policy Committee, 6 August 1975, USX.

8. US Steel, "Annual Report for 1980; and US Steel Corporate Policy Committee, USX.

9. US Steel Corporate Policy Committee, 18 August 1980, USX.

10. US Steel Corporate Policy Committee, 15 September 1978, USX.

11. *Pittsburgh Press,* 28 October 1981.

12. *Financial Times,* 8 April 1997.

13. US Steel, 1st quarter letter, 1985, USX.

14. W. T. Hogan, *Global Steel in the 1990s* (Lexington, Mass.: D.C. Heath, 1991), 11.

15. S. Brubaker, *Trends in the World Aluminum Industry* (Washington, D.C.: Resources for the Future, 1967), 5.

16. Kaiser Steel, "The Western Steel Market, 1970," and ibid., 1977–1978.

17. *American Metal Market,* 13 October 1966, 1; 6 February 1969, 9; and 1 March 1972, 1; *US Steel Quarterly,* November 1976.

18. US Steel Corporate Policy Committee, 16 June 1980, USX.

19. J. P. Hoerr, *And the Wolf Finally Came: The Decline of the American Steel Industry* (Pittsburgh: University of Pittsburgh Press, 1988), 489.

20. *Mining Annual Review,* 1984.

21. US Steel, reports on 4th quarters of 1985 and 1986, USX; *Iron and Steel Engineer,* April 1986; USS–POSCO, Pittsburg Plant Modernization Plan, February 1987, USX.

22. *Pittsburgh Post Gazette,* 6 June 1972; *American Metal Market,* 7 June 1972.

23. Federal Reserve Bank of Dallas, *Voice,* February 1978, 6.

24. US Steel Corporate Policy Committee, 18 August 1980, USX.

25. *Houston Chronicle,* 12 October 1981.

26. US Steel, "Annual Report for 1988."

27. D. Roderick, statement to Congressional Steel Caucus, 19 April 1983, USX.

28. US Steel, "1st Quarter Report for 1983," USX, 7, 9.

29. *US Steel Quarterly,* November 1974, 4; US Steel, "Annual Report for 1981," 7.

30. *Pittsburgh Post Gazette,* 3 January 1988.

31. *Financial Times,* 9 July 1984.

32. *Fortune,* 13 February 1978, 127.

33. *American Metal Market,* 29 June 1972, 4.

34. *American Metal Market,* 31 August 1972, 7.

35. *Pittsburgh Press,* 23 November 1973.

36. *Financial Times,* 7 October 1977, 5 January 1978.

37. US Steel, "1st Quarter Report for 1979," USX.

38. US Steel Corporate Policy Committee, 7 January 1980, USX.

39. USWA undated note, USX.

40. US Steel Corporate Policy Committee, 7 January 1980, USX.

41. Ibid.

42. US Steel Corporate Policy Committee, September 1985, USX.

43. US Steel Corporate Policy Committee, 3 December 1979, 3 January 1979, 7 January 1980, USX.

44. National Works, untitled booklet, c.1979, USX.

45. *American Metal Market,* 27 September 1966.

46. *Iron and Steel Engineer,* January 1972, 80–82.

47. US Steel Corporate Policy Committee, 6 December 1982, USX.

48. Ibid.

49. US Steel, "Annual Report for 1982," 4.

50. *Financial Times,* 7 October 1986; *Pittsburgh Post Gazette,* 3 April 1994, 11 August 1994; US Steel Corporate Policy Committee, 20 August 1979, USX; Hoerr, *And the Wolf Finally Came,* 339.

51. *Iron and Steel Engineer,* 1993.

52. *Forbes,* 13 July 1987, 75.

53. J. Heitman, "T. C. Graham," in Seeley, *Iron and Steel in the Twentieth Century.*

54. *Financial Times,* 11 December 1985.

55. Dun and Bradstreet, *Million Dollar Directory: America's Leading Public and Private Corporations* (1997).

23. Labor on the Defensive during the Rationalization of the 1980s and 1990s

1. SCCC, press release, 30 July 1982, USX.

2. B. Johnston to "Each US Steel Employee," 1 August 1986, USX.

3. T. C. Graham to "Valued Customers and Friends of USS," 7 October 1986, USX.

4. B. Johnston to "All Steel Employees," 2 December 1986, USX.

5. T. C. Graham to US Steel Corporate Policy Committee, 27 January 1987, USX; US Steel, "Annual Report for 1987," 4, 51, 56.

6. US Steel Corporate Policy Committee, 25 March 1981, USX.

7. D. M. Roderick to C. H. Percy, 24 August 1983, USX.

8. Notes of press conference, 10 November 1983, USX.

9. USWA, notice, 27 December 1983, USX.

10. Locker-Albrecht, "Report on Duquesene Works for USWA," 1985, 30.

11. "Glimmers of Hope at Braddock," *New Steel*, September 1996, 37.

Conclusion. United States Steel in the Long View

1. On world context, see U.S. Department of Commerce, *Historical Statistics of the United States* (Washington, D.C.: GPO, 1998).

2. Steel Net <www.steelnet.org>, 20 October 1998.

3. *World Steel Dynamics*, December 1998.

4. *New Steel*, April 1999.

Bibliography

Archives

American Sheet and Tin Plate Company Papers. US Steel Corporation, Pittsburgh, Pennsylvania.

Bethlehem Papers. Hagley Library, Wilmington, Delaware.

Carnegie Papers. Library of Congress, Washington, D.C.

Carnegie Steel Company Papers. US Steel Corporation, Pittsburgh, Pennsylvania.

Frick Papers. Frick Art and Historical Center, Pittsburgh, Pennsylvania.

Illinois Steel Company Papers. US Steel Corporation, Pittsburgh, Pennsylvania.

Steel Workers Organizing Committee Papers. Pattee Library, Pennsylvania State University, University Park.

US Steel Corporation Papers. US Steel Corporation, Pittsburgh, Pennsylvania.

United Steel Workers of America Papers. Pattee Library, Pennsylvania State University, University Park.

Whipple Papers. Hagley Library, Wilmington, Delaware.

William Dickson Papers. Pattee Library, Pennsylvania State University, University Park.

Government Documents

Proceedings of a Conference of Governors at the White House, May 1908. Washington, D.C.: GPO, 1909.

United States Bureau of Census. *1954 Census of Manufactures: General Summary.* Washington, D.C.: GPO, 1957.

United States Commissioner of Corporations. *Report on the Steel Industry.* 2 vols. in 3 pts. Washington, D.C.: GPO, 1911, 1913.

United States Congress. *Hearings before the Joint Economic Committee on Steel Prices, Unit Costs, Profits, and Foreign Competition.* Washington, D.C.: GPO, 1963.

———. Transport Investigation and Research Board. *Economics of Iron and Steel Transportation.* Washington, D.C.: GPO, 1945.

———. House. *Hearings before the Committee on Investigation of U.S. Steel [Stanley Committee].* Washington, D.C.: GPO, 1912.

———. House. *Subcommittee on the Study of Monopoly Power.* Washington, D.C.: GPO, 1950.

United States Department of Commerce. *Historical Statistics of the United States.* Washington, D.C.: GPO, annually.

———. Senate Judiciary Committee. *Hearings before the Subcommittee on Antitrust and Monopoly.* Washington, D.C.: GPO, 1957–58.

————. *United States Industry and Trade Outlook '98.* 1998.

United States Department of the Interior. *Minerals Yearbook.* Washington, D.C.: GPO, annually.

United States Federal Trade Commission. *In the Matter of the United States Steel Corporation, et al.: Complaint, Findings, and Order.* Washington, D.C.: GPO, 1924.

————. *The United States Steel Industry and Its International Rivals.* Washington, D.C: GPO, 1977.

United States Industrial Commission. *Reports.* 13 vols. Washington, D.C.: GPO, 1899–1901.

United States Temporary National Economic Committee. *Investigation of Concentration of Economic Power.* Washington, D.C.: GPO, 1939–1940.

Books and Articles

Adams, W., ed. *The Structure of American Industry.* 1950. 6th edition. New York: Macmillan, 1992.

AISA. *Bulletin (BAISA).* Philadelphia: AISI, 1866–1912.

AISI. *Statistical Report.* New York and Washington, D.C.: AISI, annually.

AISI. *Yearbook.* Philadelphia: AISI, annually.

AISI. *The Steel Import Problem.* New York: AISI, 1968.

AISI. *Steel at the Crossroads: The American Steel Industry in the 1980s.* Washington, D.C.: AISI, 1980.

AISI. *Iron and Steel Works Directory of the United States and Canada.* New York: AISI, triennially.

Aldefer, E. B., and H. E. Michl. *The Economics of American Industry.* New York: McGraw Hill, 1942.

Allen, F. L. *Only Yesterday.* 1931. Reprint, London: Penguin, 1938.

————. *The Lords of Creation.* London: Hamish Hamilton, 1935.

American Iron Ore Association. *Iron Ore.* Cleveland: American Iron Ore Association, annually.

American Steel and Wire. *Thirtieth Anniversary.* N.p.: American Steel and Wire of New Jersey, c.1930.

Appleton, J. B. *The Iron and Steel Industry of the Calumet District.* University of Illinois Studies in the Social Sciences, no. 13. Chicago: University of Illinois, 1925.

Armes, E. *The Story of Coal and Iron in Alabama.* Birmingham, Ala.: Birmingham Chamber of Commerce, 1910.

Armstrong, G. S., et al. *The Iron and Steel Industry.* Vol. 10 of *An Engineering Interpretation of the Economic and Financial Aspects of American Industry.* N.p., 1952.

Association of Iron and Steel Engineers. *The Modern Strip Mill.* Pittsburgh: AISE, 1941.

Auerbach, P. *Competition: the Economics of Industrial Change.* Oxford: Basil Blackwell, 1988.

Baker, R. S. "What the Steel Corporation Really Is and How It Works." *McClure's Magazine* 18, no. 1 (November 1901).

Barloon, M. J. "Steel, the Great Retreat." *Harpers Monthly Magazine.* August 1947.

————. "Institutional Foundations of Pricing Policy in the Steel Industry." *Business History Review* 28 (1954).

Barron, C. W. *They Told Barron.* New York: Harper, 1930.

Berglund, A. *The United States Steel Corporation.* New York: Columbia University Press, 1907.

BISRA. *The Fairless Works.* London: British Iron and Steel Research Association, 1953.

Borrus, M. "The Politics of Competitive Economies in the United States Steel Industry." In *American Industry in International Competition: Government Policies and Corporate Strategies,* edited by J. Zysman and L. Tyson. Ithaca, N.Y.: Cornell University Press, 1983.

Bowie, R. R. *Government Regulation of Business.* Cases from the National Reporter System. N.p., 1949.

Bowron, J. "Steel Making in Alabama." *TAIMME* 81 (1925).

Bridge, J. H. *The Inside Story of the Carnegie Steel Company.* New York: Aldine, 1903.

Brody, D. *Steelworkers in America: The Nonunion Era.* Cambridge, Mass.: Harvard University Press, 1960.

Brooks, R. R. *As Steel Goes.* New Haven, Conn.: Yale University Press, 1940.

Broude, H. W. *Steel Decisions and the National Economy.* New Haven, Conn.: Yale University Press, 1963.

Brubaker, S. *Trends in the World Aluminum Industry.* Washington, D.C.: Resources for the Future, 1967.

Burns, A. R. *The Decline of Competition.* New York: McGraw Hill, 1936.

Burnham, T. H., and G. O. Hoskins. *Iron and Steel in Britain, 1870–1930.* London: Allen and Unwin, 1943.

Butler, J. G. *Fifty Years of Iron and Steel.* Cleveland: Penton Press, 1923.

———. *Recollections of Men and Events.* New York: Putnams, 1927.

Campbell, H. H. *The Manufacture and Properties of Structural Steel.* New York: Scientific Publishing, 1896.

———. *The Manufacture and Properties of Iron and Steel.* New York: Hill Publishing, 1907.

Carnegie, A. "Steel." In *The Empire of Business.* New York: Doubleday Page, 1912. First published in "Review of the Century" issue, *New York Evening Post,* 1900.

Casson, H. N. *The Romance of Steel.* New York: A. S. Barnes, 1907.

Chandler, A. D. *Scale and Scope: The Dynamics of Industrial Capitalism.* Cambridge, Mass.: Harvard University Press, 1990.

Clark, V. S. *History of Manufactures in the United States.* 3 vols. New York: McGraw-Hill, 1929.

Columbia Steel Company. *A History of Columbia Steel Corporation.* US Steel, n.d.

Cordero, H. G. *Iron and Steel Works of the World.* 2d edition. London: Quinn Press, 1957.

Cordero, R. *Iron and Steel Works of the World.* 6th edition. London: Metal Bulletin, 1974.

Cotter, A. *United States Steel: A Corporation with a Soul.* New York: Doubleday Page, 1921.

———. *The Gary I Knew.* Boston: Stratford, 1928.

Daugherty, C. R.; M. G. de Chazeau; and S. S. Stratton. *Economics of the Iron and Steel Industry.* New York: McGraw Hill, 1937.

Diamond, T. "This New Round of Steel Expansion." *Harvard Business Review* 34, no. 3 (May–June 1956).

Dickson, W. B. *History of the Carnegie Veterans Association.* Montclair, N.J: Mountain Press, 1938.

Dun and Bradstreet. *Million Dollar Directory: America's Leading Public and Private Corporations.* 1997.

Edwards, C. D. "The Effect of Recent Basing Point Decisions upon Business Practices." *American Economic Review* 38 (1948).

Eppelsheimer, D. "The Development of the Continuous Strip Mill." *JISI* (1938).

Evans, H. O. *Iron Pioneer: Henry W. Oliver, 1840–1904.* New York: E. P. Dutton, 1942.

Fairless, B. *It Could Only Happen in the United States.* N.p., 1957.

Fifty Years in Iron and Steel. Youngstown, Ohio: Youngstown Sheet and Tube Company, 1950.

Fisher, D. A. *Steel Serves the Nation: The Fifty Year Story of United States Steel.* New York: US Steel, 1951.

———. *The Epic of Steel.* New York: Harper and Row, 1963.

Fitch, J. A. "Old Age at Forty." *American Magazine* 71 (March 1911).

Foner, P. S. *History of the Labor Movement in the United States.* New York: International Publishers, 1955.

Foster, W. Z. *The Great Steel Strike and Its Lessons.* New York: B. W. Huebsch, 1920.

Garraty, J. *Right Hand Man: The Life of George W. Perkins.* New York: Harper, 1957.

Goodale, S. L. *Chronology of Iron and Steel.* Cleveland: Penton Publishing, 1931.

Gulick, C. A. *Labor Policy of the United States Steel Corporation.* New York: Columbia University Press, 1924.

Hall, J. W. *Mechanical Treatment.* Vol. 2 of *The Metallurgy of Steel.* London: Charles Griffin, 1918.

Harvey, G. *Henry Clay Frick: The Man.* New York: Charles Scribners, 1928.

Hauck, W. A. "Steel Expansion for War." Reprinted in *Steel,* June 1945.

Hawley, E. W. *The New Deal and the Problem of Monopoly.* Princeton, N.J.: Princeton University Press, 1966.

Hessen, R. *Steel Titan: The Life of Charles M. Schwab.* New York: Oxford, 1975. Reprinted, Pittsburgh: University of Pittsburgh Press, 1991.

Hoerr, J. P. *And the Wolf Finally Came: The Decline of the American Steel Industry.* Pittsburgh: University of Pittsburgh Press, 1988.

Hogan, W. T. *Economic History of the Iron and Steel Industry in the United States.* 5 vols. Lexington, Mass.: D. C. Heath, 1971.

Hogan, W. T. *Global Steel in the 1990s.* Lexington, Mass.: D.C. Heath, 1991.

Hotchkiss, W. O. "Iron Ore Supply of the Future." *Economic Geology* (May 1947).

Hoover, E. M. *The Location of Economic Activity.* New York: McGraw Hill, 1948.

Hubbard, E. "American Trusts and English Combinations." *Economic Journal* (June 1902).

Iron and Steel Engineers. *Directory of Iron and Steel Plants.* Pittsburgh: AISE, 1993.

Irvin, W. A. "Changing Trends in the Steel Industry." In AISI *Yearbook.* New York, 1938.

Isard, W. "Some Locational Factors in the Iron and Steel Industry since the Early Nineteenth Century." *Journal of Political Economy* 56 (1948).

Isard, W., and R. E. Kuenne. "The Impact of Steel upon the Greater New York–Philadelphia Industrial Region." *Review of Economics and Statistics* 35 (1953).

Jeans, J. S. *American Industrial Conditions and Competition.* London: British Iron Trade Association, 1902.

Jenks, J. W., and W. E. Clark. *The Trust Problem.* Garden City, N.J.: Doubleday, 1920.

Jones, E. *The Trust Problem in the United States.* New York: Macmillan, 1922.

King, C. D. *Seventy-five Years of Progress in Iron and Steel: Manufacture of Coke, Pig Iron, and Steel Ingots.* New York: AIME, 1948.

Lamoreaux, N. R. *The Great Merger Movement in American Business, 1895–1904.* Cambridge: Cambridge University Press, 1985.

Lorant, S. *Pittsburgh: The Story of an American City.* Lenox, Mass.: S. Lorant, 1988.

McCallum, E. D. *The Iron and Steel Industry in the United States.* London: P. S. King, 1931.

McCarty, H. H. *The Geographic Basis of American Economic Life.* New York: Harper, 1940.

McLaughlin, G. E. *Growth of American Manufacturing Areas.* Pittsburgh: University of Pittsburgh Press, 1938.

Manners, G. *The Changing World Market for Iron Ore, 1950–1980: An Economic Geography.* Baltimore: Johns Hopkins University Press, 1971.

Meade, E. S. "The Genesis of the United States Steel Corporation." *Quarterly Journal of Economics* 15 (August 1901).

———. "The Steel Corporation Bond Conversion." *Quarterly Journal of Economics* 18 (1904).

Minnesota, State of. "Iron Ore Beneficiation." *Federal Reserve Bank of Minnesota Review,* October 1957.

Moore, A. P., ed. *The Book of Prominent Pennsylvanians.* Philadelphia: Leader Publishing, 1913.

Morison, S. E.; H. S. Commager; and W. E. Leuchtenburg. *The Growth of the American Republic.* New York: Oxford University Press, 1969.

Mussey, H. R. *Combinations in the Mining Industry: A Study of Combination in Lake Superior Iron Ore Production.* Columbia University Studies in History, Economics, and Public Law 23, no. 3. New York: Columbia University Press, 1905.

National Tube. *A History of National Tube.* New York: US Steel, 1953.

Nourse, E. G., et al. *America's Capacity to Produce.* Washington, D.C.: Brookings Institution, 1934.

Paskoff, P., ed. *The Iron and Steel Industry in the Nineteenth Century: Encyclopedia of American Business History and Biography.* New York: Facts on File, 1989.

Perkins, F. *The Roosevelt I Knew.* New York: Viking Press, 1946.

Perloff, H. S., et al. *Regions, Resources, and Economic Growth.* Baltimore: Johns Hopkins University Press, 1960.

Popplewell, F. *Some Modern Conditions and Recent Developments in Iron and Steel Production in America.* Manchester: Manchester University Press, 1906.

Randall, C. B. "Mining Taxation in the Lake Superior District." In AISI *Yearbook,* 1932.

Rodgers, A. L. "Industrial Inertia: A Major Factor in the Location of the Steel Industry in the United States." *Geographical Review* 42 (1952).

Rostow, W. W. *The Stages of Economic Growth: A Non-Communist Manifesto.* Cambridge: Cambridge University Press, 1971.

Schroeder, G. G. *The Growth of Major Steel Companies, 1900–1950.* Baltimore: Johns Hopkins University Press, 1953.

Seager, H. R., and C. A. Gulick. *Trust and Corporation Problems.* New York: Harper and Brothers, 1929.

Seeley, B. E., ed. *Iron and Steel in the Twentieth Century: Encyclopedia of American Business History and Biography.* New York: Facts on File, 1994.

Shadwell, A. *Industrial Efficiency: A Comparative Study of Industrial Life in England, Germany, and America.* London: Longmans, Green, 1906.

Skinner, T. et al. *The Stock Exchange Year Book.* London: Thomas Skinner, 1931.

Stocking, G. W. *Basing Point Pricing and Regional Development: A Case Study of the Iron and Steel Industry.* Chapel Hill: University of North Carolina Press, 1954.

Swank, J. M. *Introduction to a History of Iron Making and Coal Mining in Pennsylvania.* Philadelphia: AISA, 1878.

———. *History of the Manufacture of Iron in All Ages.* Philadelphia: AISA, 1892.

Tarbell, I. M. *The Life of Elbert H. Gary: The Story of Steel.* New York: Appleton, 1925.

Taussig, F. "The Iron Industry in the United States." *Quarterly Journal of Economics* 14 (February 1900).

———. *Some Aspects of the Tariff Question.* Cambridge, Mass.: Harvard University Press, 1915.

Temin, P. *Iron and Steel in Nineteenth Century America: An Economic Enquiry.* Cambridge, Mass.: MIT Press, 1964.

Tiffany, P. *The Decline of American Steel: How Management, Labor and Government Went Wrong.* New York: Oxford University Press, 1988.

USWA. *Foreign Competition and Steel Wages.* N.p., 1959.

USWA. *A Brief History of the United Steelworkers of America.* N.p., 1999.

Vangermeers, R. "The Capitalization of Fixed Assets in the Birth, Life, and Death of US Steel, 1901–1986." N.p., n.d.

Van Hise, C. R. *The Conservation of the Natural Resources of the United States.* New York: Macmillan, 1910.

Wall, J. F. *Andrew Carnegie.* New York: Oxford University Press, 1970.

Warren, K. *The American Steel Industry, 1850–1970.* Pittsburgh: University of Pittsburgh Press, 1988.

———. *Triumphant Capitalism: Henry Clay Frick and the Industrial Transformation of America.* Pittsburgh: University of Pittsburgh Press, 1996.

White, C. M. "Technological Advances in Steel Production." In *AISI Yearbook,* 1937.

Whitney, S. N. *Anti-Trust Policies.* New York: Twentieth Century Fund, 1958.

"Why Are America's Steel Plants Closing?" *US Steel News* special issue. July 1982.

Wiebel, A. *Biography of a Business.* Birmingham, Ala.: TCI, 1960.

Wilgus, H. L. *The United States Steel Corporation in Its Industrial and Legal Aspects.* Chicago: Callaghan, 1901.

Williams, C. E. "Beneficiation of Iron Ore." *TAIMME* (1931).

Williamson, H. F., ed. *The Growth of the American Economy.* Englewood, N.J.: Prentice Hall, 1944.

Worthing, M. "Comparative Assembly Costs in the Manufacture of Pig Iron." *Pittsburgh Business Review,* January 1938.

Yellen, S. *American Labor Troubles.* New York: Harcourt Brace, 1936.

Unpublished Works

Corplan Associates. "Technological Change: Primary Metals." Report for US Steel, 1964.

Craig, R. S. "A History of the United States Steel Corporation and Predecessor Companies: A Case Study of American Business." Report for US Steel, 1966.

Ford, Bacon, and Davis. "Reports on the Operations of the United States Steel Corporation." 200 vols. Special reports for US Steel, 1936–1938.

Fuller, J. "History of the Tennessee Coal Iron and Railroad Company, 1852–1907." Master's thesis, Emory University, 1958.

Gregg, R. "Origin and Development of the Tennessee Coal, Iron, and Railroad Company." Report for US Steel, n.d.

Hill, Knowlton. "Account of the 1959 USWA strike for the S.C.C.C." 1959–1960.

Leffler, R. L. "The Future of the Delaware Valley." Address to Philadelphia Mortgage Bankers Association, 14 November 1955.

Locker-Albrecht. "Report on Duquesene Works for USWA," 1985.

North, W. H., and W. H. King. "Report of a Visit to the United States." Prepared for Dorman, Long of Middlesbrough, Eng., 1946.

Olds, I. "Address to the 3rd Annual Conference of the Greater Philadelphia–South Jersey Council." N.p., 5 October 1951.

Rodgers, A. L. "The Iron and Steel Industry of the Mahoning Shenango Valleys." Ph.D. diss., University of Wisconsin, 1949.

Russell, G. A. V. "Report on Observations Made during a Tour of Representative Steel Works in the United States, March–April 1938." Report to United Steel Companies, Sheffield, Eng., 1938.

———. "Notes on Visit to the United States and Canada, February–April 1941." Report to United Steel Companies, Sheffield, Eng., 1941.

Taylor, M. "Address to the Conference of Major Industries." New York, October 1928.

———. "Ten Years of Steel." Statement at the annual meeting of US Steel stockholders, Hoboken, New Jersey, 4 April 1938.

Wolf, G. W. "General World Steel Situation." Special statement to the Joint Committee on the Economic Report, Washington, D.C., January 1950.

Index